Emergence, Complexity and Computation

Volume 27

The Emergence, Complexity and Computation (ECC) series publishes new developments, advancements and selected topics in the fields of complexity, computation and emergence. The series focuses on all aspects of reality-based computation approaches from an interdisciplinary point of view especially from applied sciences, biology, physics, or chemistry. It presents new ideas and interdisciplinary insight on the mutual intersection of subareas of computation, complexity and emergence and its impact and limits to any computing based on physical limits (thermodynamic and quantum limits, Bremermann's limit, Seth Lloyd limits…) as well as algorithmic limits (Gödel's proof and its impact on calculation, algorithmic complexity, the Chaitin's Omega number and Kolmogorov complexity, non-traditional calculations like Turing machine process and its consequences,…) and limitations arising in artificial intelligence field. The topics are (but not limited to) membrane computing, DNA computing, immune computing, quantum computing, swarm computing, analogic computing, chaos computing and computing on the edge of chaos, computational aspects of dynamics of complex systems (systems with self-organization, multiagent systems, cellular automata, artificial life,…), emergence of complex systems and its computational aspects, and agent based computation. The main aim of this series it to discuss the above mentioned topics from an interdisciplinary point of view and present new ideas coming from mutual intersection of classical as well as modern methods of computation. Within the scope of the series are monographs, lecture notes, selected contributions from specialized conferences and workshops, special contribution from international experts.

More information about this series at http://www.springer.com/series/10624

Pierre-Yves Louis · Francesca R. Nardi
Editors

Probabilistic Cellular Automata

Theory, Applications and Future Perspectives

Springer

Editors
Pierre-Yves Louis
Laboratoire de Mathématiques et Applications UMR 7348
Université de Poitiers, CNRS
Poitiers
France

Francesca R. Nardi
University of Technology
Eindhoven, Noord-Brabant
The Netherlands

ISSN 2194-7287 ISSN 2194-7295 (electronic)
Emergence, Complexity and Computation
ISBN 978-3-030-09742-4 ISBN 978-3-319-65558-1 (eBook)
https://doi.org/10.1007/978-3-319-65558-1

Softcover re-print of the Hardcover 1st edition 2018

Printed on acid-free paper

This Springer imprint is published by Springer Nature
The registered company is Springer International Publishing AG
The registered company address is: Gewerbestrasse 11, 6330 Cham, Switzerland

Editorial Board

Preface

Probabilistic cellular automata (PCA) are interacting discrete stochastic dynamical systems used as a modelling tool for a wide range of natural and societal phenomena. On the applied side, PCA constitute an attractive computational framework for high-performance computing, distributed computing and simulations. Indeed, PCA have been put to good use as part of multiscale simulation frameworks for studying natural systems or large interconnected network structures. On the mathematical side, PCA have a rich mathematical theory that leads to a better understanding of the role of randomness and synchronicity in the evolution of large systems.

This book attempts to present a wide panorama of the current status of PCA theory and applications. These various contributions cover important subjects seen from probabilistic, statistical mechanical, computational and dynamical system points of view, and illustrated with applications to computer science and natural sciences. It *gathers contributions* from scientists with different perspectives, expectations, backgrounds and techniques. By bringing together works from international experts, this book intends to disseminate terminologies, common knowledge, tools, references and challenges.

This project started with a workshop organised in June 2013 at EURANDOM[1], TU Eindhoven. The organising committee wants to thank O. Boxma, F. den Hollander and R. van der Hofstad for scientific support. P. Koorn is warmly acknowledged for organisational and administrative help regarding this meeting. The workshop was very well attended by international participants. Presentations produced lively discussions both at theoretical and at applied levels. Feedback was very positive about the event. The opportunity to publish associated papers in a scientific journal was considered. Nevertheless, the strong interdisciplinary aspects were not compatible with the specific scope of a journal. Motivated by the nice atmosphere and proven interest, the organising committee enthusiastically became the editorial board of a *contributed book*. This choice granted more freedom and allowed a more

[1]http://www.eurandom.nl/events/workshops/2013/PCA/PCA.html

introductory style of the contributions, requiring them to be understandable and useful for scientists in different communities. The board worked on a collective basis sharing the different tasks and applying the respective areas of expertise to develop the three parts. An effort was made to select contributors from different academic status to foster interaction among generations. We are grateful to all senior and junior researchers who responded to this effort.

It is a great pleasure for the editorial board to see this project finalised. We want to thank all the colleagues for their contributions that helped concreting our vision. We acknowledge many referees whose help was decisive. Finally, the main editors Pierre-Yves Louis and Francesca R. Nardi are grateful to the other board members E. Cirillo, N. Fatès, Roberto Fernández, R. Merks, W. Ruszel and C. Spitoni for their competence and involvement in carrying out this project.

This book is aimed to researchers and motivated students who want to gain more insight into this broad topic. Our purpose is to stimulate cross-fertilisation both at theoretical and at applied levels. We think that the introductory level of this book could be a good starting point for non-specialists wishing to enter the field. The book should also be of interest as a source of challenges and open issues. As emphasised, probabilistic cellular automata is a lively topic, and this book does not claim to be either self-contained or exhaustive. Finally, despite our efforts, some errors, omissions or amendments may have escape our attention. We are grateful for feedback, corrections and comments.

Eindhoven, The Netherlands
July 2016

Pierre-Yves Louis
Francesca R. Nardi

Example of probabilistic and deterministic cellular automata in the visual arts. See Chap. 2. Breed 1.2 #e365 (3D print), Driessens & Verstappen, 2007, in private collection

Acknowledgements

The editorial board warmly thanks the authors and referees whose contribution made this book possible. This project is based on a workshop[2] organised in 2013 at EURANDOM[3], TU Eindhoven. The board thanks this institution for the kind hospitality and support. Furthermore, the board gratefully acknowledges the following programs that contributed to the funding of the meeting:

- European Science Foundation, Program *Random Geometry of Large Interacting Systems and Statistical Physics* (RGLIS),
- NWO—*Nederlandse Organisatie voor Wetenschappelijk Onderzoek* (program *Incidentele steun*),
- STAR Research Cluster–Stochastics Theoretical and Applied Research,
- 3TU.AMI Applied Mathematics Institute, *Mathematics For Innovation*.

[2]http://www.eurandom.nl/events/workshops/2013/PCA/PCA.html

[3]http://www.eurandom.tue.nl/

Contents

Editors, Associate Editors and Contributors

About the Editor

Pierre-Yves Louis is an associate professor at the 'Laboratoire de Mathématiques et Applications' (University of Poitiers and CNRS, UMR 7348, France), where he teaches applied probability, statistics and stochastic modelling. His research interests include interacting systems of stochastic processes and algorithms, and stochastic models from mathematical, statistical, computational and applied perspectives (in particular in biology and medicine). He holds a PhD in mathematics from Université Lille 1 (France) and Politecnico di Milano (Italy), and his dissertation focused on a study of probabilistic cellular automata from a stochastic processes and statistical mechanics viewpoint. He previously served as an assistant at the University of Potsdam and post-doc fellow at the TU Berlin (Germany). He was trained in both pure mathematics and physics at the University Lille 1. While he was visiting scholar at Eurandom (TU Eindhoven and CNRS-UMI 3022), he was the main organiser of a workshop dedicated to PCA. P.-Y. Louis is a regularly invited researcher in the Netherlands, Germany and Italy, where he maintains several ongoing research collaborations. He supervises master students and PhD candidates.

Francesca R. Nardi is an associate professor at the University of Florence's Department of Mathematics and Computer Science since 2016. Previously, she had a tenure track position (2006–2016) in the Stochastics Section of the Department of Mathematics and Computer Science at Eindhoven University of Technology (the Netherlands). She received her MSc degree and PhD degree in mathematics from the University of Rome "Tor Vergata", Italy, and subsequently held post-doc and teaching positions at Eurandom (Eindhoven University of Technology), the University of Rome "La Sapienza", and the University of Rome "Roma tre". Her main research interests are in probability theory, statistical mechanics, and mathematical physics. Applications that motivate her research are statistical physics of equilibrium: phase diagrams and phase transitions; the statistical physics of non-equilibrium: meta-stability; the spread of epidemics over networks for moving populations; random walks in random environments; cut-off phenomena and random graphs. Francesca was the recipient (as PI) of a STAR grant and an Aspasia grant from the Netherlands Organization for Scientific Research (NWO). Francesca is regularly visiting The Netherlands where she has several ongoing projects and research collaboration. Francesca has more than 15 years experience teaching courses in basic probability, stochastic processes, large deviations, metastability and random graphs. She supervised and coached master students, PhD students and post docs.

Associate Editors

Emilio N. M. Cirillo università di Roma: la Sapienza (Italy)
Nazim Fatès inria Nancy (France)
Roberto Fernández Universiteit Utrecht (The Netherlands)
Roeland M. H. Merks Centrum Wiskunde & Informatica, Amsterdam and Mathematical Institute, Leiden University (The Netherlands)
Wioletta M. Ruszel TU Delft (The Netherlands)
Cristian Spitoni Universiteit Utrecht (The Netherlands)

Contributors

Franco Bagnoli Department of Physics and Astronomy and CSDC, University of Florence, Florence, Italy; INFN, sez. Firenze, Sesto Fiorentino, Italy

Sonja E.M. Boas Centrum Wiskunde & Informatica (CWI), Amsterdam, The Netherlands

Olivier Bouré Université de Lorraine, CNRS, LORIA, Nancy, France

Vincent Chevrier Université de Lorraine, CNRS, LORIA, Nancy, France

Emilio N.M. Cirillo Dipartimento di Scienze di Base e Applicate per l'Ingegneria, Università di Roma: la Sapienza, Roma, Italy

Daan Crommelin Centrum Wiskunde & Informatica (CWI), Amsterdam, The Netherlands; Korteweg-de Vries Institute for Mathematics, University of Amsterdam, Amsterdam, The Netherlands

Paolo Dai Pra Dipartimento di Matematica, Università di Padova, Padova, Italy

Nazim Fatès inria, Université de Lorraine, CNRS, LORIA, Nancy, France

Roberto Fernández Universiteit Utrecht, Utrecht, The Netherlands

Henryk Fukś Department of Mathematics and Statistics, Brock University, St. Catharines, ON, Canada

Lucas Gerin CMAP, École Polytechnique, Palaiseau Cedex, France

Antal A. Járai Department of Mathematical Sciences, University of Bath, Bath, UK

Yi Jiang Department of Mathematics and Statistics, Georgia State University, Atlanta, GA, USA

Carlo Lancia Department of Medical Statistics and Bioinformatics, Leiden University Medical Center, Leiden, The Netherlands

Kerry A. Landman School of Mathematics and Statistics, University of Melbourne, Parkville, VIC, Australia

Pierre-Yves Louis Laboratoire de Mathématiques et Applications UMR 7348, Université de Poitiers, CNRS, Poitiers, France

Christian Maes Instituut voor Theoretische Fysica, KU Leuven, Leuven, Belgium

Roeland M.H. Merks Centrum Wiskunde & Informatica, Amsterdam, The Netherlands; Mathematical Institute, Leiden University, Leiden, The Netherlands

Ida G. Minelli Dipartimento di Ingegneria e Scienze dell'Informazione e Matematica, Università degli Studi dell'Aquila, L'Aquila, Italy

Francesca R. Nardi Department of Mathematics and Computer Science, Eindhoven University of Technology, Eindhoven, The Netherlands; Eurandom, Eindhoven, The Netherlands; Dipartmento di Matematica e Informatica, Universitá di Firenze, Firenze, Italy

Donald F. Newgreen Murdoch Children's Research Institute, Royal Children's Hospital, Parkville, VIC, Australia

Sotiris A. Prokopiou Department of Integrated Mathematical Oncology, H. Lee Moffitt Cancer Center and Research Institute, Tampa, FL, USA

Raúl Rechtman Instituto de Energías Renovables, Universidad Nacional Autónoma de México, Temixco, MOR, Mexico

Elisabeth G. Rens Centrum Wiskunde & Informatica, Amsterdam, The Netherlands; Mathematical Institute, Leiden University, Leiden, The Netherlands

Elena Sartori Dipartimento di Matematica, Università di Padova, Padova, Italy

Benedetto Scoppola Dipartimento di Matematica, Università di Roma: Tor Vergata, Roma, Italy

Piotr Słowiński College of Engineering, Mathematics and Physical Sciences, Harrison Building, Streatham Campus, University of Exeter, Exeter, United Kingdom

Cristian Spitoni Institute of Mathematics, University of Utrecht, Utrecht, The Netherlands

Siamak Taati University of British Columbia, Vancouver, Canada

Lorenzo Taggi Max Planck Institute for Mathematics in the Sciences, Leipzig, Germany; TU Darmstadt, Darmstadt, Deutschland

Marco Tolotti Dipartimento di Management, Università Ca' Foscari, Venezia, Italy

Aernout C.D. van Enter Johann Bernoulli Institute for Mathematics and Computer Science, University of Groningen, Groningen, The Netherlands

Anja Voss-Böhme Hochschule für Technik und Wirtschaft Dresden, University of Applied Sciences, Dresden, Germany; Zentrum für Informationsdienste und Hochleistungsrechnen, Technische Universität Dresden, Dresden, Germany

Chapter 1
Overview: PCA Models and Issues

Roberto Fernández, Pierre-Yves Louis and Francesca R. Nardi

Abstract Probabilistic cellular automata (PCA) are interacting discrete stochastic dynamical systems used as a modeling tool for a wide range of natural and societal phenomena. Their key features are: (i) a stochastic component that distinguishes them from the well-known cellular automata (CA) algorithms and (ii) an underlying parallelism that sets them apart from purely asynchronous simulation dynamics in statistical mechanics, such as interacting particle systems and Glauber dynamics. On the applied side, these features make PCA an attractive computational framework for high-performance computing, distributed computing, and simulation. Indeed, PCA have been put to good use as part of multiscale simulation frameworks for studying natural systems or large interconnected network structures. On the mathematical side, PCA have a rich mathematical theory that leads to a better understanding of the role of randomness and synchronicity in the evolution of large systems. This book is an attempt to present a wide panorama of the current status of PCA theory and applications. Contributions cover important issues and applications in probability, statistical mechanics, computer science, natural sciences, and dynamical systems. This initial chapter is intended both as a guide and an introduction to the issues discussed in the book. The chapter starts with a general overview of PCA modeling, followed by a presentation of conspicuous applications in different contexts. It closes with a discussion of the links between approaches and perspectives for future developments.

R. Fernández
Universiteit Utrecht, Utrecht, The Netherlands
e-mail: R.Fernandez1@uu.nl

P.-Y. Louis (✉)
Laboratoire de Mathématiques et Applications UMR 7348, Université de Poitiers, CNRS,
11 Boulevard Marie et Pierre Curie, Téléport 2 - BP 30179, Poitiers, France
e-mail: pierre-yves.louis@math.cnrs.fr

F.R. Nardi
Technische Universiteit Eindhoven, Eindhoven, The Netherlands
e-mail: F.R.Nardi@tue.nl; francescaromana.nardi@unifi.it

F.R. Nardi
Università di Firenze, Florence, Italy

F.R. Nardi
Dipartmento di Matematica e Informatica, Universitá di Firenze, Firenze, Italy

P.-Y. Louis and F.R. Nardi (eds.), *Probabilistic Cellular Automata*, Emergence, Complexity and Computation 27, https://doi.org/10.1007/978-3-319-65558-1_1

1.1 Introduction

Cellular Automata (CA) are lattices of interconnected finite-state automata (also called cells) which evolve synchronously in discrete time steps according to deterministic rules involving the states of adjacent automata. That is, at each time step each of the automata is updated independently of the others to a new value which is a function of the value of the automata in a suitably defined neighborhood. Their genesis is usually traced back to the 1948 paper by von Neumann and Ulam [174, 208] who introduced them as computational devices. An earlier, 1946 paper, by Wiener and Rosenbluth [171]—modeling impulse conduction in cardiac systems—should also be considered a precursor. These papers already setup the scene for two important areas of application: *cybernetics* and *excitable media*. Interest in these systems was boosted in the 70s by Conway's Game of Life—a two-dimensional cellular automaton—and in the 80s by Wolfram's classification of one-dimensional automata.

CA are surprising computational tools whose dynamics, despite being defined through rather simple local rules, lead to a rich zoo of patterns and structures that emerge without being designed a priori. These structures can be transient, oscillating or stable and can exhibit order at different levels of complexity or downright chaotic features. This richness has been exploited in a number of applications in different areas, of exact, natural, and social sciences. They have even been proposed as an alternative discrete way to express physical laws using present computational tools [211, 223].

Probabilistic Cellular Automata (PCA) are the extension obtained when the rules for updating are allowed to be random: New values of each automaton are chosen according to probability distributions determined by the configuration of its neighborhood. Usually, this updating is parallel or *synchronous* —all cells are simultaneously updated independently of each other— and neighborhoods are finite sets. At present, however, the notion of PCA is understood in a rather general sense that includes (partially) asynchronous dynamics and not necessarily finite neighborhoods (see, e.g., Chaps. 16 and 18 below). In this book, we adopt this general point of view.

The probabilistic component turns PCA into flexible computing tools for complex numerical constructions, and realistic simulation tools for phenomena driven by interactions among a large number of neighboring structures. PCA are, therefore, useful for the study of key issues of statistical mechanical and mathematical physics, such as phase transitions, metastability, percolation, and transport theory. But they are also naturally adapted to the study of systems and processes in life and social sciences involving systems characterized by high levels of complexity and low level of reproducibility, even under extremely controlled conditions, due to inherent randomness or experimental limitations.

Mathematically, PCA are systems of Markov chains interconnected through a network which typically is a lattice or a finite subpart of it. These Markov chains evolve in a parallel but coupled fashion, in which the distribution of future states of each chain depends on present states of neighboring chains. This coupling of transition probabilities is, however, local, and this makes PCA appealing as algorithms for high-performance computing, distributed computing, and simulations. Indeed, this locality makes the design of parallel implementations relatively straightforward, both on distributed architectures (e.g., computing clusters) and on massively parallel architectures (e.g., GPUs).

It is difficult to establish priorities and summarize the history of PCA developments. Their study was initiated by soviet mathematicians interested in artificial intelligence and cybernetics. Initially, PCA were studied to determine the robustness of CA dynamics subject to noise perturbation [92, 203]. In this setting, (non-)reliability is related to (non-)ergodicity [66]. The well-known North-East-Center PCA rule (see Sect. 1.3.1 below) was introduced in 1978 by Toom [204, for English translation] to provide a first PCA with a non-trivial instance of rigorously proven lack of ergodicity. Early applications also included models of neuronal networks [197], biological systems [206], and large systems of interacting automata [209]. In a somewhat independent way, PCA were studied in the 70's and 80's by probabilists [53, 54, 89, 123, 140, 141]—interested in their properties as stochastic processes—and by statistical physicists [102, 104, 112, 146, 214] interested in the study of equilibrium and non-equilibrium statistical mechanical distributions on lattices. The interdisciplinary nature of PCA studies has led to a convoluted history of independent rediscoveries and alternative terminology. Initially, the automata were termed *locally interacting Markov chains*; other names include *stochastic CA* or *random CA*. Some references on applications are presented in Sect. 1.3.3. We refer to [185] and Part 1.4 in [159] for surveys on historical aspects.

This introductory chapter presents some general types of phenomena that have been represented through PCA, emphasizing open issues and challenges that will be discussed in the remainder of this book. Our goal here is to exhibit the main common ideas —that often traverse scientific boundaries—, leaving specific analyses for relevant chapters. In doing so we have preferred to follow our personal, hence subjective, viewpoints, avoiding exhaustiveness. We therefore apologize to the many important actors of the field whose work we have failed to cite.

The rest of this chapter is organized in four sections. They deal, respectively, with the following aspects: (1) paradigmatic examples of PCA and their mathematical issues, (2) the three faces of the current interest in PCA: mathematical, computational, and scientific modeling, (3) useful links and future directions of research, and (4) summary of the structure of this book.

1.2 Phenomena Addressed by PCA Modeling

PCA dynamics belong to the category of *non-equilibrium lattice models*. In modeling circles, a *lattice* is a graph defined by a countable set of vertices (called sites or nodes) and a set of links. The latter are pairs of vertices usually visualized as a segment joining them. A popular lattice is, for instance, $\mathbb{Z}^d$. The cells or elementary components of the automata sit in the vertices, and the links are interpreted as vehicles for interactions or communications between cells. Informally, the lattice is a network interconnecting the cells. The strength of an interaction between cells is expected to decrease with the number of mediating links (graph distance). Thus, the definition of the PCA usually involves a notion of *neighborhood* defined as vertices separated by a maximum prescribed number of links. In particular, two vertices are *nearest neighbors* if they are the end point of a link. The qualifier *non-equilibrium* refers to the type of questions addressed by the theory. To be sure, the issues of the existence, number, nature, and basin of attraction of invariant (equilibrium) measures remain as important as in the theory of any stochastic process. Nevertheless, PCA theory focuses, particularly, in phenomena taking place *during the evolution* toward equilibrium. See for instance Chap. 9 in this book.

In this section, we describe three scenarios that lead to typical non-equilibrium issues addressed through PCA. The first one—metastability—refers to the appearance of traps and barriers delaying convergence. In some instances, these barriers are related to the emergence of non-trivial collective behavior manifested as phase transitions. These statistical mechanics phenomena are also related to some highly challenging optimization issues [1, 160]. The second scenario (epidemiology) addresses the issue of survival vs extinction in large interacting populations. The third scenario (wildfires) illustrates the study of dynamic percolation phenomena. The mathematical treatment of the latter presents some differences with the better known theory of equilibrium percolation models [110, 134, 212].

1.2.1 Metastability and Traps

Phase transitions are one of nature's more surprising phenomena. They refer to the sudden change on physical properties upon alterations of one or a few key parameters, for instance temperature or presence of a field: Liquids solidify at the freezing point and magnets acquire a nonzero magnetization even when the field is removed. In particular, *first-order phase transitions* are characterized by the presence of *coexistence curves*, that is manifolds of parameter values where the system presents several possible *stable phases*, i.e., extremal equilibrium measures. Examples are water that at the right combinations of pressure and (low) temperature can be either in liquid or solid form, and magnets that at zero field and low enough temperature can be magnetized in different directions. In such conditions, the actual state of the system depends on how it is prepared: Water will remain solid if the coexistence curve is reached from the high pressure side and liquid if reached by increasing the pressure; the remaining magnetization of a magnet at low temperature will remember the direction the field had when turned off.

Phase transitions are equilibrium phenomena whose description involves no obvious reference to any dynamics. In contrast, the more mysterious phenomenon of *metastability* can only be understood through (stochastic) dynamical consideration. Physical systems exhibit metastable behavior in the vicinity of first-order phase transitions, for instance in supercooled vapors and liquids, in supersaturated solutions, and in ferromagnets undergoing *hysteresis*. The common feature is the persistence of the systems in a state that resembles one of the coexistent states in the transition curve, but is different from the true equilibrium associated to the actual value of the parameters: Supercooled water remains liquid at temperatures (slightly) below the freezing point; magnets during hysteresis point in a direction different from that of the (small) field present.

Typically, such state of affairs results from small temperature or field changes, performed in an extremely smooth fashion (by nature or by the laboratory technician). As a result the system finds that, in order to achieve the equilibrium corresponding to the new parameter values, it must overcome a "barrier" that is a remnant of the initial coexistence situation. The height of this "barrier" causes the system to remain for *extremely long times* in an apparent equilibrium—the *metastable state*—from which it will abruptly jump into the actual equilibrium as a result of some external perturbation or some internal random fluctuation. The observation of these metastability phenomena extends well beyond physics and includes processes in chemistry, biology, climatology, economics.

A similar phenomenon takes place in numerical algorithms and simulation protocols whose convergence is often impaired by "traps" that retain the system for very long times. This type of metastability is of different nature than the one described above, as it is not due to a slight changes in parameters. Rather, the "traps" are an inherent feature of the dynamics, and the resulting evolution is closer to glassy transitions than to the neighborhood of first-order ones.

All types of metastability manifestations share a number of attributes that point to the existence of a general theory. Such a theory should elucidate the following questions:

1. Distribution of the exit time from the metastable to the stable state. Typically this time is exponentially distributed with a rate that depends of the value of relevant parameters (temperature, magnetic field, type of "trap").
2. Determination of the nature of the "metastable trap." In gases and magnetic systems—and many asynchronous dynamics such as Metropolis or Glauber—the trap is associated to an "energy well." This is far from universal, however; in some instances, the trap is of purely entropic nature or, more generally, due to a "free-energy well".
3. Details of the typical trajectories that the system takes when exiting the metastable state. This requires an understanding of the mechanisms behind this exit. In general, exit happens when a large atypical fluctuation allows the system to overcome the probabilistic barrier protecting the metastable situation. It is important to understand the nature of such a fluctuation. In many physical systems, exit is due to *nucleation*, that is the emergence of a sufficiently large region looking as the stable state. In general processes, however, no such appealing mechanism seems to be available.

These three questions have been largely elucidated for many ferromagnetic systems—see [23, 32, 173] for finite-volume models (i.e., finite-size systems) and

[22, 190] for models in infinite volume. Superheated water and supersaturated gas have been described through lattice-gas models subjected to Kawasaki dynamics — see [100, 118, 119, 172] in finite volume and [98, 99, 119] for infinite volume. General overviews of these types of metastability phenomena are presented, for instance, in [176, Chap. 7] and [21].

Parallel dynamics open tantalizing perspectives for the understanding of metastability, because they exhibit metastable traps and exit mechanisms that differ from those of asynchronous dynamics and processes. In PCA, these differences stem from the observation that *a large number of cells may change in a single time step*, leading to metastability mechanisms different from those of asynchronous dynamics. These facts make PCA metastability studies both a challenging [61, 112] and revealing component of the quest for a general theory of metastability. Comparative studies with asynchronous dynamics are particularly interesting, as they may lead to faster convergent simulation algorithms. See Chap. 3 in this book. For some pioneer studies on PCA metastability, we refer to [40–42, 44, 45, 173]. A more general theory, that applies also to PCA, has been developed in [43].

1.2.2 Epidemics and Extinction

A natural model in the epidemiology context is to consider a population of susceptible individuals sitting in the vertices of a lattice whose links determine the possibility of direct communication (interaction). The definition of the model includes:

- A set of possible states for each individual. Simple models assume three possibilities: sane/susceptible, ill/infecting, and recovered with or without immunity.
- A neighborhood of individuals that can pass the infection to a given one. Often, but not always, only nearest neighbors are considered.
- A rule deciding when an infection is passed to an individual from his/her neighbors.
- A rule specifying how one individual can recover and either become susceptible again or stay healthy forever due to acquired immunity.

The infection rule is stochastic in nature—exposition does not imply automatic contagion—, and so is the recovery rule. Furthermore, both rules should act on all individuals at the same time. PCA are, therefore, the model *par excellence* for epidemiological processes.

The rules depend on parameters that can be empirically estimated. For instance, the probability that the individual at site k gets the infection at the n-th time step can be of the form $1 - (1 - p)^{N_k(n)}$, where $N_k(n)$ is the number of neighbors of k that are infected at time n. Here $p \in [0, 1]$ is a parameter that measures susceptibility to infection. The capacity to overcome the infection is, on the other hand, an attribute of each individual. Hence, the probability of recovery of each individual at a given step is often assumed to be equal to another parameter $q \in [0, 1]$ independent of the rest of the population.

The main questions addressed to each epidemiology model refer to the range of parameters that will prevent the illness to become an epidemic. This is a risk management strategic question involving a number of measures: vaccines decrease p, isolation decreases N_k, general health situation increases q, *etc.*. In relation to this, the order of magnitude of the *spreading time* is an important piece of information, as it determines optimal vaccination strategies or, if there is no time, the need for quarantine.

Whole families of *individual-based lattice models* have been introduced in the last decades. They are known by their acronyms, e.g., SIR (Susceptible, Infectious, and Recovered), SIS (Susceptible, Infectious, and Susceptible) (see for instance [19, 188, 213]). The propagation of computer viruses through technological networks is, of course, another natural area of application of epidemiological models [182]. This issue is also analyzed in Chap. 12 below.

1.2.3 Wildfire and Percolation Phenomena

Mathematical models can be an important aid in establishing policies to limit the damage caused by wildfires in forests. The model should answer, for a given spatial distribution of trees, questions such as "Will the whole tree population eventually burn?" or "What will be the shape of the frontline of burning trees?" More generally the model should gauge the influence of the network structure and, hopefully, lead to the design of tree distributions for which the propagation to the whole forest is unlikely.

Wildfire models seem, conceptually, related to epidemiological models. Nevertheless, there is an important mathematical difference. Models in epidemiology deal with a finite population and focus on the persistence of the pathology *in time*. Wildfire models, on the other hand, consider a potentially infinite forest and study the *spatial* extension of the fire. The latter is, therefore, directly related to percolation models, as mentioned above. See [120] for recent related work as well as Chaps. 5 and 14 in this book.

The mathematical ingredients of a wildfire model are the following. As usual, there is a lattice, for instance $\mathbb{Z}^2$, with trees (potentially) sitting at its sites. The model is defined by the following choices.

- The possible states of each tree. In its simplest version, it must include three possibilities: non-burning tree, burning tree, and no tree (e.g., because it has burnt)
- Rule for the beginning of the fire. Possibilities are: the fire starts at a uniformly chosen tree (finite lattice), to simulate accidents, or ignition instances are (space–time) Poisson distributed to simulate lightning.
- Rule for the fire to pass from tree to tree. This should be a stochastic rule involving neighborhoods whose shape depends on actual conditions in the terrain. Examples include, but are not limited to, the nearest-neighbor CA. In general, neighborhoods are assumed to be uniformly finite.

Let us, as example, detail the rules for the Drossel–Schwabl model. The situation at each site is represented by three possibilities: 0 (no tree), 1 (burning tree), and 2 (non-burning tree). All the trees are simultaneously updated according to the following rules. Let us denote $\sigma_k(n)$ the state of the site k at time step n.

- A burning tree disappears at the next time step:

$$\sigma_k(n) = 1 \longmapsto \sigma_k(n+1) = 0 \qquad \text{with probability } 1 \ .$$

- The growth of a new tree by chance at an empty node k is tuned through a parameter $p \in [0, 1]$

$$\sigma_k(n) = 0 \longmapsto \sigma_k(n+1) = 2 \qquad \text{with probability } p \ .$$

- A tree starts to burn at node k, either by ignition from another burning neighboring tree or by chance. The later probability is given by a parameter f:

$$\begin{aligned} &\sigma_k(n) = 2 \longmapsto \sigma_k(n+1) = 1 \\ &\quad \text{with probability} \begin{cases} 1 & \text{if at least one neighboring tree is burning;} \\ f & \text{if no neighboring tree is burning.} \end{cases} \end{aligned}$$

Main general references on this specific model are [69, 70, 106]. An up-to-date reference for practitioners is [221]. Different critical behaviors have been studied in [107] through simulations and numerical analysis. Some theoretical results are given, for instance, in [48, 88]; see also [5].

1.3 The Multiple Faces of the PCA Paradigm

CA and PCA were initially introduced both as theoretical models for *decentralized systems* and as computational tools. Eventually, however, they were seized by different scientific communities which exploited them in a number of directions, ranging from purely mathematical studies to practical modeling of natural structures. At present, PCA can be considered a code word for a "parallelization paradigm" that allows to clarify and deepen the understanding of fundamental issues in mathematics and physics while, at the same time, leading to efficient computational procedures and simulation algorithms. In Sect. 1.2, we discussed how the PCA paradigm contributes to the understanding of some key non-equilibrium phenomena. Here we focus, instead, on issues and features pertaining to the PCA dynamics in itself. These aspects are crucial for the design and trust of PCA as modeling and computational tools.

1.3.1 Mathematical Issues

1.3.1.1 The Mathematical Setup

Unlike asynchronous dynamics—e.g., coupled differential equations, spin-flip dynamics, or interacting particle systems—CA and PCA can be directly defined for infinite lattices of cells. Indeed, its parallel character ensures existence of the corresponding process without involving some finite-region limit or otherwise conditions on the parameters. We present in this section the main aspects of the mathematical definition of the automata, starting with a list of the main ingredients of the setup:

The network G: This is a graph $G = (V(G), E(G))$ in which the set of vertices $V(G)$ marks the location of the automata (cells), and the set of edges $E(G)$ corresponds to interaction (or communication) channels between pairs of automata.
The alphabet S: Also called *local space* or *spin space*, describes the possible settings each automata may take. In most CA (for instance in this book), S is a finite set and hence it is endowed with the discrete σ-algebra and topology, and the uniform measure.
The configuration space $S^{V(G)}$: It represents the situation of the whole network of automata. It is endowed with the product topological and measure structure. Below we denote configurations as $\sigma = \{\sigma_k : k \in V(G)\}$, where σ_k is the configuration of the automaton at k.
The neighborhoods V_k: Each $V_k \subset V(G)$ represents the automata that can interact with the automaton sitting at $k \in V(G)$. For instance, a popular choice in $G = \mathbb{Z}^2$ is $V_k = \{k, k \pm e_1, k \pm e_2\}$ where (e_1, e_2) is the canonical basis of $\mathbb{Z}^2$ (north/south, east/west, and center neighbors). This is the so-called *von Neumann* neighborhood.

CA are Discrete-Time Dynamical Systems

Deterministic cellular automata are defined by the iteration of a transformation (global update) of the form

$$\begin{aligned} F : S^{V(G)} &\longrightarrow S^{V(G)} \\ \big(F(\sigma)\big)_k &= f_k(\sigma_{V_k}) \end{aligned}$$

for some single-updating functions $f_k : S^{V_k} \to S$. In many CA, $V(G)$ admits the action of the group $\mathbb{Z}^d$. These actions are called *translations*, and special emphasis is placed in translation-invariant CA, that is those whose updating rules commute with these translations.

As an illustration, let us consider the already mathematically rich CA in which $G = \mathbb{Z}$ (*one-dimensional PCA*). In this network, translations are generated by the *shift* map

$$\begin{aligned} T : S^{\mathbb{Z}} &\longrightarrow S^{\mathbb{Z}} \\ T(\sigma)_k &= \sigma_{k+1}. \end{aligned}$$

Translation invariance amounts, then, to homogeneity of neighborhood's $V_k = V_0 + k$ and single-updating functions $f_k(\sigma) = f_0(T^{-k}\sigma)$ for some function $f : S^{V_0} \mapsto S$.

The Curtis–Hedlund–Lyndon theorem [115, Theorem 3.1] characterizes translation-invariant CA transformations within the framework of dynamical systems. It states that a map F from $S^{\mathbb{Z}}$ to itself is a translation-invariant CA if and only if it is continuous (in the sense of the product topology) and commutes with the shift map. Moreover, if S has only two values, the map F is surjective if and only if it leaves invariant the uniform Bernoulli measure $\otimes_{k\in\mathbb{Z}}\mathcal{B}(1/2)$. We refer the reader to [155] for additional developments, to [130, 156] for recent results and to [142] for a recent survey of one-dimensional CA in the framework of topological dynamics. Readers interested in general introductions to CA are referred, for instance, to [129, 131]. See the Chaps. 11 and 6 in this book.

PCA are Interacting Families of Markov Stochastic Processes

Stochastic updating rules are defined by Markovian transition-probability kernels. If $V(G)$ and the alphabet S are finite sets, these kernels are defined by functions of the form $P(\sigma \mid \eta)$, interpreted as the probability that a configuration η at time step t ($t \in \mathbb{N}$) will be updated into the configuration σ at time step $t+1$. These functions must, therefore, satisfy the normalization condition

$$\sum_{\sigma} P(\sigma \mid \eta) = 1 \,. \tag{1.1}$$

The Markovianness stems from the fact that the distribution of the new configuration σ is independent of preceding configurations other than immediately preceding η. In this finite setting, a PCA corresponds to transitions of the form

$$P(\sigma \mid \eta) = \prod_{k\in G} p_k(\sigma_k \mid \eta_{V_k}) \tag{1.2}$$

where $\{p_k(\cdot \mid \eta_{V_k}),\ k \in G,\ \eta_{V_k} \in S^{V_k}\}$ is a family of probabilities on S. The product (1.2) corresponds to a family of Markov processes, one at each site. Nevertheless, the processes interact with each other because they share a common past configuration η.

PCA, however, can be directly defined on infinite (but countable!) graphs G. In these CA, transition probabilities are defined by *probability kernels* on $S^{V(G)} \times S^{V(G)}$. These are functions $P(\cdot \mid \cdot)$ whose two arguments are of a different nature. Indeed, the kernel is a *probability measure* with respect to the first argument and a measurable function with respect to the second one. Explicitly, the requirements are:

(i) $P(\cdot \mid \eta)$ is a probability measure on $S^{V(G)}$ for each $\eta \in S^{V(G)}$.
(ii) $P(A \mid \cdot)$ is a measurable function for each measurable $A \subset S^{V(G)}$.

PCA stochastic dynamics correspond to kernels of the form

$$P(d\sigma \mid \eta) = \prod_{k\in G} p_k(d\sigma_k \mid \eta_{V_k}) \tag{1.3}$$

where each p_k is a probability kernel on $S \times S^{V_k}$. These product measures exist and are uniquely defined due to Kolmogorov existence theorem. CA correspond to the particular CA in which the measures $p_k(\cdot \mid \eta_{V_k})$ are *delta*-like.

It is important to distinguish PCA from interacting particle systems (IPS) [62, 149, 150, 196]. Both, PCA and IPS are Markovian processes defined by families of interacting stochastic processes. The difference lies in the level of (a)synchronism. IPSs update one spin per time step (or a few per unit time in the continuous-time version). Furthermore, the interactions in IPS models are not only due to a common past but also to constraints and penalties imposed at the arrival time. These aspects lead to delicate construction processes, involving limits of dynamics in finite parts of the graph, that may be feasible only under supplementary conditions on transition probabilities and rates. The study of processes on infinite graphs is not just mathematical entertaining, but it is made necessary by the huge number of entities composing real-life systems (10^{25} molecules in a cubic inch of fluid, 10^{11} neurons in the human brain).

Despite their differences, often PCA and IPSs offer alternative approaches to the study of the same type of phenomena. One instance is the transport mechanisms and phenomena studied through the *Totally Asymmetric Exclusion Process* that have similar manifestations in some PCA dynamics; see [55, 139, 147], [155, part 4.3] and Chap. 16 in this book.

The degree of synchronism also distinguishes PCA from Glauber-like spin-flip dynamics such as the ones used to simulate equilibrium spin models. PCA are specially suited for models in complete graphs, such as mean-field (e.g., Curie–Weiss) models, but can also provide efficient alternatives to study, with a controllable margin of error, the Ising and similar models [51, 52, 143, 179]. This issue is discussed in Chap. 7 below.

Let us also mention the results in [64, 162] where the cardinality of *directed animals* on the lattice is related to properties of some PCA dynamics. Exact solutions —that is, solutions determined by a closed system of a few analytic equations— have been found for some PCA. See, for instance, [127] for an exactly solvable non-reversible PCA.

1.3.1.2 Ergodicity and Phase Transitions

PCA in infinite graphs exhibit a rich taxonomy of invariant measures. These include product measures [156] and Markov random chains or fields [31, 49]. Further insight is achieved by studying space–time distributions [146, 151, 156] which, not surprisingly, are found to be related to statistical mechanical distributions in one further dimension [104, 146]. This relation is particularly fruitful for *reversible* PCA, that is for stochastic dynamics invariant under time reversal [61, 112, 137, 141, 151]. The connection between PCA and space–time statistical mechanics links the lack of ergodicity in the former with phase transitions in the latter. A PCA is *ergodic* if whichever its initial condition, it asymptotically converges in distribution to a unique invariant measure. In space–time picture, lack of ergodicity can often be related to

Table 1.1 Elementary cellular automata rule 192

V_k's pattern at time $n-1$	111	110	101	100	011	010	001	000
New state in k at time n	1	1	0	0	0	0	0	0

a statistical mechanical phase transition triggered by boundary conditions on the initial space boundary of the space–time domain. Such lack of ergodicity was first rigorously exhibited in the simple ("toy") models presented below. Let us point out that, for instance if all transition probabilities are strictly positive, the dynamics is ergodic for a finite number of automata but looses this property when the number of automata becomes infinite. This is a remarkable example of *global effect emerging* when infinitely many sites interact.

Phase transitions are associated to multiple invariant measures. It is natural to wonder whether when the invariant measure is unique the automata is necessarily ergodic (i.e., this measure is attained for all initial configurations). The answer is negative [36, 125]. Transition probabilities usually depend on one or several parameters. The catalog of invariant measures for different values of those is called a *phase diagram*, in analogy with the statistical mechanical nomenclature. The rigorous determination of phase diagrams is often a difficult task, and numerical studies are the only available option. See [195] for a numerical analysis of the phase diagram of some *majority voter* PCA. See also Chap. 15 below.

The Stavskaya Model

This is the first model in which lack of ergodicity was rigorously proven. The model, to be considered in Chap. 13, depends on a *noise* parameter $\varepsilon > 0$. Its definition is as follows:

- (S1) Graph $G = \mathbb{Z}$.
- (S2) Alphabet $S = \{0, 1\}$.
- (S3) Neighborhoods $V_k = \{k-1, k\}$.
- (S4) Updating rule: For $\varepsilon \in [0, 1]$ fixed

$$p_k(1 \mid \eta_{\{k-1,k\}}) = \begin{cases} 1 \text{ if } \eta_k = \eta_{k-1} = 1 \\ \varepsilon \text{ otherwise .} \end{cases} \tag{1.4}$$

For $\varepsilon = 0$, this is the elementary CA with rule 192 (Wolfram's denomination [214])—see Table 1.1. Such rule is not symmetric under the interchange $0 \leftrightarrow 1$ but has both configurations $\underline{1}$ ("all ones") and $\underline{0}$ ("all 0") as fixed points. If $0 < \varepsilon < 1$, the dynamics can be thought as Rule 192 followed by a non-symmetric noise—*error mechanism*—which flips "0" into "1" independently with probability ε while leaving "1" unaltered. The resulting PCA has the following properties:

- (SN1) The configuration $\underline{1}$ is absorbing, already in finite volume.
- (SN2) The rates are not all (strictly) positive.
- (SN3) The dynamics is not reversible.

Fig. 1.1 Sample of a space–time configuration of the Stavskaya model. The parameter value is $\varepsilon = 0.28$. Boundary conditions were chosen periodic. The initial configuration is a central activated site (1) and the other sites are inactivated (0). *Black* (resp. *white*) *squares* represent activated (resp. inactivated) sites. Time is running downward. Simulated using FiatLux software [77]

The Stavskaya PCA is ergodic in its finite-graph version, but the ergodicity is lost for the full lattice for small values of ε. The precise result is as follows.

Theorem 1 *For the PCA defined by (S1)–(S4) above there exists a critical value* $\varepsilon^* > 0$ *such that:*

(i) If $\varepsilon > \varepsilon^*$ *the dynamics is ergodic with* $\lim_{n\to\infty} \mathbb{P}_\rho(\sigma(n) = \cdot) = \delta_{\underline{+1}}(\cdot)$ *for any initial distribution* ρ.

(ii) If $0 < \varepsilon < \varepsilon^*$ *there is a second invariant measure in which the value* 0 *survives:*

$$\lim_{n\to\infty} \mathbb{P}_{\delta_{\underline{0}}}(\sigma(n) = \cdot) = \mu_\varepsilon(\cdot) \neq \delta_{\underline{+1}} . \tag{1.5}$$

Furthermore, every translation-invariant stationary distribution is a convex combination of μ_ε *and* $\delta_{\underline{+1}}$.

The proof was first described in [191] and spelled out in [210]. More recent developments are given in [59, 164, 200] and Chap. 13. The exact value ε^* is still unknown. Numerical simulations and estimation give $\varepsilon^* \sim 0.29450$. Figure 1.1 shows a sample of a space–time diagram for ε slightly subcritical.

The North–East-Centre PCA Model and the Erosion Property

This very celebrated PCA is defined as follows:

- (N1) Graph $G = \mathbb{Z}^2$.
- (N2) Alphabet $S = \{0, 1\}$.
- (N3) Neighborhoods $V_k := \{k, k - e_1, k + e_2\}$.
- (N4) Updating rule: For $\varepsilon \in [0, 1]$ fixed

$$p_k(0 \mid \eta_{V_k}) = (1 - \varepsilon)\big(1 - \mathrm{Maj}(\eta_{V_k})\big) \tag{1.6}$$

where $\mathrm{Maj}(\sigma_x, \sigma_y, \sigma_z)$ is the value adopted by the majority of the three arguments.

For $\varepsilon \in]0, 1[$, the NEC PCA is a noisy perturbation of the *majority* CA, defined by

$$f_k(\sigma_{V_k}) = \mathrm{Maj}(\sigma_{V_k}) . \tag{1.7}$$

The error mechanism is, in fact, identical to that of the Stavskaya PCA and, as a consequence, the NEC shares with the latter the features (SN1)–(SN2) listed above. The NEC PCA, however, has many more absorbing configurations, for instance those formed by an arbitrary number of vertical and/or horizontal lines filled with "1".

The NEC PCA has, however, two additional properties that act in opposed direction and which are responsible for very eventful space–time diagrams:

The erosion property: The associated CA (1.7) is such that finite sets of "1" sites in an otherwise all "0" configuration disappear in a finite time, and similarly for islands of "0" inside a "1" sea.
Alignment-suppression property: There exist "spiders" formed by a few segments of sites such that, once they are filled with "1", the dynamics propagates these "1" to the interior of a sphere. As discussed in [84], this means that presence of a sphere of "1" is penalized by the invariant measures only as a subvolume exponential. This contradicts well-known Gibbsian properties and implies that *all* invariant measures for the NEC are *non-Gibbsian*.

The loss of ergodicity for small ε was proven by Toom [203] introducing carefully defined space–time contours and has since become a model argument to prove non-ergodicity in PCA. There has been a number of rewriting, reinterpretation, refining, and extensions of this pioneer proof. Interested readers can consult, for instance, Chap. 13 and [15, 65, 111, 153, 178].

The Positive Rate Conjecture

There exist different sets of sufficient conditions ensuring ergodicity of PCA [46, 85, 152, 154]. Many of those require local updating with *positive rates*:

$$p_k(s \mid \eta_{V_k}) > 0 , \qquad \forall k \in G,\ s \in S\ \eta_{V_k} \in S^{V_k} . \tag{1.8}$$

This property implies that all pair of configurations have positive probabilities of being mutually reachable. In particular no absorbing states are possible. The long-standing *positive-rate conjecture* stated that all positive-rate PCA on $G = \mathbb{Z}$ are ergodic. This conjecture was proved to be false through a complicated example in [93, 108]. The design of an understandable counterexample with $S = \{0, 1\}$ is still an open problem.

1.3.1.3 Random Perturbations of CA

One of the motivations of introducing PCA was to study the stability of CA under random perturbations. In this regard, the non-ergodicity results of the Stavskaya and NEC PCA reported above point in the direction of stability of the Rule 192 (Table 1.1)

Table 1.2 Elementary cellular automata rule 90

V_k's pattern at time $n-1$	111	110	101	100	011	010	001	000
New state in k at time n	0	1	0	1	1	0	1	0

Table 1.3 Transition's rules of elementary cellular automata rule 110

V_k's pattern at time $n-1$	111	110	101	100	011	010	001	000
New state in k at time n	0	1	1	0	1	1	1	0

and majority [rates (1.7)] CA under an asymmetric noise that flips "0" into "1" with probability ε. Indeed, the perturbation is proven to preserve both CA fixed points — $\delta_{\underline{1}}$ and $\delta_{\underline{0}}$— though the latter becomes a non-trivial probability measure supported in configurations that include infinitely many "1" (due to the erosion property). The question arises as to how dense these "1" are.

Kinzel [136] studied this question for stochastic perturbations of the CA defined by rule 90 in Wolfram's denomination [214] (Table 1.2). Performing a non-rigorous extrapolation from finite-size scaling, Kinzel concluded that the invariant measure obtained by perturbing $\delta_{\underline{0}}$ has zero density of "1" for small noise. This conclusion was shown to be false in [24], for arbitrarily small asymmetric noise. As soon as $\varepsilon > 0$, the probability of survival of "1" is strictly positive uniformly in the size of the system. This shows that even a very small noise can change properties drastically and, incidentally, that to formalize arguments based on finite-size scaling is a delicate task.

The proof in [24] is based on a connection with a process of *oriented percolation*. This type of connections has been later exploited, for instance, in [67]. The connection of PCA long-time behavior with processes of directed percolation is part of the epidemiology scenario discussed above and has been developed in [71, 132, 133, 165, 175, 181].

1.3.2 Computational Issues

Synchronous or quasi-synchronous updating turn PCA into natural tools for efficient parallel computing algorithms [56, 57, 217]. Here we present the main issues addressed by the theory of CA and PCA as computational tools.

CA as Universal Computational Systems

A computational system is said *universal* if it can run any program or, equivalently, execute any algorithm. Such attribute can be exhibited, for instance, by proving that the system can run programs equivalent to any program run by a Turing machine. The first CA proven to be computationally universal was Rule 110 (Table 1.3), analyzed in [47]. Recently, there have been exhaustive studies [161, 198] on the computational properties of all the elementary cellular automata (ECA) in Wolfram's classification [214].

CA and PCA as Decentralized Computational Systems

Both CA and PCA are archetypical models of *decentralized computing*. Each cell has its own resources and operates at each time step exchanging information and results only with neighboring units. Global features of the system emerge as collective results of these local interactions, without being driven by any external rule (or only partially ruled by external factors, as could be argued for global magnetic fields in Ising-like systems). The attributes of the computational approach offered by cellular automata are appropriately summarized in the "formula" stated in [193]:

$$\text{simple} + \text{parallel} + \text{local} = \text{cellular computing}\,.$$

Outputs of CA and PCA provide instances of self-organizing behavior and constitute a natural framework to relate this with the theory of formal languages and measures of algorithmic complexity. These issues are discussed in [216]. See also [34] for an introduction to algorithmic complexity.

PCA as Stochastic Algorithms

Stochastic algorithms tend to have better convergence properties than deterministic ones. Perhaps the main reason is that the former incorporate mechanisms to avoid or escape the drift toward local minima that constitute terminal traps for the latter. Nevertheless, in the presence of randomness, these traps become metastable states from which escape times, though always finite, can become excruciatingly long. At this point, the mathematical issue of ergodicity, discussed above, has a direct relevance.

In finite networks, most PCA dynamics can be proven to be ergodic, guaranteeing the eventual convergence of the associated algorithm to a well-defined final law. This is a consequence of general results in the theory of Markov chains, which apply for instance to positive rates PCA dynamics. Nevertheless, finite systems are expected to exhibit metastable behavior if the associated infinite system undergoes a phase transition/loss of ergodicity phenomenon. This is a well-known fact for MCMC practitioners [74, 128]. The issue was rigorously studied for a parallel implementation of the Gibbs sampler associated to the Ising Hamiltonian [40, 49, 61, 112] and in connection with the simulated annealing approach for stochastic algorithms [86].

PCA and Robustness with Respect to Errors

PCA constitute, on the one hand, the natural framework to study sensitivity to round off and other sources of errors in CA computations and, on the other hand, an excellent laboratory for the design of robust and trustworthy computational approaches. This perspective has been analyzed in [92, 111] and, more recently, in [16, 83, 183].

Ergodicity, in particular, corresponds to (extreme) robustness with respect to the initial conditions. The ergodicity for a PCA dynamics on $G = \mathbb{Z}$ is proven to be undecidable from an algorithmic point of view. See [28, 205] and [155, Sect. 3.1]. In a complementary way, the sensitivity to starting conditions implied by the absence of ergodicity has also been put to good use through PCA computations. A conspicuous example are the PCA dynamics that solve the *majority* or *density-classification* problem, namely to determine, on the basis of large-time outputs, whether there was a majority of some spin value in the starting condition [4, 80, 90, 163]; see Chap. 10 below.

Synchronicity and Updating Schemes

While parallel (synchronous) updating has obvious mathematical and computational advantages, sequential (asynchronous) updating has also important favorable features. For instance, the latter is well adapted to simulations of short-range spin models and, on the practical front, does not require the existence of a universal clock to which all the automata must synchronize. Furthermore, sequential sampling can be fine-tuned by adopting an appropriate updating scheme (e.g., uniformly at random, random with respect to the last updated node or deterministic).

The question arises whether efficiency can ultimately be improved by adjusting *both* the degree of synchronicity and the updating scheme. This issue has been actively investigated in the last two decades [2, 18, 35, 58, 82, 184]. A promising alternative are the so-called α-asynchronous PCA in which, when updating time arrives, the node is updated with probability α and otherwise left invariant. See the survey paper [79] and the references therein for more details.

1.3.3 Applications of PCA

PCA models have found applications in a diversity of fields from exact, natural and social sciences. We offer in this section an overview of the reach of these applications.

PCA as a Flexible Modeling and Simulation Framework in a Variety of Applied Contexts

The seminal work of Vichniac [211] placed CA as an *exact* modeling alternative to differential equations and not only as an approximation scheme [159, 201]. See for instance [124] for a comprehensive development. Here is a comparative list of advantages of cellular automata approaches:

- PCA models are simple to define if rules are given on a context-dependent basis. They provide a complete description of the evolution of the system even at the level of individual agents (cells) or of clusters of few individuals ("low-level" description).
- Models based on differential equations are suitable, in general, only for large space and timescales and if there is some level of homogeneity or some "steering global influence" that justifies a description in terms of densities.
- PCA belong to so-called *individual-based models*, that is:
  - They are based on information given at the individual level and, therefore, have a number of variables proportional to the number of individual cells.

- Simple low-level rules are the only source of the complex global phenomena or *collective behavior* that may eventually emerge at the level of the whole population or of a high fraction of its individuals ("high-level" phenomena).

- Unlike differential equations, PCA can incorporate specific individual attributes. The framework is particularly appropriated for systems which can be decomposed into interconnected elementary entities and where there is lack of homogeneity like in biological systems.
- PCA approaches can describe fluctuations which integro-differential approaches in general smooth out, average or neglect.
- PCA modeling applies at scales where no averaging is reasonable and therefore not amenable to analysis through differential equations.

Some general references on the connection between low- and high-level scales are [12, 60, 177, 187, 192] for modeling considerations. Of particular interest is the recent development of *hybrid models* involving different time or space scales. We cite for instance [167] where a CA approach is used with an environment governed by a partial differential equation. Another original contribution is the reverse engineering approach developed in [122] to find out PCA rules able to generate some fixed experimental patterns.

As a preliminary glimpse for interested readers, here is a (very incomplete!) overview of modeling applications of PCA in different scientific and technological areas.

PCA as Models for Complex Systems

PCA dynamics have two distinctive features:

- *Emergence:* Complex collective behavior appears solely as a result of local rules. We refer to [109, 145] for a general presentation of the *emergence* concept in physics, life sciences, and economical and other social phenomena.
- *Multiscale behavior:* This emergence acts at different scales with different levels of complexity. This multiscale feature is, in fact, a trademark of complex phenomena. See, for instance, [158] and the recent book [116] for a discussion of this aspect of the automata.

These attributes make PCA one of the most used class of models to analyze complex systems [11, 13, 38, 215]. PCA simulations are useful to understand complex behavior and to make predictive analysis. These predictions can often be rather surprising and counterintuitive, as illustrated in Chap. 18 of this book.

PCA as Models for Life Sciences

PCA systems are of great value in biological modeling, due to their sensitivity to space heterogeneity and their capacity to give rise to self-organized global structures. See [95] for a survey of life science applications and [10, 116, 166] for general modeling considerations. More specifically, PCA have been successful tools to describe pattern formation in cell development (*morphogenesis*) [63, 206] and cell biology [39,

96], specially in relation to multiscale phenomena [3, 60, 114, 189]. Other applications include immunology [26, 202, 219, 222]; neurosciences [94, 105, 135, 138, 157] and Chap. 17 of this book; oncology [60, 126, 199], and epidemiology [19, 75, 91, 148, 168, 194, 213].

In ecology, PCA models have long being proposed both as a paradigm [117] and to describe or simulate concrete issues. These include, for instance, evolution [180] and population dynamics [29, 87]. Most of the studies were of numerical character, but some rigorous results are also available [73, 169]

To conclude, let us mention, as part of the life sciences applications, a stochastic extension of the famous *game of life* (deterministic) CA, which was studied through computer simulations [170].

PCA as Models for Social Sciences

Opinion dynamics have been modeled through PCA in which spin values correspond to voting opinions [8]. The use of PCA systems has been advocated [14, 76] to model the evolution of markets driven by economic or financial agents. Such models have been applied to study *crisis propagation* in a network of companies, discriminating among different regimes that range from almost independent entities to strongly interconnected markets [14, 50, 103, 121]. See Chap. 4 below.

PCA as Models for Exact Sciences

PCA have been proposed as a general model for physical phenomena [37]. At present, the literature involving PCA in physics and chemistry is immense. Readers can consult the proceedings of the 2014 International Conference on Cellular Automata for Research and Industry [113] to have an idea of the state of the art. Recent notorious contributions range from an analogy with chemical reactions [186] and traffic models [220]. See also Chap. 16 of this book for a study of transport in lattice gases.

PCA as Models for Art

Perhaps not surprisingly given the plethora of patterns cellular automata can produce, PCA have also found applications in the visual arts. See [81] and Chap. 2 of this book.

1.4 Future Perspectives

As a result of the intense activity in PCA modeling, both theory and applications are moving into new directions that require extensions and refinement of the present conceptual framework and available techniques. Here are some pressing issues.

General Alphabets

All the automata studied so far have finite discrete alphabets. The need to consider more general alphabets— including, perhaps, some global constraints—relies both on mathematical [30] and modeling reasons (see e.g., Chaps. 18 and 19 of this book). Here are further examples that justify the development of a more general theory:

- Cellular Potts models for biological tissues require very large alphabets.
- The swarming model considered in Chap. 16 is a fully synchronous PCA with some conserved quantities.
- Countable spin spaces are considered in an ecological context in [17].
- In Chap. 8, the interacting Pólya urn model is seen as a PCA with $S = [0, 1]$. Automata with continuous alphabets are sometimes called *continuous automata* [33].

General Interactions

Historically, the interactions among automata have been of finite range. This is consistent with the interest in computer science to settle questions relating local and global transfer of information. Nevertheless, the modeling of dynamics of complex systems or the simulation of non-equilibrium statistical mechanical systems requires more general types of interactions. A few examples:

- Models with global constraints [78], Chaps. 16, 18 in this book and [7] (swarming models; silicon cells as in the CPM model, and models for glioma cell migration).
- Mean-field interactions, modeled in versions used for equilibrium spin models [9, 25, 144].
- Theoretical generalizations of PCA for simulation purposes [6].

Disorder

The architecture and the parameters of real-life networks are themselves subject to errors and fluctuations. Appropriate models require, then, the introduction of disorder either in interaction parameters—like in random field Ising models [20]—or in the underlying graph where the automata sit—like the power-law random graphs describing social and electronic networks [68, 218]. Yet another type of disorder of interest is the one characterizing the *stochastic spatial models* [72], in which sites can "mutate" their state. Updating rules can also change in a disorderly fashion. This possibility can be related to the issue of using observed data to statistically infer updating rules as in Chap. 20 of this book.

Implementation as Computational Schemes

The exceptional potential of PCA for high-performance computations has not yet been fully exploited for research-oriented simulations and computations. Some pioneer examples that point the way for future developments are the following.

- The implementation [143] (and Chap. 7) of a toy PCA on parallel architectures (like GPU units) for theoretical research purposes.
- The implementation on parallel architecture of classical stochastic algorithms (MCMC, Gibbs sampling, stochastic approximation). It can be done in a synchronous or quasi-synchronous way [97].
- The use of PCA in [101] to find large cliques in Erdös random graphs.
- The connection to statistics and machine learning algorithms recently presented in [207].

1.5 Structure of the Book

PCA have deservedly gained widespread recognition as versatile and efficient computational and simulation tools. They are presently been used in many areas of knowledge ranging from pure probability to social studies and including a wealth of scientific and technological applications. Furthermore, they constitute interesting mathematical objects on their own, whose theory lies at the crossroad of probability, statistical mechanics, and theoretical computer science. This situation has led to a highly diversified pool of theoreticians, developers, and practitioners whose interaction is highly desirable but can be hampered by differences in jargon and focus.

This book—as the workshop in which it is based—is an attempt to approach these different research communities by offering a tribune for them to present achievements, pressing issues, and future directions. The book is not intended as a treatise, but rather as a gentle introduction, for a general readership, of the role and relevance of PCA technology. The goal is to foster interest of newcomers and interaction between the different community-dependent perspectives, hopefully promoting new syntheses and applications. Each chapter can be read independently, in particular it carries its own bibliography section. Notation and formal aspects vary, according to standard usage in each research area, but differences are not dramatic and transitions should be straightforward for the reader.

The remaining of the book is divided into three parts oriented toward different families of applications:

Part I: Probability and statistical mechanics. Its seven chapters deal with probabilistic issues arising from the use of PCA as statistical mechanical models. These models share properties—and have contrasting attributes— with the standard sequential stochastic models used in out-of-equilibrium statistical mechanics.

Part II: Computer science and discrete dynamical systems. The six chapters of this part are devoted to central questions regarding robustness and computational aspects of PCA. Issues include comparisons with deterministic CA, general mathematical properties (e.g., convergence to a fixed point, phase transitions, existence of invariant measures) and determination of computations best suited to the use of PCA (e.g., density classification [27, 80]). Here, the term "computation" is intended in a general sense which includes, for instance, pattern formation [63] and classification of initial conditions. A particularly interesting question is to which extent randomness helps to speed up computations.

Part III: Applications to natural sciences and computational (cell) biology. It is formed by five chapters with applications to cell functions (e.g., Cellular Potts Model and stability of emerging patterns), challenging aspects of computational biology (Chaps. 17, 18, and 19) in particular weakened parallel with CPM, and multiscale modeling of atmospheric or oceanic circulation (Chap. 20). Chapter 16 introduces a model of swarming.

Acknowledgements The authors of this overview chapter thank the full editorial board for their help to set up this book. [We thank, in particular, R. Merks and N. Fatés for their comments and suggestions] about this chapter.

References

1. Achlioptas, D., Naor, A., Peres, Y.: Rigorous location of phase transitions in hard optimization problems. Nature **435**(June), 759–764 (2005)
2. Adachi, S., Peper, F., Lee, J.: Computation by asynchronously updating cellular automata. J. Stat. Phys. **114**(January), 261–289 (2004)
3. Alber, M.S., Kiskowski, M.A., Glazier, J.A., Jiang, Y.I.: On cellular automaton approaches to modeling biological cells. In: J. Rosenthal, D.S. Gilliam (eds.) Mathematical Systems Theory in Biology, Communications, Computation, and Finance. The IMA Volumes in Mathematics and its Applications. The IMA Volumes in Mathematics and its Applications, vol. 7105, p. 12. Springer, New York (2002)
4. Alippi, C., Polycarpou, M., Panayiotou, C., Ellinas, G. (eds.): Computing with Probabilistic Cellular Automata. Lecture Notes in Computer Science, vol. 1. Springer, Berlin (2009)
5. Almeida, R.M., Macau, E.E.N.: Stochastic cellular automata model for wildland fire spread dynamics. J. Phys. Conf. Ser. **285**(1), 12,038 (2011)
6. Arrighi, P., Schabanel, N., Theyssier, G.: Stochastic cellular automata: correlations, decidability and simulations. Fundamenta Informaticae **126**(2–3), 121–156 (2013)
7. Aubert, M., Badoual, M., Féreol, S., Christov, C., Grammaticos, B.: A cellular automaton model for the migration of glioma cells. Phys. Biol. **3**(2), 93 (2006)
8. Bahr, D., Passerini, E.: Statistical Mechanics of Collective Behavior: Macro-Sociology. J. Math. Sociol. **23**(1), 29–49 (1998)
9. Balister, P., Bollobás, B., Kozma, R.: Large deviations for mean field models of probabilistic cellular automata. Random Struct. Algorithms **29**(3), 399–415 (2006)
10. Bandini, S., Manzoni, S., Vizzari, G.: Agent Based Modeling and Simulation. In: Computational Complexity. Theory, Techniques, and Applications, pp. 105–117 (2012)
11. Bandini, S., Mauri, G., Serra, R.: Cellular automata: from a theoretical parallel computational model to its application to complex systems. Parallel Comput. **27**, 539–553 (2001)
12. Bandman, O.: Simulating spatial dynamics by probabilistic cellular automata. Lect. Notes Comput. Sci. **2493**, 10–19 (2002)
13. Barrat, A., Barthelemy, M., Vespignani, A.: Dynamical Processes on Complex Networks. Cambridge University Press, Cambridge (2008)
14. Bartolozzi, M., Thomas, A.W.: Stochastic cellular automata model for stock market dynamics. Phys. Rev. E **046112**(4), 1–17 (2004)
15. Berezner, S.A., Krutina, M., Malyshev, V.A.: Exponential convergence of Toom's probabilistic cellular automata. J. Stat. Phys. **73**(5–6), 927–944 (1993)
16. Berry, H., Fatès, N.: Robustness of the critical behaviour in the stochastic Greenberg–Hastings cellular automaton model. Int. J. Unconv. Comput. **7**(1), 65–85 (2011)
17. Birkner, M., Depperschmidt, A.: Survival and complete convergence for a spatial branching system with local regulation. Ann. Appl. Probab. **17**(5/6), 1777–1807 (2007)
18. Blok, H.J., Bergersen, B.: Synchronous versus asynchronous updating in the game of Life. Phys. Rev. E **59**(4), 1–16 (1999)
19. Boccara, N., Cheong, K.: Critical behaviour of a probabilistic automata network SIS model for the spread of an infectious disease in a population of moving individuals. J. Phys. A **26**, 3707–3717 (1999)
20. Bouchaud, J.P.: Crises and collective socio-economic phenomena: simple models and challenges. J. Stat. Phys. **151**(3–4), 567–606 (2013)

21. Bovier, A., den Hollander, F.: Metastability, Grundlehren der mathematischen Wissenschaften, vol. 351. Springer International Publishing, Cham (2015)
22. Bovier, A., den Hollander, F., Spitoni, C.: Homogeneous nucleation for Glauber and Kawasaki dynamics in large volumes at low temperatures. Ann. Probab. **38**(2), 661–713 (2010)
23. Bovier, A., Manzo, F.: Metastability in Glauber dynamics in the low-temperature limit: beyond exponential asymptotics. J. Stat. Phys. **107**(3–4), 757–779 (2002)
24. Bramson, M., Neuhauser, C.: Survival of one-dimensional cellular automata under random perturbations. Ann. Probab. **22**(1), 244–263 (1994)
25. Bricmont, J., Bosch, H.V.D.: Intermediate model between majority voter PCA and its mean field model. J. Stat. Phys. **158**(5), 1090–1099 (2014)
26. Burkhead, E.G., Hawkins, J.M., Molinek, D.K.: A dynamical study of a cellular automata model of the spread of HIV in a lymph node. Bull. Math. Biol. **71**(1), 25–74 (2009)
27. Bušić, A., Fatès, N., Mairesse, J., Marcovici, I.: Density classification on infinite lattices and trees. Electron. J. Probab. **18**, 109–120 (2013)
28. Bušić, A., Mairesse, J., Marcovici, I.I.: Probabilistic cellular automata, invariant measures, and perfect sampling. Adv. Appl. Probab. **980**(July 2012), 960–980 (2011)
29. Carvalho, K.C.D., Tomé, T.: Anisotropic probabilistic cellular automaton for a predator-prey system. Braz. J. Phys. **37**(2a), 466–471 (2007)
30. Casse, J.: Probabilistic cellular automata with general alphabets letting a Markov chain invariant. Adv. Appl. Probab. **48**(2) (2016)
31. Casse, J., Marckert, J.F.: Markovianity of the invariant distribution of probabilistic cellular automata on the line. Stochastic Process. Appl. **1**(9), 1–29 (2014)
32. Cerf, R., Manzo, F.: Nucleation and growth for the Ising model in d dimensions at very low temperatures. Ann. Probab. **41**(6), 3697–3785 (2013)
33. Cervelle, J.: Constructing continuous systems from discrete cellular automata. In: P. Bonizzoni, V. Brattka, B. Loewe (eds.) The Nature of Computation. Logic, Algorithms, Applications. Lecture Notes in Computer Science, vol. 7921, pp. 55–64. Springer, Berlin (2013)
34. Cervelle, J., Formenti, E.: Algorithmic Complexity and Cellular Automata. Computational Complexity. Theory, Techniques, and Applications pp. 132–146 (2012)
35. Chassaing, P., Gerin, L.: Asynchronous cellular automata and Brownian motion. In: 2007 Conference on Analysis of Algorithms, Discrete Math. Theor. Comput. Sci. Proc., AH, pp. 385–401. Assoc. Discrete Math. Theor. Comput. Sci., Nancy (2007)
36. Chassaing, P., Mairesse, J.: A non-ergodic probabilistic cellular automaton with a unique invariant measure. Stochastic Process. Appl. **121**(11), 2474–2487 (2010)
37. Chopard, B., Droz, M.: Cellular automata modeling of physical systems. Comput. Compl. Theory Tech. Appl. **122**, 407–433 (1998)
38. Chopard, B., Dupuis, A., Masselot, A., Luthi, P.: Cellular automata and lattice boltzmann techniques: an approach to model and simulate complex systems. Adv. Complex Syst. **05**(02n03), 103–246 (2002)
39. Chopard, B., Ouared, R., Deutsch, A., Hatzikirou, H., Wolf gladrow, D.: Lattice-gas cellular automaton models for biology: from fluids to cells. Acta biotheoretica **58**(4), 329–340 (2010)
40. Cirillo, E.N.M., Nardi, F.R.: Metastability for a stochastic dynamics with a parallel heat bath updating rule. J. Stat. Phys. **110**(1–2), 183–217 (2003)
41. Cirillo, E.N.M., Nardi, F.R.: Relaxation height in energy landscapes: an application to multiple metastable states. J. Stat. Phys. **150**(6), 1080–1114 (2013)
42. Cirillo, E.N.M., Nardi, F.R., Polosa, A.D.: Magnetic order in the Ising model with parallel dynamics. Phys. Rev. E **64**, 057,103 (2001)
43. Cirillo, E.N.M., Nardi, F.R., Sohier, J.: Metastability for general dynamics with rare transitions: escape time and critical configurations. J. Stat. Phys. **161**(2), 365–403 (2015)
44. Cirillo, E.N.M., Nardi, F.R., Spitoni, C.: Competitive nucleation in reversible probabilistic cellular automata. Phys. Rev. E **78** (2008)
45. Cirillo, E.N.M., Nardi, F.R., Spitoni, C.: Metastability for reversible probabilistic cellular automata with self-interaction. J. Stat. Phys. **132**(3), 431–471 (2008)

46. Coletti, C.F., Tisseur, P.: Invariant measures and decay of correlations for a class of ergodic probabilistic cellular automata. J. Stat. Phys. **140**(1), 103–121 (2010)
47. Cook, M.: Universality in Elementary Cellular Automata. Complex Syst. **15**, 1–40 (2004)
48. Cox, J.T., Durrett, R.: Limit theorems for the spread of epidemics and forest fires. Stochastic Process. Appl. **30**, 171–191 (1988)
49. Dai Pra, P., Louis, P.Y., Roelly, S.: Stationary measures and phase transition for a class of probabilistic cellular automata. ESAIM Probab. Stat. **6**, 89–104 (2002)
50. Dai Pra, P., Runggaldier, W., Sartori, E., Tolotti, M.: Large portfolio losses; a dynamic contagion model. Ann. Appl. Probab. **19**(1), 1–38 (2007)
51. Dai Pra, P., Scoppola, B., Scoppola, E.: Sampling from a Gibbs measure with pair interaction by means of PCA. J. Stat. Phys. **149**(4), 722–737 (2012)
52. Dai Pra, P., Scoppola, B., Scoppola, E.: Fast mixing for the low temperature 2d Ising model through irreversible parallel dynamics. J. Stat. Phys. **159**(1), 1–20 (2015)
53. Dawson, D.A.: Synchronous and asynchronous reversible Markov systems. Canad. Math. Bull. **17**(5), 633–649 (1974)
54. Dawson, D.A.: Stable states of probabilistic cellular automata. Inf. Control **34**(2), 93–106 (1977)
55. De Masi, A., Esposito, R., Lebowitz, J.L., Presutti, E.: Hydrodynamics of stochastic cellular automata. Commun. Math. Phys. **125**(1), 127–145 (1989)
56. Delorme, M., Mazoyer, J. (eds.): Cellular Automata: A Parallel Model. Springer, Berlin (1999)
57. Demongeot, J., Goles, E., Tchuente, M. (eds.): Dynamical Systems and Cellular Automata. Academic Press, New York (1985)
58. Dennunzio, A., Formenti, E., Fatès, N.: Foreword: cellular automata and applications. Nat. Comput. **12**(3), 305–305 (2013)
59. Depoorter, J., Maes, C.: Stavskaya's measure is weakly Gibbsian. Markov Process. Related Fields **12**(4), 791–804 (2006)
60. Deroulers, C., Aubert, M., Badoual, M., Grammaticos, B.: Modeling tumor cell migration: from microscopic to macroscopic models. Phys. Rev. E **79**(3), 1–14 (2009)
61. Derrida, B.: Dynamical phase transitions in spin models and automata. In: Fundamental Problems in Statistical Mechanics VII (Altenberg, 1989), pp. 273–309. North-Holland, Amsterdam (1990)
62. Deuschel, J.D., Greven, A. (eds.): Interacting Stochastic Systems. Springer, Berlin (2005)
63. Deutsch, A., Dormann, S., Maini, P.K.: Cellular Automaton Modeling of Biological Pattern Formation: Characterization, Applications, and Analysis. Modeling and Simulation in Science Engineering and Technology. Birkhäuser (2005)
64. Dhar, D.: Equivalence of the two-dimensional directed-site animal problem to Baxter's hard-square lattice-gas model. Phys. Rev. Lett. **49**(14), 959–962 (1982)
65. Diakonova, M., Mackay, R.S.: Mathematical examples of space-time phases. Int. J. Bifurc. Chaos **21**(08), 1–8 (2011)
66. Dobrushin, R.L.: Markov processes with a large number of locally interacting components: existence of a limit process and its ergodicity. Problemy Peredači Informacii **7**(2), 70–87 (1971)
67. Domany, E., Kinzel, W.: Equivalence of cellular automata to Ising models and directed percolation. Phys. Rev. Lett. **53**(4), 311–314 (1984)
68. Dommers, S., Giardinà, C., van der Hofstad, R.: Ising models on power-law random graphs. J. Stat. Phys. **141**(4), 638–660 (2010)
69. Drossel, B., Clar, S., Schwabl, F.: Exact results for the one-dimensional self-organized critical forest-fire model. Phys. Rev. Lett. **71**(23), 3739 (1993)
70. Drossel, B., Schwabl, F.: Self-organized critical forest-fire model. Phys. Rev. Lett. **69**(11), 1629–1632 (1992)
71. Durrett, R.: Oriented percolation in two dimension. Ann. Probab. **12**(4), 929–1227 (1984)
72. Durrett, R.: Stochastic spatial models. SIAM Rev. **41**(4), 677–718 (1999)
73. Durrett, R., Levin, S.A.: Stochastic spatial models: a user's guide to ecological applications. Philos. Trans. B **343**, 329–350 (1994)

74. Earl, D.J., Deem, M.W.: Parallel tempering: theory, applications, and new perspectives. Phys. Chem. Chem. Phys. PCCP **7**(23), 3910–3916 (2005)
75. Elsayed, W.M., El Bassiouny, A.H., Radwan, E.F.: Applying inhomogeneous probabilistic cellular automata rules on epidemic model. Int. J. Adv. Res. Artif. Intell. **2**(4) (2013)
76. Farmer, J.D., Foley, D.: The economy needs agent-based modelling. Nature **460**(7256), 685–686 (2009)
77. Fatès, N.: FiatLux: a simulation program in Java for cellular automata and discrete dynamical systems. http://fiatlux.loria.fr (Cecill licence) APP `IDDN.FR.001.300004.000.S.P.2013.000.10000`
78. Fatès, N.: Solving the decentralised gathering problem with a reaction-diffusion-chemotaxis scheme. Swarm Intell. **4**(2), 91–115 (2010)
79. Fatès, N.: A guided tour of asynchronous cellular automata. Lect. Notes Comput. Sci. **8155**, 15–30 (2013)
80. Fatès, N.: Stochastic cellular automata solutions to the density classification problem. Theory Comput. Syst. **53**(2), 223–242 (2013)
81. Fatès, N.: Aesthetics and Randomness in Cellular Automata. pp. 137–139. Springer International Publishing, Berlin (2016)
82. Fatès, N., Gerin, L.: Examples of fast and slow convergence of 2D asynchronous cellular systems. In: Umeo, H., Morishita, S., Nishinari, K., Komatsuzaki, T., Bandini, S. (eds.) Cellular Automata. Lecture Notes in Computer Science, vol. 5191, pp. 184–191. Springer, Berlin (2008)
83. Fatès, N., Morvan, M.: An experimental study of robustness to asynchronism for elementary cellular automata. Complex Syst. **16**(1), 1–27 (2005)
84. Fernández, R., Toom, A.: Non-gibbsianness of the invariant measure of non-reversible cellular automata with totally asymmetric noise. Astérisque **287**, 71–87 (2003)
85. Ferrari, P.: Ergodicity for a class of probabilistic cellular automata. Rev. Mat. Apl. **12**(2), 93–102 (1991)
86. Ferrari, P., Frigessi, A., Schonmann, R.H.: Convergence of some partially parallel gibbs samplers with annealing. Ann. Appl. Probab. **3**(1), 137–153 (1993)
87. Soares Filho, B.S., Coutinho Cerqueira, G., Lopes Pennachin, C.: Dinamica–a stochastic cellular automata model designed to simulate the landscape dynamics in an Amazonian colonization frontier. Ecol. Model. **154**(3), 217–235 (2002)
88. Fisch, R., Gravner, J., Griffeath, D.: Metastability in the Greenberg-Hastings model. Ann. Appl. Probab. **3**(4), 935–967 (1993)
89. Föllmer, H.: Tail structure of Markov chains on infinite product spaces. Z. Wahrsch. Verw. Gebiete **285**(3), 273–285 (1979)
90. Fukś, H.: Non-deterministic density classification with diffusive probabilistic cellular automata. Phys. Rev. E **66**(6), 1–4 (2002)
91. Fukś, H., Lawniczak, A.T.: Individual-based lattice model for spatial spread of epidemics. Discret. Dyn. Nat. Soc. **6**(3), 191–200 (2001)
92. Gács, P.: Reliable computation with cellular automata. J. Comput. Syst. Sci. **32**, 15–78 (1986)
93. Gács, P.: Reliable cellular automata with self-organization. J. Stat. Phys. **103**(1–2), 45–267 (2001)
94. Galves, A., Löcherbach, E.: Modeling networks of spiking neurons as interacting processes with memory of variable length. Journal de la Société Française de Statistique **157**(1), 17–32 (2016)
95. Ganguly, N., Sikdar, B.K., Deutsch, A., Canright, G., Chaudhuri, P.P.: A survey on cellular automata. Engineering pp. 1–30 (2003)
96. Garijo, N., Manzano, R., Osta, R., Perez, M.A.: Stochastic cellular automata model of cell migration, proliferation and differentiation: validation with in vitro cultures of muscle satellite cells. J. Theor. Biol. **314**, 1–9 (2012)
97. Garzon, M.: Models of Massive Paralellism: Analysis of Cellular Automata and Neural Networks. Springer, Berlin (1985)

98. Gaudillière, A., den Hollander, F., Nardi, F., Olivieri, E., Scoppola, E.: Ideal gas approximation for a two-dimensional rarefied gas under Kawasaki dynamics. Stochastic Process. Appl. **119**(3), 737–774 (2009)
99. Gaudillière, A., Nardi, F.R.: An upper bound for front propagation velocities inside moving populations. Braz. J. Probab. Stat. **24**(2), 256–278 (2010)
100. Gaudillière, A., Olivieri, E., Scoppola, E.: Nucleation pattern at low temperature for local Kawasaki dynamics in two dimensions. Markov Process. Related Fields **11**, 553–628 (2005)
101. Gaudillière, A., Scoppola, B., Scoppola, E., Viale, M.: Phase transitions for the cavity approach to the clique problem on random graphs. J. Stat. Phys. **145**(5), 1127–1155 (2011)
102. Georges, A., Le Doussal, P.: From equilibrium spin models to probabilistic cellular automata. J. Stat. Phys. **54**(3–4), 1011–1064 (1989)
103. Giesecke, K., Weber, S.: Credit contagion and aggregate losses. J. Econom. Dyn. Control **30**(5), 741–767 (2006)
104. Goldstein, S., Kuik, R., Lebowitz, J.L., Maes, C.: From PCA to equilibrium systems and back. Commun. Math. Phys. **125**(1), 71–79 (1989)
105. Goltsev, A.V., de Abreu, F.V., Dorogovtsev, S.N., Mendes, J.F.F.: Stochastic cellular automata model of neural networks. Phys. Rev. E **81**(6), 61,921 (2010)
106. Grassberger, P.: On a self-organized critical forest-fire model. J. Phys. A **26**(9), 2081 (1993)
107. Grassberger, P.: Critical behaviour of the Drossel-Schwabl forest fire model. New J. Phys. **4**(1), 17 (2002)
108. Gray, L.F.: A reader's guide to Gacs's "positive rates" paper. J. Stat. Phys **103**(1–2), 1–44 (2001)
109. Griffeath, D.: Self-organization of random cellular automata: four snapshots. In: G.R. Grimmett (ed.) Probability and Phase Transition, NATO ASI Series, vol. 420, pp. 49–67. Springer, Berlin (1994)
110. Grimmett, G.: Percolation. Springer, Berlin (1999)
111. Grinstein, G.: Can complex structures be generically stable in a noisy world? IBM J. Res. Develop. **48**(1), 5–12 (2004)
112. Grinstein, G., Jayaprakash, C., He, Y.: Statistical mechanics of probabilistic cellular automata. Phys. Rev. Lett. **55**(23), 2527–2530 (1985)
113. Habel, L., Schreckenberg, M.: Cellular Automata: 11th International Conference on Cellular Automata for Research and Industry, ACRI 2014, Krakow, Poland, September 22–25, 2014. Proceedings. In: J. Was, G.C. Sirakoulis, S. Bandini (eds.) Lecture Notes in Computer Science, vol. 8751, pp. 620–629 (2014)
114. Hatzikirou, H., Deutsch, A.: Cellular automata as microscopic models of cell migration in heterogeneous environments. Curr. Top. Dev. Biol. **81**(07), 401–434 (2008)
115. Hedlund, G.A.: Endomorphisms and automorphisms of the shift dynamical system. Math. Syst. Theory **3**(4), 320–375 (1969)
116. Hoekstra, A., Kroc, J., Sloot, P.: Introduction to modeling of complex systems using cellular automata. In: Simulating Complex Systems by Cellular Automata, pp. 1–16 (2010)
117. Hogeweg, P.: Cellular automata as a paradigm for ecological modeling. Appl. Math. Comput. **27**(1), 81–100 (1988)
118. den Hollander, F., Nardi, F., Olivieri, E., Scoppola, E.: Droplet growth for three-dimensional Kawasaki dynamics. Probab. Theory Related Fields **125**(2), 153–194 (2003)
119. den Hollander, F., Olivieri, E., Scoppola, E.: Metastability and nucleation for conservative dynamics. J. Math. Phys. **41**(3), 1424–1498 (2000)
120. Holroyd, A.E., Marcovici, I., Martin, J.B.: Percolation games, probabilistic cellular automata, and the hard-core model. arXiv:1503.05614 [math] (2015)
121. Horst, U.: Stochastic CAcade, credit contagion, and large portfolio losses. J. Econ. Behav. Organ. **63**(1), 25–54 (2007)
122. Ichise, Y., Ishida, Y.: Reverse engineering of spatial patterns in cellular automata. Artif. Life Robot. **13**(1), 172–175 (2008)
123. Ignatyuk, I.A., Malyshev, V.A.: Processes with local interaction, and communication networks. Zap. Nauchn. Sem. S.-Peterburg. Otdel. Mat. Inst. Steklov. (POMI) **25**(1), 65–77 (1989)

124. Ilachinski, A.: Cellular Automata; A Discrete Universe. World Scientific, Singapore (2001)
125. Jahnel, B., Külske, C.: A class of non-ergodic probabilistic cellular automata with unique invariant measure and quasi-periodic orbit. Stoch. Process. Appl. **125**(6), 2427–2450 (2015)
126. Jiang, Y.: Understanding a killer: a predictive model for tumor development. Contemp. Math. **410**, 173–185 (2006)
127. Just, W.: Toom's model with Glauber rates: an exact solution. J. Stat. Phys. **139**(6), 985–990 (2010)
128. Karafyllidis, I., Thanailakis, A.: A model for predicting forest fire spreading using cellular automata. Ecol. Model. **99**, 87–97 (1997)
129. Kari, J.: Theory of cellular automata: a survey. Theoret. Comput. Sci. **334**(1–3), 3–33 (2005)
130. Kari, J., Taati, S.: Conservation laws and invariant measures in surjective cellular automata. In: AUTOMATA-2011, 2, pp. 113–122. DMTCS Proceedings (2012)
131. Kari, J.J.: Basic concepts of cellular automata. In: Rozenberg, G., Bäck, T., Kok, J.N. (eds.) Handbook of Natural Computing, pp. 3–24. Springer, Berlin (2012)
132. Katori, M., Konno, N., Tanemura, H.: Limit theorems for the nonattractive Domany-Kinzel model. Ann. Probab. **30**(2), 933–947 (2002)
133. Katori, M., Tsukahara, H.: Two-neighbour stochastic cellular automata and their planar lattice duals. J. Phys. A **28**(14), 3935 (1995)
134. Kesten, H.: What is Percolation? Notices of the AMS (2006)
135. Kinouchi, O., Copelli, M.: Optimal dynamical range of excitable networks at criticality. Nat. Phys. **2**(5), 348–351 (2006)
136. Kinzel, W.: Phase transitions of cellular automata. Zeitschrift für Physik B Condensed Matter **58**(3), 229–244 (1985)
137. Kozlov, O., Vasilyev, N.: Reversible Markov chains with local interaction. Multicompon. Random Syst. **6**, 451–469 (1980)
138. Kozma, R., Puljic, M., Balister, P., Bollobas, B., Freeman, W.J.: Neuropercolation: a random cellular automata approach to spatio-temporal neurodynamics. In: Sloot, P.M.A., Chopard, B., Hoekstra, A.G. (eds.) Lecture Notes in Computer Science: Cellular Automata, Proceedings, vol. 3305, pp. 435–443. Springer, Berlin (2004)
139. Kriecherbauer, T., Krug, J.: A pedestrian's view on interacting particle systems, KPZ universality, and random matrices. J. Phys. **43**(40) (2010)
140. Künsch, H.: Nonreversible stationary measures for infinite interacting particle systems. Z. Wahrsch. Verw. Gebiete **66**(3), 407–424 (1984)
141. Künsch, H.: Time reversal and stationary Gibbs measures. Stoch. Process. Appl. **17**(1), 159–166 (1984)
142. Kurka, P.: Topological dynamics of one-dimensional cellular automata. Technical report
143. Lancia, C., Scoppola, B.: Equilibrium and Non-equilibrium ising models by means of PCA. J. Stat. Phys. **153**(4), 641–653 (2013)
144. Le Ny, A.: (Non-) Gibbs Description of Mean-field Models, vol. 60, chap. 21, pp. 463–480. Birkhäuser (2008)
145. Lebowitz, J.L.: Emergent Phenomena. Physik J. **6**(8/9) (2007)
146. Lebowitz, J.L., Maes, C., Speer, E.R.: Statistical mechanics of probabilistic cellular automata. J. Stat. Phys. **59**(1–2), 117–170 (1990)
147. Lebowitz, J.L., Orlandi, E., Presutti, E.: Convergence of stochastic cellular automation to Burgers' equation: fluctuations and stability. Physica D: Nonlinear Phenomena **33**(1–3), 165–188 (1988)
148. Levin, S.A., Durrett, R.: From individuals to epidemics. Philos. Trans. B **351**(1347), 1615–1621 (1996)
149. Liggett, T.M.: Stochastic models of interacting systems. Ann. Probab. **25**(1), 1–29 (1997)
150. Liggett, T.M.: Stochastic models for large interacting systems and related correlation inequalities. Proc. Natl. Acad. Sci. U.S.A. **107**, 16413–16419 (2010)
151. Louis, P.-Y.: Automates Cellulaires Probabilistes: mesures stationnaires, mesures de Gibbs associées et ergodicité. Ph.D. thesis, Politecnico di Milano, Italy and Université Lille 1, France (2002)

152. Louis, P.-Y.: Ergodicity of PCA: equivalence between spatial and temporal mixing conditions. Electron. Commun. Probab. **9**, 119–131 (2004)
153. de Maere, A., Ponselet, L.: Exponential decay of correlations for strongly coupled toom probabilistic cellular automata. J. Stat. Phys. **147**(3), 634–652 (2012)
154. Maes, C., Shlosman, S.B.: Ergodicity of probabilistic cellular automata: a constructive criterion. Commun. Math. Phys. **135**(2), 233–251 (1991)
155. Mairesse, J., Marcovici, I.: Around probabilistic cellular automata. Theor. Comput, Sci. (2014)
156. Mairesse, J., Marcovici, I.: Probabilistic cellular automata and random fields with i.i.d. directions. Annales de l'Institut Henri Poincaré, Probabilités et Statistiques **50**(2), 455–475 (2014)
157. Manchanda, K., Yadav, A.C., Ramaswamy, R.: Scaling behavior in probabilistic neuronal cellular automata. Phys. Rev. E **87**(1), 12,704 (2013)
158. Manneville, P., Boccara, N., Vichniac, G., Bidaux, R.: Cellular Automata and the Modeling of Complex Physical Systems. Springer, Berlin (1989)
159. Margolus, N., Toffoli, T.: Cellular Automata Machines: A New Environment for Modeling, vol. 1. MIT Press (1987)
160. Martin, O.C., Monasson, R., Zecchina, R.: Statistical mechanics methods and phase transitions in optimization problems. Theor. Comput. Sci. **265**(1–2), 3–67 (2001)
161. Martínez, G.J., Seck-Tuoh Mora, J.C., Zenil, H.: Wolfram's classification and computation in cellular automata classes III and IV. In: Zenil, H. (ed.) Irreducibility and Computational Equivalence, vol. 2, pp. 237–259. Springer, Berlin (2013)
162. Bousquet Mélou, M.: New enumerative results on two-dimensional directed animals. Discret. Math. **180**(1–3), 73–106 (1998)
163. Mendonça, J.: Sensitivity to noise and ergodicity of an assembly line of cellular automata that classifies density. Phys. Rev. E **83**(3) (2011)
164. Mendonça, J.R.G.: A Monte Carlo investigation of the critical behavior of Stavskaya's probabilistic cellular automaton. Phys. Rev. E **83**(1), 18–21 (2011)
165. Mendonça, J.R.G.: The inactive-active phase transition in the noisy additive (exclusive-or) probabilistic cellular automaton. Int. J. Mod. Phys. C **27**(2) (2016)
166. de Menibus, B.H., Sablik, M.: Self-organization in cellular automata: a particle-based approach. In: G. Mauri, A. Leporati (eds.) Developments in Language Theory. Lecture Notes in Computer Sciences, vol. 6795, pp. 251–263. Springer, Berlin (2011)
167. Merks, R.M.H., Perryn, E.D., Shirinifard, A., Glazier, J.a.: Contact-inhibited chemotaxis in de novo and sprouting blood-vessel growth. PLoS Comput. Biol. **4**(9), e1000, 163 (2008)
168. Mikler, A.R., Venkatachalam, S., Abbas, K.: Modeling infectious diseases using global stochastic cellular automata. J. Biol. Syst. **13**(4), 421–439 (2005)
169. Molofsky, J., Bever, J.D.: A New Kind of Ecology? BioScience **54**(5), 440 (2004)
170. Monetti, R.A., Albano, E.V.: On the emergence of large-scale complex behavior in the dynamics of a society of living individuals: the stochastic game of life. J. Theor. Biol. **187**(2), 183–194 (1997)
171. N., W., A., R.: The mathematical formulation of the problem of conduction of impulses in a network of connected excitable elements, specifically in cardiac muscle. Arch Inst Cardiol Mex (1946)
172. Nardi, F.R., Olivieri, E., Scoppola, E.: Anisotropy effects in nucleation for conservative dynamics. J. Stat. Phys. **119**(3), 539–595 (2005)
173. Nardi, F.R., Spitoni, C.: Sharp asymptotics for stochastic dynamics with parallel updating rule with self-interaction. J. Stat. Phys **4**(146), 701–718 (2012)
174. von Neumann, J.: The theory of self-reproducing automata. In: Burks, A.W. (ed.) The Theory of Self-Reproducing Automata. University of Illinois Press, Urbana (1966)
175. Ódor, G., Szolnoki, A.: Directed-percolation conjecture for cellular automata. Phys. Rev. E **53**, 2231–2238 (1996)
176. Olivieri, E., Vares, M.E.: Large deviations and metastability, Encyclopedia of Mathematics and its Applications, vol. 100. Cambridge University Press, Cambridge (2005)
177. Penington, C., Hughes, B., Landman, K.: Building macroscale models from microscale probabilistic models: A general probabilistic approach for nonlinear diffusion and multispecies phenomena. Phys. Rev. E **84**(4), 41,120 (2011)

178. Ponselet, L.: Phase transitions in probabilistic cellular automata. Ph.D. thesis, Université catholique de Louvain (2013)
179. Procacci, A., Scoppola, B., Scoppola, E.: Probabilistic Cellular Automata for low temperature Ising model. J. Stat. Phys. **165**, 991–1005 (2016)
180. Rajewsky, N., Schreckenberg, M.: A probabilistic cellular automaton for evolution. J. Phys. **5**(9), 1129–1134 (1995)
181. Regnault, D.: Directed percolation arising in stochastic cellular automata analysis. In: Ochmanski, E., Tyszkiewicz, J. (eds.) Mathematical Foundations of Computer Science 2008. Lecture Notes in Computer Science, vol. 5162, pp. 563–574. Springer, Berlin (2008)
182. del Rey, A.M.: A computer virus spread model based on cellular automata on graphs. In: Omatu, S., Rocha, M.P., Bravo, J., Fernández, F., Corchado, E., Bustillo, A., Corchado, J.M. (eds.) LNCS : Distributed Computing, Artificial Intelligence, Bioinformatics, Soft Computing, and Ambient Assisted Living, vol. 5518, pp. 503–506. Springer, Berlin (2009)
183. Rouquier, J.-B.: Robustesse et émergence dans les systèmes complexes: le modèle des automates cellulaires. Ph.D. thesis, ÉNS Lyon (2008)
184. Rouquier, J.-B., Morvan, M.: Coalescing cellular automata: synchronization by common random source for asynchronous updating. J. Cell. Automata **4**, 55–78 (2009)
185. Sarkar, P.: A brief history of cellular automata. ACM Comput. Surv. **32**(1), 80–107 (2000)
186. Scalise, D., Schulman, R.: Emulating cellular automata in chemical reaction-diffusion networks. Nat. Comput. **15**(2), 197–214 (2016)
187. Schiff, J.L.: Cellular Automata: A Discrete View of the World. Wiley (2012)
188. Schneckenreither, G., Popper, N., Zauner, G., Breitenecker, F.: Modelling SIR-type epidemics by ODEs, PDEs, difference equations and cellular automata - A comparative study. Simul. Model. Pract. Theory **16**(8), 1014–1023 (2008)
189. Schnell, S., Grima, R., Maini, P.: Multiscale modeling in biology. Am. Sci. **95**, 134–142 (2007)
190. Schonmann, H.R., Shlosman, B.S.: Wulff droplets and the metastable relaxation of kinetic isingmodels. Commun. Math. Phys. **194**(2), 389–462 (1998)
191. Shnirman, M.: On the problem of ergodicity of a Markov chain with infinite set of states. Probl. Kibern. **20**, 115–124 (1968)
192. Simpson, M.J., Merrifield, A., Landman, K.A., Hughes, B.D.: Simulating invasion with cellular automata: connecting cell-scale and population-scale properties. Phys. Rev. E **76**(2), 021918 (2007)
193. Sipper, M.: Simple + parallel + local = Cellular computing. In: M. Schwefel, A. Eiben, E. Bäck, T. Schoenauer (ed.) Fifth International Conference on Parallel Problem Solving from Nature (PPSN V), vol. 1498, Lecture Notes in Computer Science, pp. 653–662. Springer, Heidelberg (1998)
194. Slimi, R., El Yacoubi, S.: Spreadable Probabilistic Cellular Automata model: an application in epidemiology. In: ACRI 2006. Lecture Notes in Computer Science, vol. 4173, pp. 330–336. Springer, Berlin (2006)
195. Słowiński, P., MacKay, R.S.: Phase diagrams of majority voter probabilistic cellular automata. J. Stat. Phys. **159**(1), 43–61 (2015)
196. Spitzer, F.: Interaction of Markov processes. Adv. Math. **5**, 246–290 (1970)
197. Stavskaya, O.N., Piatetsky shapiro, G.: Homogeneous networks of spontaneously active elements. Problemy Kibernet **20**, 91–106 (1968)
198. Sutner, K.: Computational classification of cellular automata. Int. J. Gen. Syst. **41**(6), 595–607 (2012)
199. Szabó, A., Merks, R.M.H.: Cellular potts modeling of tumor growth, tumor invasion, and tumor evolution. Front. Oncol. **3**, 87 (2013)
200. Taggi, L.: Critical probabilities and convergence time of percolation probabilistic cellular automata. J. Stat. Phys. **159**(4), 853–892 (2015)
201. Toffoli, T.: Cellular automata as an alternative to (rather than an approximation of) differential equations in modeling physics. Phys. D **10**, 117–127 (1984)

202. Tomé, T., de Felício, J.R.D.: Probabilistic cellular automaton describing a biological immune system. Phys. Rev. E **53**(4), 3976–3981 (1996)
203. Toom, A.: Nonergodic multidimensional systems of automata. Probl. Inf. Trans. **10**, 239–246 (1974)
204. Toom, A.: Multicomponent Random Systems, chap. Stable and Attractive Trajectories in Multicomponent Systems, pp. 549–575. Marcel Dekker Inc (1980)
205. Toom, A.: Algorithmical unsolvability of the ergodicity problem for binary cellular automata. Markov process. Related Fields **6**(4), 569–577 (2000)
206. Toom, A.L., Vasilyev, N.B., Stavskaya, O.N., Mityushin, L.G., Kurdyumov, G.L., Pirogov, S.A.: Locally interacting systems and their application in biology. In: Dobrushin, Kryukov, Toom (eds.) Stochastic Cellular Systems: Ergodicity, Memory, Morphogenesis, pp. 1–182. Springer, Berlin (1978)
207. Tristan, J.B., Zaheer, M., Steel, G.L.J., Green, S.J.: Learning topics with stochastic cellular automata dynamics. In: Neural Information Processing Systems (2015)
208. Ulam, S.: Random processes and transformations. In: Proceedings of the International Congress of Mathematicians, Cambridge, Mass., 1950, pp. 264–275. American Mathematical Society, Providence, R. I. (1952)
209. Vaserstein, L.N.: Markov processes over denumerable products of spaces describing large system of automata. Problemy Peredači Informacii **5**(3), 64–72 (1969)
210. Vaserstein, L.N., Leontovich, A.M.: Invariant measures of certain Markov operators that describe a homogeneous random medium. Zap. Nauchn. Sem. S.-Peterburg. Otdel. Mat. Inst. Steklov. (POMI) **6**(1), 71–80 (1970)
211. Vichniac, G.Y.: Simulating physics with cellular automata. Phys. D **10**, 96–116 (1984)
212. Werner, W.: Percolation et modèle d'Ising. Collection SMF. Société Mathématique de France, Paris (2009)
213. White, S.H., Del Rey, A.M., Sanchez, G.R.: Using cellular automata to simulate epidemic diseases. Appl. Math. Sci. **3**, 959–968 (2009)
214. Wolfram, S.: Statistical mechanics of cellular automata. Rev. Mod. Phys. **55**(3), 601–644 (1983)
215. Wolfram, S.: Cellular automata as models of complexity. Nature **311**(5985), 419–424 (1984)
216. Wolfram, S.: Computation theory of cellular automata. Commun. Math. Phys. **96**(1), 15–57 (1984)
217. Worsch, T.: Cellular automata as models of parallel computation. In: Computational Complexity Theory, Techniques, and Applications, pp. 298–311 (2012)
218. Wu, A.C., Xu, X.J., Wang, Y.H.: Excitable greenberg-hastings cellular automaton model on scale-free networks. Phys. Rev. E **75**, 032,901 (2007)
219. Xiao, X., Shao, S.H., Chou, K.C.: A probability cellular automaton model for hepatitis B viral infections. Biochem. Biophys. Res. Commun. **342**(2), 605–610 (2006)
220. Zamith, M., Leal-Toledo, R.C.P., Clua, E., Toledo, E.M., Magales, G.V.d.: A new stochastic cellular automata model for traffic flow simulation with drivers' behavior prediction. J. Comput. Sci. **9**, 51–56 (2015)
221. Zinck, R.D., Johst, K., Grimm, V.: Wildfire, landscape diversity and the Drossel-Schwabl model. Ecol. Model. **221**(1), 98–105 (2010)
222. Zorzenon dos Santos, R., Coutinho, S.: Dynamics of HIV infection: a cellular automata approach. Phys. Rev. Lett. **87**(16), 168,102 (2001)
223. Zuse, K.: The computing universe. Int. J. Theor. Phys. **21**(6–7), 589–600 (1982)

Chapter 2
Probabilistic Cellular Automata in the Visual Arts

Roeland M.H. Merks

In January 1970, computer scientist Leo Geurts walked into Swart Gallery in Amsterdam, The Netherlands, to see the solo exhibition by Dutch artist Peter Struycken (The Netherlands, 1939). He was struck by Struycken's black and white works "Computerstructuren" (1969), which were painted after grid patterns generated by algorithms. Geurts assumed that they must have been produced using cellular automata. He started working with Lambert Meertens at *Mathematisch Centrum* (now CWI) in Amsterdam to make a similar work. This led to what is possibly the first example of the use of probabilistic cellular automata (PCA), entitled *Kristalstructuren* [4, 5, 7]. Struycken did not know both scientists, but "their assumptions about [my] algorithm were too highly fetched. As non-mathematician, I had thought up a much less advanced algorithm for producing my paintings. Their approach made their results more elegant and varied" [18].

The work by Geurts and Meertens was produced using variants of the majority voting rule with asynchronous updating (also related to the Ising spin model (see Chap. 11) and to Potts models (see Chap. 8)). Each lattice site $\mathbf{x}$ had one of two states, $\sigma(\mathbf{x}) \in \{\text{black, white}\}$. The new state was either the majority state (Fig. 2.1c) or the opposite of the majority state (Fig. 2.1d) in the Von Neumann neighborhood $\{\mathbf{x} \pm (1, 0), \mathbf{x} \pm (0, 1)\}$, or, alternatively, in the neighborhood formed by the four diagonal neighbors $\{\mathbf{x} \pm (1, 1), \mathbf{x} \pm (1, -1)\}$. The simulations were initialized with random configurations of $P(\sigma = \text{black}) = 1/2$ and they were updated until the patterns stabilized or entered into oscillation. By means of boundary conditions, the

R.M.H. Merks (✉)
Centrum Wiskunde & Informatica,
Science Park 123, 1098 XG Amsterdam, The Netherlands
e-mail: merks@cwi.nl

R.M.H. Merks
Mathematical Institute, Leiden University,
Niels Bohrweg 1, 2333 CA Leiden, The Netherlands

P.-Y. Louis and F.R. Nardi (eds.), *Probabilistic Cellular Automata*, Emergence, Complexity and Computation 27, https://doi.org/10.1007/978-3-319-65558-1_2

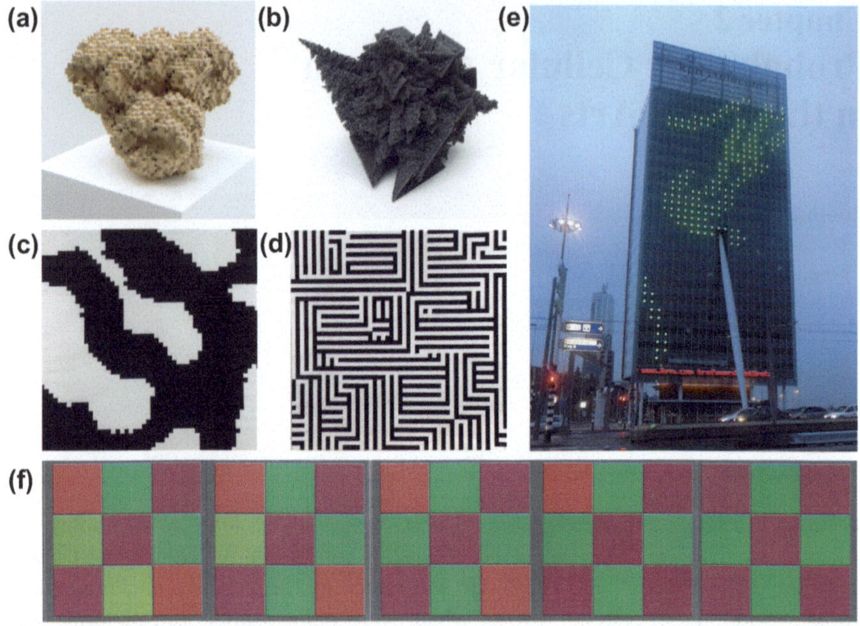

Fig. 2.1 Examples of probabilistic and deterministic cellular automata in the visual arts. **a** *Breed 0.1 #1*, Driessens and Verstappen, 1995; **b** *Accretor #2777-4*, Driessens and Verstappen, 2013, courtesy DAM gallery Berlin; **c,d** *Kristalstructuren* (1970) Geurts and Meertens. **c** Voting rule with Von Neumann neighborhood; **d** opposite of voting rule with Von Neumann neighborhood; Swart Gallery, Amsterdam; **e** *Pixelsex* (2005) courtesy Tim Otto Roth. **f** SPLASH 1972/1974 (1972–1974) Peter Struycken. Stage 24–28 in a series of 28; scan of leporello [16]

boundary rows and columns were initialized randomly like the rest of the lattice and left unchanged during the simulations [5].

Although *Kristalstructuren* found its way into the art market via *Swart Gallery*, this artistic work was a side-branch of Geurts' and Meertens' main line of work in computer science; apart from a few follow-up projects, including a bag for the Dutch mobile supermarket chain "SRV" they concentrated on their scientific work. Struycken's own first use of CA-like algorithms were in his works SPLASH 1972/1974 (Fig. 2.1f) [2, 16, 17]—in which color patterns evolved from an initial pattern towards a preset, final pattern—and later in FIELDS 1979/1980 [14].

Given the attractive patterns that cellular automata can produce [1] and the conceptual interest in the use of algorithms for art, it is perhaps not surprising that other visual artists have also applied cellular automata in their work. Page ix shown an example, entitled *Breed 1.2 #e365* (2007) by the Dutch artist duo Driessens and Verstappen (The Netherlands, 1963, 1964). *Breed* are a series of plywood and 3D printed sculptures (Fig. 2.1a). Not based on PCA in the strict sense of the word, these sculptures were generated by three-dimensional, recursive CA-like rules, generated at random using evolutionary algorithms. To simulate cell division, the 3D lattice was refined after each iteration. Their later series *Accretor* (Fig. 2.1b) are sculptures generated using multi-material 3D printing. Using a three-dimensional accretive growth model [19], similar to a deterministic version of the Eden growth rule [3], randomly selected, deterministic CA-rules determine at which surface positions new particles are added.

In the art project *Pixelsex* by Tim Otto Roth [11] (Germany, 1974), simulations of probabilistic cellular automata were displayed on Renzo Piano's KPN Telecom Tower in Rotterdam in 2005 and 2006 [10]. These PCA have biological application: they are a simulation of the collective behavior of self-propelled myxobacteria ("slime bacteria") [15] using the Cellular Potts model (Chap. 8). Further contemporary professional artists who have used probabilistic cellular automata, include Paul Brown (UK, 1947) and John F. Simon Jr. (USA, 1963); also see Ref. [6, 14].

Despite the attractive patterns they produce, for conceptual reasons many artists are hesitant with respect to the stochasticity of PCA. Driessens and Verstappen deliberately apply deterministic CA-rules, using randomness only for generating initial conditions or sets of deterministic rules. For them the challenge is to 'breed' complex shapes using entirely deterministic rules: "The use of stochasticity in a generative process is a 'trick' that is often used to make the system look more lively" [20]. Tim Otto Roth shares this artistic viewpoint on generative art, contrasting deterministic CA with earlier probability-based art: "I like the contrast that these [deterministic] CA are emergent dynamical systems, but not accidental at all." [12] However, he adds that his "CA based performances with people are in a certain way probabilistic as the actors cannot behave perfectly." [13]

Hopefully this book will help to show that, despite their "accidental" nature, probabilistic cellular automata are more than a 'trick'. In statistical models of natural systems, the probabilistic rules capture the stochastic fluctuations that are a key component of living systems [9] and of many non-living systems [8]. They can drive 'accidental' behavior in some cases, and practically deterministic behavior in others.

References

1. Adamatzky, A., Martnez, G.J. (eds.): Designing Beauty: The Art of Cellular Automata. Emergence, Complexity and Computation. Springer, Berlin (2016)
2. Dietrich, F.: Visual intelligence: The first decade of computer art (1965-1975). IEEE Comput. Gr. Appl. **5**(7), 33–45 (1985). https://doi.org/doi.ieeecomputersociety.org/10.1109/MCG.1985.276440
3. Eden, M.: A Two-Dimensional Growth Process. Proc. Fourth Berkeley Symp. Math. Stat. Probab. **4**, 223–239 (1961)
4. Geurts, L., Meertens, L.: Crystallization. In: Eighth annual computer art contest. Comput. Autom. **19**(8), 13–24 (1970)
5. Geurts, L.J.M.: Kristalstrukturen, een experiment in computer-kunst (in Dutch). In: Vakantiecursus Abstracte Informatica, pp. VI–1 – VI–21 (1973)
6. Javid, M., al Rifaie, M.M., Zimmer, R.: An informational model for cellular automata aesthetic measure. In: Proceedings of AISB 2015's second international symposium on computational creativity, pp. 9–15 (2015)

7. Meertens, L., Geurts, L.: Kristalstructuren (crystal structures) (1970) In: V & A Collection. E.81-2008. http://collections.vam.ac.uk/item/O158415/kristalstructuren-crystal-structures-print-meertens-lambert
8. Newman, M.E.J., Barkema, G.T.: Monte Carlo Methods in Statistical Physics. Clarendon Press, Oxford (1999)
9. Noble, D.: Dance to the Tune of Life. Biological Relativity. Cambridge University Press, Cambridge (2016)
10. Roth, T.O., Deutsch, A.: Universal Synthesizer and Window: Cellular Automata as a New Kind of Cybernetic Image. In: O. Grau, T. Veigl (eds.) Imagery in the 21st Century, pp. 269–288. Cambridge/Mass (MIT Press) (2011)
11. Roth, T.O.: http://pixelsex.org (Accessed on May 29th, 2017)
12. Roth, T.O.: Presentation at CWI, October 3rd, 2016: Personal communication by e-mail, May 17th, 2017
13. Roth, T.O.: Personal communication by e-mail, June 2nd, 2017
14. Scha, R.: Kunstmatige kunst (in dutch). De Connectie **2**(1), 4–7 (2006)
15. Starruß, J., Bley, T., Søgaard-Andersen, L., Deutsch, A.: A new mechanism for collective migration in Myxococcus xanthus. J. Stat. Phys. **128**(1–2), 269–286 (2007)
16. Struycken, P.: Plons 1972/1974; Splash 1972/1974 Verfindustrie Jac Eyck bv, Heerlen, Holland, and Sikkens Service Centrum voor Limburg with preface by Carel Blotkamp (in Dutch). Includes leporello of 28 colour folds
17. Struycken, P.: Splash 1972/1974. In: R. Leavitt (ed.) Artist and Computer, pp. 30–31. Harmony Books (1976)
18. Struycken, P.: Personal communication by e-mail, May 17th, 2017
19. Whitelaw, M.: Accretor: Generative materiality in the work of driessens and verstappen. Artif. Life **21**, 307–312 (2015). https://doi.org/10.1162/ARTL_a_00171
20. Verstappen, M.: Personal communication by e-mail, May 17th, 2017

Part I
Probability and Statistical Mechanics

Chapter 3
Basic Ideas to Approach Metastability in Probabilistic Cellular Automata

Emilio N.M. Cirillo, Francesca R. Nardi and Cristian Spitoni

Abstract Cellular Automata are discrete-time dynamical systems on a spatially extended discrete space, which provide paradigmatic examples of nonlinear phenomena. Their stochastic generalizations, i.e., Probabilistic Cellular Automata, are discrete-time Markov chains on lattice with finite single-cell states whose distinguishing feature is the *parallel* character of the updating rule. We review the some of the results obtained about the metastable behavior of Probabilistic Cellular Automata, and we try to point out difficulties and peculiarities with respect to standard Statistical Mechanics Lattice models.

3.1 Introduction

Cellular Automata are discrete-time dynamical systems on a spatially extended discrete space. They are well known for being easy to implement and for exhibiting a rich and complex nonlinear behavior as emphasized for instance in [22] for Cel-

E.N.M. Cirillo
Dipartimento di Scienze di Base e Applicate per l'Ingegneria,
Università di Roma: la Sapienza, via A. Scarpa 16, 00161 Roma, Italy
e-mail: emilio.cirillo@uniroma1.it

F.R. Nardi
Department of Mathematics and Computer Science, Eindhoven University of Technology, P.O. Box 513, 5600 MB Eindhoven, The Netherlands
e-mail: F.R.Nardi@tue.nl

F.R. Nardi
Eurandom, P.O. Box 513, 5600 MB Eindhoven, The Netherlands

F.R. Nardi
Dipartmento di Matematica e Informatica, Universitá di Firenze, Firenze, Italy

C. Spitoni (✉)
Institute of Mathematics, University of Utrecht, Budapestlaan 6, 3584 CD Utrecht, The Netherlands
e-mail: C.Spitoni@uu.nl

P.-Y. Louis and F.R. Nardi (eds.), *Probabilistic Cellular Automata*, Emergence, Complexity and Computation 27, https://doi.org/10.1007/978-3-319-65558-1_3

lular Automata on one-dimensional lattice. For the general theory of deterministic Cellular Automata, we refer to the recent paper [12] and references therein

Probabilistic Cellular Automata (PCA) are Cellular Automata straightforward generalization where the updating rule is stochastic. They are used as models in a wide range of applications. From a theoretic perspective, the main challenges concern the nonergodicity of these dynamics for an infinite collection of interacting cells.

Strong relations exist between PCA and the general equilibrium statistical mechanics framework [14, 21]. Important issues are related to the interplay between disordered global states and ordered phases (*emergence of organized global states, phase transition*) [19]. Although PCA initial interest arose in the framework of Statistical Physics, in the recent literature many different applications of PCA have been proposed. In particular, it is notable to remark that a natural context in which the PCA main ideas are of interest is that of evolutionary games [20].

In this paper, we shall consider a particular class of PCA, called *reversible* PCA, which are reversible with respect to a Gibbs-like measure defined via a translation invariant multi-body potential. In this framework, we shall pose the problem of metastability and show its peculiarities in the PCA world.

Metastable states are ubiquitous in nature and are characterized by the following phenomenological properties: (i) The system exhibits a single phase different from the equilibrium predicted by thermodynamics. The system obeys the usual laws of thermodynamics if small variations of the thermodynamical parameters (pressure, temperature, ...) are considered. (ii) If the system is isolated, the equilibrium state is reached after a very large random time; the lifetime of the metastable state is practically infinite. The exit from the metastable state can be made easier by forcing the appearance large fluctuations of the stable state (droplets of liquid inside the super-cooled vapor, ...). (iii) The exit from the metastable phase is irreversible.

The problem of the rigorous mathematical description of metastable states has long history which started in the 70s, blew up in the 90s, and is still an important topic of mathematical literature. Different theories have been proposed and developed, and the pertaining literature is huge. We refer the interested reader to the monograph [18]. In this paper, we shall focus on the study of metastability in the framework of PCA.

In [1, 5, 8, 9, 16], the metastable behavior of a certain class of reversible PCA has been analyzed. In this framework, it has been pointed out the remarkable interest of a particular reversible PCA (see Sect. 3.3) characterized by the fact that the updating rule of a cell depends on the status of the five cells forming a cross centered at the cell itself. In this model, the future state of the spin at a given cell depends also on the present value of such a spin. This effect will be called *self-interaction*, and its weight in the updating rule will be called *self-interaction intensity*.

The paper is organized as follows. In Sect. 3.2, we introduce reversible Probabilistic Cellular Automata and discuss some general properties. In Sect. 3.3, we introduce the model that will be studied in this paper, namely the nearest neighbor and the cross PCA, and discuss its Hamiltonian. In Sect. 3.4, we pose the problem of metastability in the framework of Probabilistic Cellular Automata and describe the main ingredients that are necessary for a full description of this phenomenon. In Sect. 3.5, we finally state our results.

3.2 Reversible Probabilistic Cellular Automata

We shall first briefly recall the definition of Probabilistic Cellular Automata and then introduce the so-called *Reversible Probabilistic Cellular Automata.*

Let $\Lambda \subset \mathbb{Z}^d$ be a finite cube with periodic boundary conditions. Associate with each site $i \in \Lambda$ (also called *cell*) the state variable $\sigma_i \in X_0$, where X_0 is a finite single-site space and denote by $X := X_0^\Lambda$ the *state space*. Any $\sigma \in X$ is called a *state* or *configuration* of the system.

We introduce the shift Θ_i on the torus, for any $i \in \Lambda$, defined as the map $\Theta_i : X \to X$ shifting a configuration in X so that the site i is mapped onto the origin 0, more precisely such that (see Fig. 3.1)

$$(\Theta_i \sigma)_j = \sigma_{i+j}. \tag{3.1}$$

The configuration σ at site j shifted by i is equal to the configuration at site $i + j$. For example, (see Fig. 3.1) set $j = 0$, then the value of the spin at the origin 0 will be mapped onto site i.

We consider a probability distribution $f_\sigma : X_0 \to [0, 1]$ depending on the state σ restricted to $I \subset \Lambda$. A Probabilistic Cellular Automata are the Markov chain $\sigma(0), \sigma(1), \ldots, \sigma(t)$ on X with transition matrix

$$p(\sigma, \eta) = \prod_{i \in \Lambda} f_{\Theta_i \sigma}(\eta_i) \tag{3.2}$$

for $\sigma, \eta \in X$. We remark that f depends on $\Theta_i \sigma$ only via the neighborhood $i + I$. Note that the character of the evolution is local and parallel: The probability that the spin at the site i assumes at time $t + 1$ the value $s \in X_0$ depends on the value of the state variables at time t (parallel evolution) associated only with the sites in $i + I$ (locality).

A class of *reversible* PCA can be obtained by choosing $X = \{-1, +1\}^\Lambda$, and probability distribution

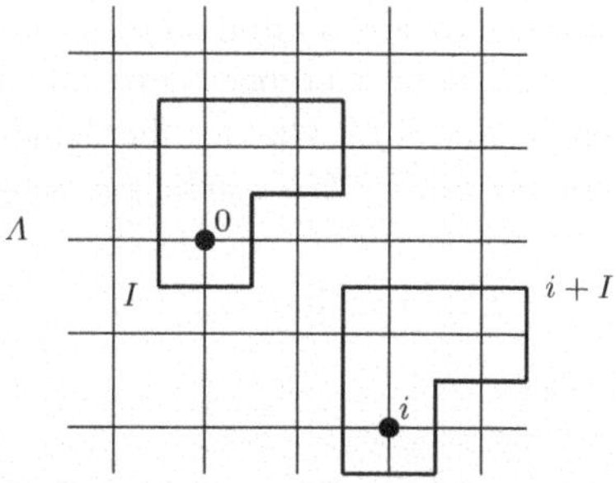

Fig. 3.1 Schematic representation of the action of the shift Θ_i defined in (3.1)

$$f_\sigma(s) = \frac{1}{2}\Big\{1 + s\tanh\Big[\beta\Big(\sum_{j\in\Lambda} k(j)\sigma_j + h\Big)\Big]\Big\} \tag{3.3}$$

for all $s \in \{-1, +1\}$ where $T \equiv 1/\beta > 0$ and $h \in \mathbb{R}$ are called *temperature* and *magnetic field*. The function $k : \mathbb{Z}^2 \to \mathbb{R}$ is such that its support[1] is a subset of Λ and $k(j) = k(j')$ whenever $j, j' \in \Lambda$ are symmetric with respect to the origin. With the notation introduced above, the set I is the support of the function k. We shall denote by $p_{\beta,h}$ the corresponding transition matrix defined by (3.2).

Recall that Λ is a finite torus, namely periodic boundary conditions are considered throughout this paper. It is not difficult to prove [10, 13] that the above-specified PCA dynamics is reversible with respect to the finite-volume Gibbs-like measure

$$\mu_{\beta,h}(\sigma) = \frac{1}{Z_{\beta,h}} e^{-\beta G_{\beta,h}(\sigma)} \tag{3.4}$$

with *Hamiltonian*

$$G_{\beta,h}(\sigma) = -h\sum_{i\in\Lambda}\sigma_i - \frac{1}{\beta}\sum_{i\in\Lambda}\log\cosh\Big[\beta\Big(\sum_{j\in\Lambda}k(j-i)\sigma_j + h\Big)\Big] \tag{3.5}$$

and *partition function* $Z_{\beta,h} = \sum_{\eta\in X}\exp\{-\beta G_{\beta,h}(\eta)\}$. In other words, in this case the detailed balance equation

$$p_{\beta,h}(\sigma,\eta)e^{-\beta G_{\beta,h}(\sigma)} = e^{-\beta G_{\beta,h}(\eta)}p_{\beta,h}(\eta,\sigma) \tag{3.6}$$

is satisfied thus the probability measure $\mu_{\beta,h}$ is stationary for the PCA.

Note that different reversible PCA models can be specified by choosing different functions k. In particular, the support I of such a function can be varied. In the next section, we shall introduce two common choices, the *nearest neighbor PCA* [5] obtained by choosing the support of k as the set of the four sites neighboring the origin and the *cross PCA* [9] obtained by choosing the support of k as the set made of the origin and its four neighboring sites (see Fig. 3.2).

The stationary measure $\mu_{\beta,h}$ introduced above looks like a finite-volume Gibbs measure with Hamiltonian $G_{\beta,h}(\sigma)$ (see (3.5)). It is worth noting that $G_{\beta,h}$ cannot be thought as a proper statistical mechanics Hamiltonian since it depends on the temperature $1/\beta$. On the other hand, the low-temperature behavior of the stationary measure of the PCA can be guessed by looking at the *energy* function

$$H_h(\sigma) = \lim_{\beta\to\infty} G_{\beta,h}(\sigma) = -h\sum_{i\in\Lambda}\sigma_i - \sum_{i\in\Lambda}\Big|\sum_{j\in\Lambda}k(j-i)\sigma_j + h\Big| \tag{3.7}$$

[1]Recall that, by definition, the support of the function k is the subset of Λ where the function k is different from zero.

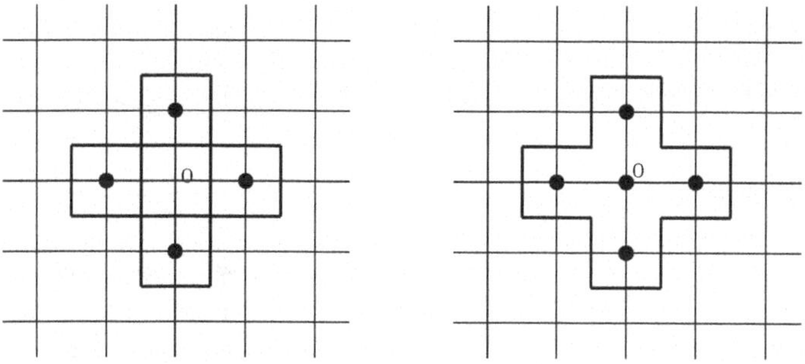

Fig. 3.2 Schematic representation of the nearest neighbor (*left*) and cross (*right*) models

The absolute minima of the function H_h are called *ground states* of the stationary measure for the reversible PCA.

3.3 The Tuned Cross PCA

We consider, now, a particular example of reversible PCA. More precisely, we set $k(j) = 0$ if j is neither the origin nor one of its nearest neighbors, i.e., it is not in the five-site cross centered at the origin, $k(0) = \kappa \in [0, 1]$, and $k(j) = 1$ if j is one of the four nearest neighbors of the origin; we shall denote by J the set of nearest neighbors of the origin. With such a choice, we have that

$$f_\sigma(s) = \frac{1}{2}\Big\{1 + s\tanh\Big[\beta\Big(\kappa\sigma_0 + \sum_{j\in J}\sigma_j + h\Big)\Big]\Big\} = \frac{1}{1 + e^{-2\beta s(\kappa\sigma_0 + \sum_{j\in J}\sigma_j + h)}} \tag{3.8}$$

We shall call this model the *tuned cross* PCA. The *self-interaction intensity* κ tunes between the *nearest neighbor* ($\kappa = 0$) and the *cross* ($\kappa = 1$) PCA.

Note that for this model, the Hamiltonian $G_{\beta,h}$ defining the stationary Gibbs-like measure is given by

$$G_{\beta,h}(\sigma) = -h\sum_{i\in\Lambda}\sigma_i - \frac{1}{\beta, h}\sum_{i\in\Lambda}\log\cosh\Big[\beta\Big(\kappa\sigma_i + \sum_{j\in i+J}\sigma_j + h\Big)\Big] \tag{3.9}$$

while the corresponding energy function, see (3.7), is

$$H_h(\sigma) = -h\sum_{i\in\Lambda}\sigma_i - \sum_{i\in\Lambda}\Big|\kappa\sigma_i + \sum_{j\in i+J}\sigma_j + h\Big| \tag{3.10}$$

In Statistical Mechanics Lattice systems, the energy of a configuration is usually written in terms of coupling constants. We could write the expansion of the energy

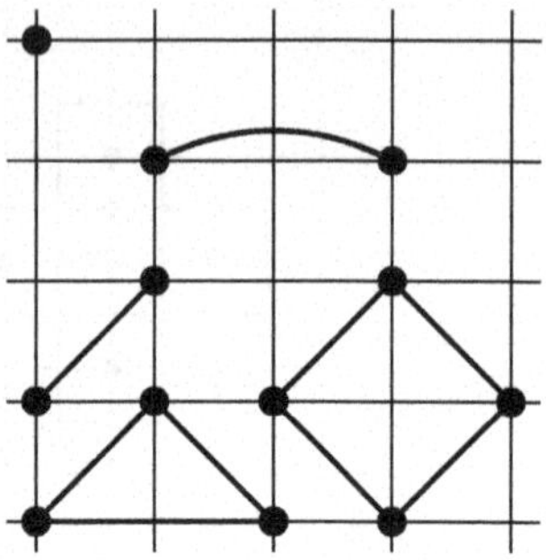

Fig. 3.3 Schematic representation of the coupling constants: from the left to the right and from the top to the bottom the couplings $J_{\cdot}$, $J_{\langle\langle\langle\rangle\rangle\rangle}$, $J_{\langle\langle\rangle\rangle}$, J_{Δ}, and $J_{\diamond}$ are depicted

H_h in (3.10), but, for the sake of simplicity, we consider the nearest neighbor PCA [5], namely we set $\kappa = 0$. We get

$$H_h(\sigma) = -J_{\cdot} \sum_{x\in\Lambda} \sigma(x) - J_{\langle\langle\rangle\rangle} \sum_{\langle\langle xy\rangle\rangle} \sigma(x)\sigma(y) - J_{\langle\langle\langle\rangle\rangle\rangle} \sum_{\langle\langle\langle xy\rangle\rangle\rangle} \sigma(x)\sigma(y)$$
$$-J_{\Delta} \sum_{\Delta_{xyz}} \sigma(x)\sigma(y)\sigma(z) - J_{\diamond} \sum_{\diamond_{xywz}} \sigma(x)\sigma(y)\sigma(w)\sigma(z)$$

where the meaning of the symbols $\cdot$, $\langle\langle\rangle\rangle$, $\langle\langle\langle\rangle\rangle\rangle$ Δ, and $\diamond$ is illustrated in Fig. 3.3 and the corresponding coupling constants are

$$J_{\cdot} = \frac{5}{2}h,\ J_{\langle\langle\rangle\rangle} = 1 - \frac{1}{4}h,\ J_{\langle\langle\langle\rangle\rangle\rangle} = \frac{1}{2} - \frac{1}{8}h,\ J_{\Delta} = -\frac{1}{8}h,\ \text{and}\ J_{\diamond} = -\frac{1}{2} + \frac{3}{8}h$$

It is interesting to note that the coupling constant $J_{\diamond}$ is negative (antiferromagnetic coupling), and this will give a physical meaning to the appearance of checkerboard configurations in the study of metastability for the nearest neighbor PCA.

The coupling constants can be computed by using [4, Eqs. (6) and (7)] (see also [11, Eqs. (3.1) and (3.2)] and [7]). More precisely, given $f : \{-1, +1\}^V \to \mathbb{R}$, with $V \subset \mathbb{Z}^2$ finite, we have that for any $\sigma \in \{-1, +1\}^V$

$$f(\sigma) = \sum_{I\subset V} C_I \prod_{i\in I} \sigma_i \tag{3.11}$$

with the coefficients C_I's given by

$$C_I = \frac{1}{2^{|V|}} \sum_{\sigma\in\{-1,+1\}^V} f(\sigma) \prod_{i\in I} \sigma_i \tag{3.12}$$

We refer to [6] for the details. We note that in that paper, the couplings have been computed for a more general model than the one discussed here.

Now, we jump back to the tuned cross PCA and we discuss the structure of the ground states, that is to say, we study the global minima of the energy function H_h

given in (3.10). Such a function can be rewritten as

$$H_h(\sigma) = \sum_{i \in \Lambda} H_{h,i}(\sigma)$$

with

$$H_{h,i}(\sigma) = -\Big[\frac{1}{5}h\Big(\sigma_i + \sum_{j \in i+J} \sigma_j\Big) + \Big|\kappa\sigma_i + \sum_{j \in i+J} \sigma_j + h\Big|\Big] \qquad (3.13)$$

We also note that

$$H_h(\sigma) = H_{-h}(-\sigma) \qquad (3.14)$$

for any $h \in \mathbb{R}$ and $\sigma \in X$, where $-\sigma$ denotes the configuration obtained by flipping the sign of all the spins of σ. By (3.14), we can bound our discussion to the case $h \geq 0$ and deduce a posteriori the structure of the ground states for $h < 0$.

The natural candidates to be ground states are the following configurations: $\mathbf{u} \in X$ such that $\mathbf{u}(i) = +1$ for all $i \in \Lambda$, $\mathbf{d} \in X$ such that $\mathbf{d}(i) = +1$ for all $i \in \Lambda$, $\mathbf{c}_e$, and $\mathbf{c}_o$ with $\mathbf{c}_e$ the checkerboard configuration with pluses on the even sub-lattice of Λ and minuses on its complement, while $\mathbf{c}_o$ is the corresponding spin-flipped configuration. Indeed, we can prove that the structure of the zero-temperature phase diagram is that depicted in Fig. 3.4.

Case $h > 0$ *and* $k_0 \geq 0$. The minimum of $H_{h,i}$ is attained at the cross configuration having all the spins equal to plus one. Hence, the unique absolute minimum of H_h is the state $\mathbf{u}$.

Case $h = 0$ *and* $k_0 > 0$. The minimum of

$$H_{0,i}(\sigma) = -\Big|\kappa\sigma_i + \sum_{j \in i+J} \sigma_j\Big|$$

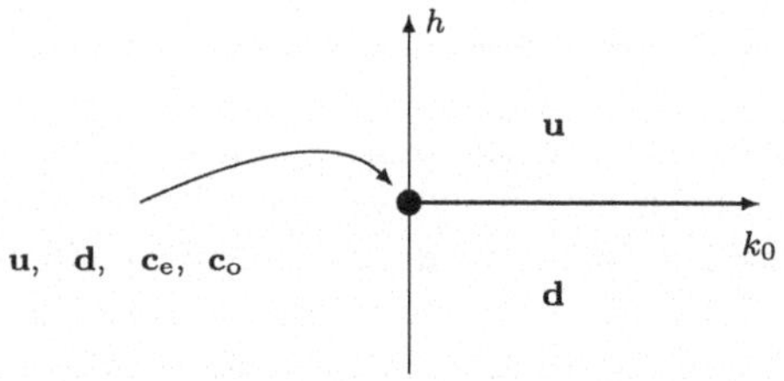

Fig. 3.4 Zero-temperature phase diagram of the stationary measure of the tuned cross PCA. On the *thick lines*, the ground states of the adjacent regions coexist. At the origin, the listed four ground states coexist

is attained at the cross configuration having all the spins equal to plus one or all equal to minus one. Hence, the set of ground states is made of the two configurations **u** and **d**.

Case $h = 0$ *and* $k_0 = 0$. The minimum of $H_{0,i}$ is attained at the cross configuration having all the spins equal to plus one or all equal to minus one on the neighbors of the center and with the spin at the center which can be, in any case, either plus or minus. Hence, the set of ground states is made of the four configurations **u**, **d**, $\mathbf{c}_e$, and $\mathbf{c}_o$.

Case $h < 0$. The set of ground states can be easily discussed as for $h > 0$ by using the property (3.14).

3.4 Main Ingredients for Metastability

At $\kappa > 0$, the zero-temperature phase diagram in Fig. 3.4 is very similar to that of the standard Ising model, which is the prototype for the description of phase transitions in Statistical Mechanics. So we expect that even in the case of the tuned cross PCA, the equilibrium behavior could be described as follows: (i) At positive magnetic field h, there exist a unique phase with positive magnetization[2]; (ii) the same it is true at negative h but with negative magnetization; (iii) at $h = 0$, the equilibrium behavior is more complicated: There exists a critical value of the temperature such that at temperatures larger than such a value there exists a unique phase with zero magnetization, while at temperatures smaller than the critical one there exists two equilibrium measures with opposite not zero magnetization, called the *residual magnetization*.

This scenario has proven to be true in the case of the two-dimensional standard Ising model, but in the context of the tuned cross PCA, the problem is much more difficult due to the complicated structure of the energy function (3.9). The validity of such a scenario has been checked via a mean-field computation in [6].

From now on, for technical reasons, we shall assume that the magnetic field satisfies the following conditions

$$0 < h < 4 \quad \text{and} \quad h \neq \kappa, 2 - \kappa, 2 + \kappa, 4 - \kappa, 4 + \kappa \tag{3.15}$$

Since $h > 0$, the equilibrium is characterized by positive magnetization. The question is: Is it possible to investigate the possibility of the existence of metastable states? In other words, is it possible to show that there exist not equilibrium phases in which the system is trapped in the sense described in the introduction (see Sect. 3.1)?

This question has a very long history: In some sense, it arose with the van der Waals theory of liquid–vapor transition and began to find some mathematically rigorous answer only in the 80s. We just quote [17] for the *pathwise approach* and [2] for

---

[2]By exploiting the translational invariance of the model, it is possible to define the magnetization as the mean value of the spin at the origin against the Gibbs-like equilibrium measure $\mu_{\beta,h}$.

the *potential theoretic* one, and we refer to [18] for the full story and for complete references.

According to the rigorous theories of metastability, the problem has to be approached from a dynamical point of view. Namely, we shall consider the evolution of the tuned cross PCA started at the initial configuration $\zeta \in X$ and study the random variable

$$\tau_{\mathbf{u}}^{\zeta} := \inf\{t > 0,\ \sigma(t) = \mathbf{u}\} \tag{3.16}$$

called the *first hitting time* to $\mathbf{u}$. The state ζ will be called metastable or not depending on the properties of the random variable $\tau_{\mathbf{u}}^{\zeta}$ in the zero-temperature limit[3] ($\beta \to \infty$). In the framework of different approaches to metastability, different definitions of metastable states have been given, but they are all related to the properties of the hitting time $\tau_{\mathbf{u}}^{\zeta}$. In particular, it has to happen that the mean value of $\tau_{\mathbf{u}}^{\zeta}$ has to be large, say diverging exponentially fast with $\beta \to \infty$.

As remarked above, for $h > 0$ small, natural candidates to be metastable states for the tuned cross PCA are the configurations $\mathbf{d}$, $\mathbf{c}_{\mathrm{e}}$, and $\mathbf{c}_{\mathrm{o}}$. But, imagine to start the chain at $\mathbf{d}$: Why should such a state be metastable? Why should the chain take a very long time to hit the "stable" state $\mathbf{u}$? The analogous question posed in the framework of the two-dimensional Ising model with Metropolis dynamics has an immediate qualitative answer: In order to reach $\mathbf{u}$ starting from $\mathbf{d}$, the system has to perform, spin by spin, a sequence of changes against the energy drift. Indeed, plus spins have to be created in the starting sea of minuses, and those transitions have a positive energy cost if the magnetic field is small enough, indeed the interaction is ferromagnetic and pairs of neighboring opposite spins have to be created.

But in the case of the tuned cross PCA, recall (3.10) and recall we assumed $h < 4$, see (3.15), the starting $\mathbf{d}$ and the final $\mathbf{u}$ configurations have energy

$$H_h(\mathbf{d}) = -|\Lambda|(4 + \kappa - 2h) \quad \text{and} \quad H_h(\mathbf{u}) = -|\Lambda|(4 + \kappa + 2h)$$

So that $H_h(\mathbf{d}) > H_h(\mathbf{u})$, as it is obvious since $\mathbf{u}$ is the ground state. Moreover, the dynamics is allowed to jump in a single step from $\mathbf{d}$ to $\mathbf{u}$ by reversing all the spins of the system. A naive (wrong) conclusion would be that $\mathbf{d}$ cannot be metastable because the jump from $\mathbf{d}$ to $\mathbf{u}$ can be performed in a single step by decreasing the energy.

The conclusion is wrong because in reversible PCA the probability to perform a jump is not controlled simply by the difference of energies of the two configurations involved in the jump. Indeed, in the example discussed above, recall (3.2) and (3.8), and we have that

$$p_{\beta,h}(\mathbf{d},\mathbf{u}) = \left[\frac{1}{1 + e^{2\beta(4+\kappa-h)}}\right]^{|\Lambda|} \overset{\beta\to\infty}{\sim} e^{-2|\Lambda|\beta(4+\kappa-h)}$$

[3]The regime outlined in this paper, i.e., finite state space and temperature tending to zero, is usually called the Wentzel–Friedlin regime. Different limits can be considered, for instance, volume tending to infinity.

which proves that the direct jump from **d** to **u** is depressed in probability when β is large.

This very simple remark shows that the behavior of the PCA cannot be analyzed by simply considering the energy difference between configurations. It is quite evident that a suitable cost function has to be introduced.

From (3.15), the local field $\kappa\sigma_0 + \sum_{j\in J}\sigma_j + h$ appearing in (3.8) is different from zero. Thus, for $\beta \to \infty$,

$$p_{\beta,h}(\sigma,\eta) \to \begin{cases} 1 & \text{if } \eta(i)\big[\kappa\sigma_i + \sum\limits_{j\in i+J}\sigma_j + h\big] > 0 \ \forall i \in \Lambda \\ 0 & \text{otherwise} \end{cases}$$

where we have used (3.2). Hence, given σ, there exists a unique configuration η such that $p_{\beta,h}(\sigma,\eta) \to 1$ for $\beta \to \infty$ and this configuration is the one such that $\eta(i)$ is aligned with the local field $\kappa\sigma_i + \sum_{j\in i+J}\sigma_j + h$ for any $i \in \Lambda$. Such a unique configuration will be called the *downhill image* of σ. This property explains well in which sense PCA are the probabilistic generalization of deterministic Cellular Automata: Indeed, in such models each configuration is changed deterministically into a unique image configuration. This property is recovered in probability in reversible PCA in the limit $\beta \to \infty$.

We now remark that if η is different from the downhill image of σ, we have that $p_{\beta,h}(\sigma,\eta)$ decays exponentially with rate

$$\Delta_h(\sigma,\eta) = -\lim_{\beta\to\infty}\frac{1}{\beta}\log p_{\beta,h}(\sigma,\eta) = \sum_{\substack{i\in\Lambda:\\ \eta(i)[\kappa\sigma_i+\sum_{j\in i+J}\sigma_j+h]<0}} 2\Big|\kappa\sigma_i + \sum_{j\in i+J}\sigma_j + h\Big| \qquad (3.17)$$

Note that if η is the downhill image of σ, then $\Delta_h(\sigma,\eta) = 0$. More precisely, we have

$$e^{-\beta\Delta_h(\sigma,\eta)-\beta\gamma(\beta)} \le p_{\beta,h}(\sigma,\eta) \le e^{-\beta\Delta_h(\sigma,\eta)+\beta\gamma(\beta)}$$

with $\gamma(\beta) \to 0$ for $\beta \to \infty$. This property is known in the literature as the Wentzel and Friedlin condition.

Since from (3.6) and (3.17), it follows that the following reversibility condition

$$H_h(\sigma) + \Delta_h(\sigma,\eta) = H_h(\eta) + \Delta_h(\eta,\sigma) \qquad (3.18)$$

is satisfied for any $\sigma, \eta \in X$; we have that the function $\Delta_h(\sigma,\eta)$ can be interpreted as the energy cost that must be paid in the transition $\sigma \to \eta$.

We are now ready to give a precise definition of metastable states in the framework of reversible Probabilistic Cellular Automata. We shall follow the approach in [15] which is based on the analysis of the energy landscape of the system. Note that in our setup, the energy landscape is not only given by the energy function H_h, but it is also decorated by the energy cost function Δ_h. It is important to remark that, for

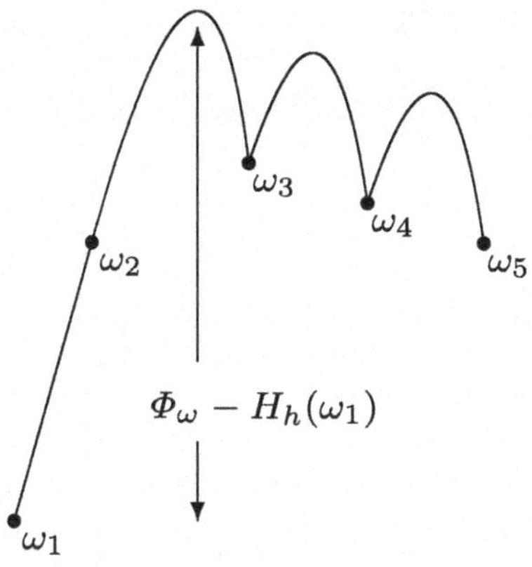

Fig. 3.5 Graphic representation of the definition of height of a path

the sake of clearness, we shall give the definition having in mind the specific case we are considering, namely the tuned cross PCA with $0 < h < \kappa$, but the definition we shall can give can be easily generalized to the broad context of reversible PCA.

A sequence of configurations $\omega = \{\omega_1, \dots, \omega_n\}$, with $\omega_i \in X$ for $i = 1, \dots, n$, is called *path*. The *height* of the path ω is defined as

$$\Phi_\omega = \max_{i=1,\dots,n-1} [H_h(\omega_i) + \Delta_h(\omega_i, \omega_{i+1})] \tag{3.19}$$

see Fig. 3.5 for a graphic illustration.

Given two sets of configurations $A, A' \subset X$, the *communication height* $\Phi(A, A')$ between A, A' is defined as

$$\Phi(A, A') = \min_{\omega: A \to A'} \Phi_\omega \tag{3.20}$$

where the minimum is taken on the set of paths starting in A and ending in A'. Given $\sigma \in X$, we define the *stability level* of σ as

$$V_\sigma = \Phi(\sigma, \{\text{states with energy smaller than } \sigma\}) - H_h(\sigma) \tag{3.21}$$

That is to say, V_σ is the height of the most convenient path that one has to follow in order to decrease the energy starting from σ.

Finally, we define the *maximal stability level* as the largest among the stability levels, i.e.,

$$\Gamma_{\mathrm{m}} = \max_{\sigma \in X \setminus \{\mathbf{u}\}} V_\sigma > 0 \tag{3.22}$$

and the set of *metastable* states

$$X_{\mathrm{m}} = \{\eta \in X \setminus \{\mathbf{u}\} : V_\eta = \Gamma_{\mathrm{m}}\} \tag{3.23}$$

This definition of metastable states is particularly nice, since it is based only on the properties of the energy landscape. In other words, in order to find the metastable states of the tuned cross PCA, one "just" has to solve some variational problems on the energy landscape of the model. This is, unfortunately, a very difficult task that has been addressed mainly in [5, 8].

Why is this definition of metastable states satisfying? Because, given $\zeta \in X_{\mathrm{m}}$, for the chain started at ζ, we can prove properties of the random variable $\tau_{\mathbf{u}}^{\zeta}$ characterizing ζ as a metastable state in the physical sense outlined in the introduction. Indeed, if we let $\mathbb{P}_\sigma$ and $\mathbb{E}_\sigma$, respectively, the probability and the average computed along the trajectories of the tuned cross PCA started at $\sigma \in X$, we can state the following theorem.

Theorem *Let $\zeta \in X_{\mathrm{m}}$. For any $\varepsilon > 0$ we have that*

$$\lim_{\beta\to\infty} \mathbb{P}_\zeta(e^{\beta(\Gamma_{\mathrm{m}}-\varepsilon)} < \tau_{\mathbf{u}}^{\zeta} < e^{\beta(\Gamma_{\mathrm{m}}+\varepsilon)}) = 1$$

Moreover,

$$\lim_{\beta\to\infty} \frac{1}{\beta} \log \mathbb{E}_\zeta[\tau_{\mathbf{u}}^{\zeta}] = \Gamma_{\mathrm{m}}$$

This theorem has been proven in [15] in the framework of Statistical Mechanics Lattice systems with Metropolis dynamics. Its generalization to the PCA case has been discussed in [8].

The physical content of the two statements in the theorem is that the first hitting time of the chain started at a metastable state $\zeta \in X_{\mathrm{m}}$ is of order $\exp\{\beta\Gamma_{\mathrm{m}}\}$. The first of the two statements ensures this convergence in probability and the second in mean.

It is important to remark that it is possible to give a more detailed description of the behavior of the chain started at a metastable state. In particular, it can be typically proven a nucleation property, that is to say, one can prove that before touching the stable state $\mathbf{u}$ the chain has to visit "necessarily" an intermediate configuration corresponding to a "critical" droplet of the stable phase (plus one) plunged in the sea of the metastable one. By necessarily, above, we mean with probability one in the limit $\beta \to \infty$. For a wide description of the results that can be proven, we refer the interested reader, for instance, to [15, 18].

3.5 Metastable Behavior of the Tuned Cross PCA

The metastable behavior of the tuned cross PCA has been studied extensively in [5] (nearest neighbor PCA, i.e., $\kappa = 0$), [1, 8] (cross PCA, i.e., $\kappa = 1$), and [9] (tuned cross PCA with $0 < \kappa < 1$). In the extreme cases, i.e., $\kappa = 0$ and $\kappa = 1$, rigorous results were proved, while in the case $0 < \kappa < 1$ only heuristic arguments have been

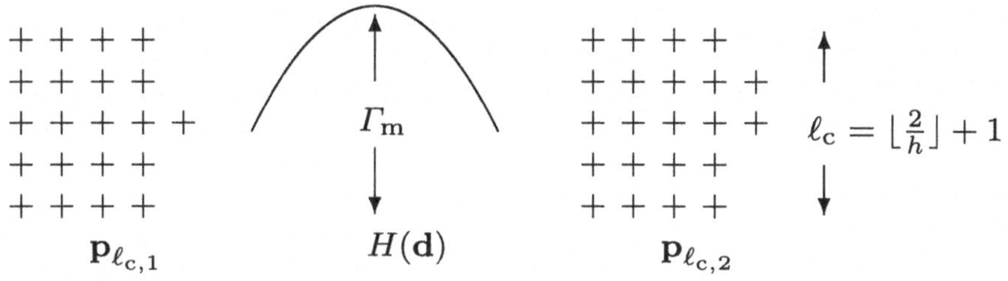

Fig. 3.6 Graphical description of Γ_m for the cross PCA

provided. In this section, we shall review briefly the main results referring the reader to the quoted papers for details. We shall always assume that h satisfies (3.15) and $2/h$ not integer; moreover, we note that the result listed below are proven for Λ large enough depending on h.

In the cross case ($\kappa = 1$), it has been proven [8] that the metastable state is unique, and more precisely, with the notation introduced above, it has been shown that $X_m = \{\mathbf{d}\}$. Moreover, it has also been proven that the maximal stability level is given by

$$\Gamma_m = H_h(\mathbf{p}_{\ell_{c,1}}) + \Delta_h(\mathbf{p}_{\ell_{c,1}}, \mathbf{p}_{\ell_{c,2}}) - H_h(\mathbf{d}) \overset{\beta\to\infty}{\sim} \frac{16}{h} \tag{3.24}$$

where[4] $\ell_c = \lfloor 2/h \rfloor + 1$ is called *critical length*, $\mathbf{p}_{\ell_{c,1}}$ is a configuration characterized by a $\ell_c \times (\ell_c - 1)$ rectangular droplet of plus spins in the sea of minuses with a single-site protuberance attached to one of the two longest sides of the rectangle, and $\mathbf{p}_{\ell_{c,2}}$ is a configuration characterized by a $\ell_c \times (\ell_c - 1)$ rectangular droplet of plus spins in the sea of minuses with a two-site protuberance attached to one of the two longest sides of the rectangle (see Fig. 3.6).

Once the model dependent problems have been solved and the metastable state found, the properties of such a state are provided by the general theorem stated in Sect. 3.4. We just want to comment that the peculiar expression of the maximal stability level that, we recall, gives the exponential asymptotic of the mean exit time has a deep physical meaning. Indeed, it is also proven that during the escape from the metastable state $\mathbf{d}$ to the stable one $\mathbf{u}$, the chain visits with probability tending to one in the limit $\beta \to \infty$ the configuration $\mathbf{p}_{\ell_{c,1}}$ and, starting from such a configuration, it performs the jump to $\mathbf{p}_{\ell_{c,2}}$. From the physical point of view, this property means that the escape from the metastable state is achieved via the nucleation of the critical droplet $\mathbf{p}_{\ell_{c,2}}$.

In the nearest neighbor case ($\kappa = 0$) it has been proven [5] that the set of metastable states is $X_m = \{\mathbf{d}, \mathbf{c}_e, \mathbf{c}_o\}$. It is important to note that the two states $\mathbf{c}_e$ and $\mathbf{c}_o$ are essentially the same metastable state; indeed, it can be easily seen that $\mathbf{c}_e$ is the downhill image of $\mathbf{c}_o$ and vice versa. So that, when the system is trapped in such a metastable state, it flip-flops between these two configurations. Moreover, it has also

[4]Given a real r we denote by $\lfloor r \rfloor$ its integer part, namely the largest integer smaller than r.

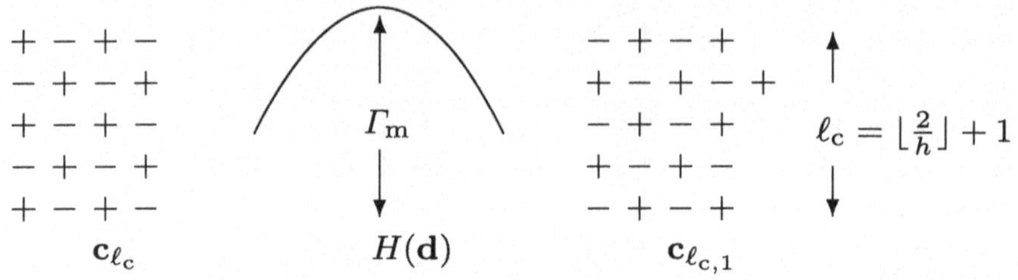

Fig. 3.7 Graphical description of $\Gamma_{\rm m}$ for the nearest neighbor PCA

been proven that the maximal stability level is given by

$$\Gamma_{\rm m} = H_h(\mathbf{c}_{\ell_c}) + \Delta_h(\mathbf{c}_{\ell_c}, \mathbf{c}_{\ell_{c,1}}) - H_h(\mathbf{d}) \overset{\beta\to\infty}{\sim} \frac{8}{h} \tag{3.25}$$

where $\ell_c = \lfloor 2/h \rfloor + 1$ is called *critical length*, $\mathbf{c}_{\ell_c}$ is a configuration characterized by a $\ell_c \times (\ell_c - 1)$ rectangular checkerboard droplet in the sea of minuses, and $\mathbf{p}_{\ell_{c,1}}$ is a configuration characterized by a $\ell_c \times (\ell_c - 1)$ rectangular checkerboard droplet in the sea of minuses with a single-site plus protuberance attached to one of the two longest sides of the rectangle (see Fig. 3.7). It is worth noting that, comparing (3.24) and (3.25), the exit from the metastable state is much slower in the case of the cross PCA with respect to the nearest neighbor one.

Even in this case, the properties of the metastable states are an immediate consequence of the theorem stated above. But also for the nearest neighbor PCA, the nucleation property is proven: During the transition, during the escape from the metastable state $\mathbf{d}$ to the stable one $\mathbf{u}$, the chain visits with probability tending to one in the limit $\beta \to \infty$ the configuration $\mathbf{c}_{\ell_c}$ and, starting from such a configuration, it performs the jump to $\mathbf{c}_{\ell_{c,1}}$. From the physical point of view, this property means that the escape from the metastable state is achieved via the nucleation of the critical checkerboard droplet $\mathbf{c}_{\ell_c}$.

Moreover, in the nearest neighbor case, it has been proven that during the escape from $\mathbf{d}$ to $\mathbf{u}$, the system has also to visit the checkerboard metastable states $\{\mathbf{c}_e, \mathbf{c}_o\}$. Starting from such a metastable state, the system performs the final escape to $\mathbf{u}$ with an exit time controlled by the same maximal stability level $\Gamma_{\rm m}$ (3.25).

Finally, we just mention the heuristic results discussed in [9] for the tuned cross PCA with $0 < \kappa < 1$. There is one single metastable state, i.e., $X_{\rm m} = \{\mathbf{d}\}$, but, depending on the ration κ/h, the system exhibits different escaping mechanisms. In particular, for $h < 2\kappa$ the systems perform a direct transition from $\mathbf{d}$ to $\mathbf{u}$, whereas for $2\kappa < h$ the system "necessarily" visits the not metastable checkerboard state before touching $\mathbf{u}$. In [9], it has been pointed out the analogies between the behavior of the tuned cross PCA and the Blume–Capel model [3]. The metastable character of the two models is very similar with the role of the self-interaction parameter κ played by that of the chemical potential in the Blume–Capel model.

References

1. Bigelis, S., Cirillo, E.N.M., Lebowitz, J.L., Speer, E.R.: Critical droplets in metastable probabilistic cellular automata. Phys. Rev. E **59**, 3935 (1999)
2. Bovier, A., Eckhoff, M., Gayrard, V., Klein, M.: Metastability and low lying spectra in reversible Markov chains. Commun. Math. Phys. **228**, 219–255 (2002)
3. Cirillo, E.N.M., Olivieri, E.: Metastability and nucleation for the Blume-Capel model. Different mechanisms of transition. J. Stat. Phys. **83**, 473–554 (1996)
4. Cirillo, E.N.M., Stramaglia, S.: Polymerization in a ferromagnetic spin model with threshold. Phys. Rev. E **54**, 1096 (1996)
5. Cirillo, E.N.M., Nardi, F.R.: Metastability for the Ising model with a parallel dynamics. J. Stat. Phys. **110**, 183–217 (2003)
6. Cirillo, E.N.M., Louis, P.-Y., Ruszel, W.M., Spitoni, C.: Effect of self-interaction on the phase diagram of a Gibbs-like measure derived by a reversible Probabilistic Cellular Automata. In press on Chaos, Solitons, and Fractals
7. Cirillo, E.N.M., Nardi, F.R., Polosa, A.D.: Magnetic order in the Ising model with parallel dynamics. Phys. Rev. E **64**, 57103 (2001)
8. Cirillo, E.N.M., Nardi, F.R., Spitoni, C.: Metastability for a reversible probabilistic cellular automata with self-interaction. J. Stat. Phys. **132**, 431–471 (2008)
9. Cirillo, E.N.M., Nardi, F.R., Spitoni, C.: Competitive nucleation in reversible probabilistic cellular automata. Phys. Rev. E **78**, 040601 (2008)
10. Grinstein, G., Jayaprakash, C., He, Y.: Statistical mechanics of probabilistic cellular automata. Phys. Rev. Lett. **55**, 2527 (1985)
11. Haller, K., Kennedy, T.: Absence of renormalization group pathologies near the critical temperature. Two examples. J. Stat. Phys. **85**, 607–637 (1996)
12. Kari, J.: Theory of cellular automata: a survey. Theor. Comput. Sci. **334**, 3–33 (2005)
13. Kozlov, V., Vasiljev, N.B.: Reversible Markov chain with local interactions. In: Multicomponent Random System. Advances in Probability & Related Topics, pp. 451–469 (1980)
14. Lebowitz, J.L., Maes, C., Speer, E.R.: Statistical mechanics of probabilistic cellular automata. J. Stat. Phys. **59**, 117–170 (1990)
15. Manzo, F., Nardi, F.R., Olivieri, E., Scoppola, E.: On the essential features of metastability: tunnelling time and critical configurations. J. Stat. Phys. **115**, 591–642 (2004)
16. Nardi, F.R., Spitoni, C.: Sharp asymptotics for stochastic dynamics with parallel updating rule. J. Stat. Phys. **146**, 701–718 (2012)
17. Olivieri, E., Scoppola, E.: Markov chains with exponentially small transition probabilities: first exit problem from a general domain. I. The reversible case. J. Stat. Phys. **79**, 613–647 (1995)
18. Olivieri, E., Vares, M.E.: Large Deviations and Metastability. Cambridge University Press, Cambridge (2005)
19. Palandi, J., de Almeida, R.M.C., Iglesias, J.R., Kiwi, M.: Cellular automaton for the order-disorder transition. Chaos Solitons Fractals **6**, 439–445 (1995)
20. Perc, M., Grigolini, P.: Collective behavior and evolutionary games - an introduction. Chaos Solitons Fractals **56**, 1–5 (2013)
21. Wolfram, S.: Statistical mechanics of cellular automata. Rev. Mod. Phys. **55**, 601–644 (1983)
22. Wolfram, S.: Cellular automata as models of complexity. Nature **311**, 419–424 (1984)

Chapter 4
Strategic Interaction in Interacting Particle Systems

Paolo Dai Pra, Elena Sartori and Marco Tolotti

Abstract In the last decades, models inspired by statistical mechanics have been vastly used in the context of social sciences to model the behavior of interacting economic actors. In particular, parallel updating models such as Probabilistic Cellular Automata have been proved to be very useful to represent rational agents aiming at maximize their utility in the presence of social externalities. What PCA do not account for is strategic interaction, i.e., the fact that, when deciding, agents forecast the action of other agents. In this contribution, we compare models that differ in the presence of strategic interaction and memory of past actions. We will show that the emergent equilibria can be very different: Fixed points, cycles of period 2, and chaotic behavior may appear and, possibly, coexist for some values of the parameters, of the model.

4.1 Introduction

The idea that principles of statistical physics could be applied to systems comprised by a large number of intelligent and rational individuals has fascinated physicists and mathematicians for several decades, besides having stimulated the imagination of science fiction writers. The number of individuals in a reasonably large community, although quite far from the Avogadro number, is large enough to separate microscopic from macroscopic scale; in more technical terms, a reasonably simple collective behavior should result from the combination of all individual behaviors through a law of large numbers, in analogy to the way thermodynamic quantities such as

P. Dai Pra (✉) · E. Sartori
Dipartimento di Matematica, Università di Padova, Padova, Italy
e-mail: daipra@math.unipd.it

E. Sartori
e-mail: esartori@math.unipd.it

M. Tolotti
Dipartimento di Management, Università Ca' Foscari, Venezia, Italy
e-mail: tolotti@unive.it

P.-Y. Louis and F.R. Nardi (eds.), *Probabilistic Cellular Automata*, Emergence, Complexity and Computation 27, https://doi.org/10.1007/978-3-319-65558-1_4

pressure and temperature result from the combination of the irregular motion of single particles in a gas. Moreover, some stylized features of the interaction between individuals (e.g., imitation) show similarities with interactions between the elements of some physical systems (e.g., spin interactions in ferromagnets).

Together with the analogies mentioned above, many differences naturally emerge. In physical systems, the *prime principles*, on which the microscopic dynamics are based, are usually well established, as they follow from fundamental laws. This is not the case in social systems: Interactions between individuals are complex and not well understood. Any stylized model is thus likely to have limited applicability. At a more technical level, some standard assumptions in models inspired by statistical physics, such as time reversibility and short range of the interaction, are often unreasonable in the dynamics of social systems.

Our view is that, despite of these difficulties, modeling of large-scale social systems is a relevant and stimulating challenge. Stylized models, though unrealistic, may reveal the key factors producing certain behaviors, allowing, for instance, the design of controls to avoid the emergence of undesirable patterns in real social systems. We remark that similar ideas have proved to be successful in other contexts, e.g., biology with applications to medicine (see, e.g., [6]).

As we mentioned already, stylized modeling for systems of interacting rational individuals has been vastly inspired by statistical mechanics. After all, elementary particles are themselves "rational," in that they aim at minimizing their contribution to the total energy of the system; temperature injects noise in the dynamics, causing entropic effects to be macroscopically relevant. In this spirit, it has been remarked in several works (see, e.g., [1, 2]) that many discrete time stochastic models for the evolution of interacting particles can be formulated as a sequence of optimizations: At each time, a particle "chooses" its next position by minimizing its contribution to the energy of the system *given* the position of all other particles, and subject to some random disturbance. As we will see later, this interpretation is particularly natural in the context of parallel updating of Probabilistic Cellular Automata (PCA). This formulation appears suitable for applications to social sciences, where "minimizing particle's contribution to the total energy of the system" is replaced by "maximizing individual utility." Since individual utilities can be arbitrary functions of positions of all individuals, one obtains a wide class of models, not necessarily time reversible.

The purpose of this paper is to propose a modification of this approach to modeling interacting systems, which takes into account one basic difference between the "rationality" of physical particles and that of human individuals. We, simply, express it as follows: in interacting with other humans, any individual tries to *forecast* what the others will be doing in the (near) future. It is easy to exhibit examples in which this forecasting plays a relevant role. An obvious example is that of agents investing in a financial market. Each agent aims at maximizing his own profit; this profit depends of the future prices of assets, which in turn depends of the future investment strategy of all agents. Naturally, the agent tries to forecast the strategies of other individuals, *well aware* that all the others will do the same.

An example in a different context is that of a car traffic jam in an intersection with a broken traffic light. A driver will decide to cross the intersection when he is reasonably confident that drivers coming from other directions will not; his decision is based on what he believes the behavior of other individuals will be. It has been observed (see, e.g., [7]) that the evolution rapidly goes to a self-organized steady state, where crossings from concurrent directions alternate in a nearly periodic way. The problem of formulating stylized models exhibiting this behavior is, to a large extent, still unsolved (see [9]).

In this paper, elaborating on ideas contained in [3], we propose a formulation of what we will refer to as *strategic interaction, i.e.*, a mechanism of interaction between individuals which includes forecasting of other individuals' behavior in the near future. Rather than aiming at generality, we illustrate our basic ideas in a very simple context, inspired by PCA. In Sect. 4.2, we illustrate the interpretation of PCA as sequential stochastic optimization problems, and we propose the version of the same models with strategic interaction. In Sect. 4.3, we analyze a simple *mean-field* model, for which the macroscopic limit is easily obtained and illustrate the effects of the strategic interaction on the steady-state behavior.

4.2 Strategic and Non-strategic Interaction

By Probabilistic Cellular Automata (PCA) we, generally, mean a discrete time Markov chain on a product space S^{Λ}, where Λ is finite or countable, whose transition probabilities are product measures; in other words, different components update simultaneously and independently. In this paper, we restrict to the case in which $S = \{-1, 1\}$ and Λ is finite. It will be clear that most ideas apply to more general choices of S, but binary models allow some peculiar explicit computation. We denote by $\sigma = (\sigma_i)_{i\in\Lambda}$ an element of $\{-1, 1\}^{\Lambda}$. The evolution $(\sigma(n))_{n\geq 0}$ of PCA is of the product form

$$\mathbb{P}(\sigma(n+1) = \sigma | \sigma(n) = \xi) = \prod_{i\in\Lambda} \mathbb{P}(\sigma_i(n+1) = \sigma_i | \sigma(n) = \xi) =: \prod_{i\in\Lambda} p_i(\sigma_i|\xi). \tag{4.1}$$

If we assume $p_i(\sigma_i|\xi) > 0$ for every i, σ, ξ, then $p_i(\sigma_i|\xi)$ can be written in the form $p_i(\sigma_i|\xi) = \exp[\sigma_i \Phi_i(\xi)]/[2\cosh(\Phi_i(\xi))]$, for some functions $\Phi_i : \{-1, 1\}^{\Lambda} \to \mathbb{R}$. Without any loss of generality, for the interpretation of the model it is convenient to introduce some parameters, writing Φ_i in the following form:

$$\Phi_i(\xi) = \beta \left[F_i(\xi) + \mu_i \xi_i\right], \quad \text{for } \beta, \mu_i \geq 0, \tag{4.2}$$

which corresponds to the *local* transition probabilities

$$p_i(\sigma_i|\xi) = \frac{\exp[\beta\sigma_i \left[F_i(\xi) + \mu_i\xi_i\right]}{2\cosh(\beta \left[F_i(\xi) + \mu_i\xi_i\right])}. \tag{4.3}$$

The function Φ in (4.2) can be interpreted as follows.

- β is an *inverse temperature* parameter that allows to tune the amount of random noise in the system. For $\beta = 0$, the system is fully random. As $\beta \to +\infty$, the dynamics converge to the (nearly) deterministic evolution

$$\sigma_i(n+1) = \text{sign}\left[F_i(\sigma(n)) + \mu_i \sigma_i(n)\right];$$

  randomness only survives in the case $F_i(\sigma(n)) + \mu_i \sigma_i(n) = 0$, where the values ± 1 for $\sigma_i(n+1)$ are equally probable.
- The function F_i describes the interaction of the ith component with the others.
- The parameter μ_i models *friction*: For large μ_i, it is unlikely for σ_i to change sign in a time step, slowing down the dynamics (see [4], where this friction plays a key role for determining a "desired" stationary distribution). Of course, the term $\mu_i \xi_i$ could be included in $F_i(\xi)$, but it will be convenient to separate this *self-interaction* term.

4.2.1 An Equivalent Optimization Problem

Suppose each $i \in \Lambda$ labels an agent that, at any time t, faces a *binary decision problem*: $\sigma_i(n) = \pm 1$ denotes the *action* of the ith agent at time n. At each time n the aim of the ith agent is to maximize a *random utility function* U_i as function of the action $s_i = \sigma_i(n)$; the function $U_i(s_i)$ is determined by the action of (possibly) all agents at time $n-1$, and by a random term $\varepsilon_i(n)$ as follows:

$$U_i(s_i; \sigma(n-1), \varepsilon_i(n)) := s_i \left[F_i(\sigma(n-1)) + \mu_i \sigma_i(n-1) + \varepsilon_i(n)\right], \tag{4.4}$$

where $(\varepsilon_i(n))_{i\in\Lambda, n\geq 1}$ are i.i.d. real random variables, having the following distribution function:

$$\eta(x) := \mathbb{P}(\varepsilon_i(n) \leq x) = \frac{1}{1+e^{-2\beta x}}. \tag{4.5}$$

All agents perform simultaneously their optimization. Note that agent i will choose $\sigma_i(n) = 1$ if and only if

$$F_i(\sigma(n-1)) + \mu\sigma_i(n-1) + \varepsilon_i(n) > 0, \tag{4.6}$$

which, *given* $\sigma(n-1)$, happens with probability

$$\begin{aligned}
\mathbb{P}&\left[F_i(\sigma(n-1)) + \mu\sigma_i(n-1) + \varepsilon_i(n) > 0\right] \\
&= \eta(F_i(\sigma(n-1)) + \mu\sigma_i(n-1)) \\
&= \frac{\exp[\beta\left[F_i(\sigma(n-1)) + \mu_i\sigma_i(n-1)\right]}{2\cosh(\beta\left[F_i(\sigma(n-1)) + \mu_i\sigma_i(n-1)\right])}.
\end{aligned} \tag{4.7}$$

Note that the case in which equality holds in (4.6), which would make the two actions equivalent, can be ignored, since it occurs with probability zero.

Comparing (4.7) with (4.3), we realize the sequence of optimization problems induces a Markov evolution with the same transition probabilities as the PCA in (4.3).

4.2.2 Strategic Interaction

In the sequence of optimization problems described above, agents update their action simultaneously and *independently*, on the basis of the past actions. We consider a modification of the model, suggested by the following considerations.

(a) The utility of agent i at time n may depend on the action of *all agents at the same time* n.
(b) Each agent is aware of the fact that all other agents are optimizing their own utility, and uses this fact to forecast their actions.
(c) Agents know all function F_i and friction parameters μ_i. The random term $\varepsilon_i(n)$ can only be observed by agent i. Agents know the distribution of $\varepsilon_i(n)$.

By (c), the action $s_i = \sigma_i(n)$ of agent i at time n may depend on $\varepsilon_i(n)$, so it is convenient to define an action s_i as a measurable function $s_i(\varepsilon_i)$ of the random term ε_i.

If $s = (s_i)_{i\in\Lambda}$ denotes the vector of actions at a given time n, we set $s^i = (s_j)_{j\neq i}$. We assume agent i aims at maximizing in s_i the following utility, obtained modifying in a simple way (4.4):

$$U_i(s_i, s^i; \sigma(n-1), \varepsilon_i(n)) = s_i \left[\mathrm{e}^i\left(F_i(s)\right) + \mu_i \sigma_i(n-1) + \varepsilon_i(n) \right], \tag{4.8}$$

where, for a vector of actions $(s_j(\varepsilon_j(n)))_{j\in\Lambda} =: s(\varepsilon(n))$, the expression $\mathrm{e}^i\left(F_i(s)\right)$ is obtained by averaging $F_i(s(\varepsilon(n)))$ over $(\varepsilon_j(n))_{j\neq i}$.

Unlike in (4.4), the utility U_i depends on the action of all agents; it is, therefore, natural to give a game-theoretic definition of an "optimal" vector of actions.

Definition 1 A vector of actions $s = (s_i(\varepsilon_i(n)))_{i\in\Lambda}$ is called a *Nash equilibrium* if for all $i \in \Lambda$

$$s_i(\varepsilon_i(n)) = \text{argmax } U_i(\cdot, s^i; \sigma(n-1), \varepsilon_i(n)). \tag{4.9}$$

In other words, in a Nash equilibrium any agent is using the best action *given* the other agents' actions. For comments and details on this notion of equilibrium, we refer the reader to [8]. It is immediate, but quite relevant, to observe that s is a Nash equilibrium if and only if it is a fixed point for the so-called *best response map* $s \mapsto \Phi(s)$ given by

$$\Phi_i(s) = \text{argmax } U_i(\cdot, s^i; \sigma(n-1), \varepsilon_i(n)). \tag{4.10}$$

In general, in games there is no guarantee that either existence or uniqueness hold for Nash equilibria. For the models above, existence is not a problem, however.

Proposition 1 *At least one Nash equilibrium exists.*

Proof By (4.10), $\Phi_i(s)(\varepsilon_i(n))$ is increasing in $\varepsilon_i(n)$. Thus, we can restrict to actions s of the form

$$s_i(\varepsilon_i) = \mathbb{1}_{(x_i,+\infty)}(\varepsilon_i),$$

with $x_i \in \overline{\mathbb{R}} := \mathbb{R} \cup \{\pm\infty\}$. Thus, Φ can be seen as a map from $\overline{\mathbb{R}}^{|\Lambda|}$ to itself. The fact that the distribution of the $\varepsilon_i(n)$ is absolutely continuous guarantees that this map is continuous. Since $\overline{\mathbb{R}}^{|\Lambda|}$ is convex and compact, the conclusion follows from a standard fixed point argument.

The uniqueness of the Nash equilibrium is, however, not guaranteed. This means that, the map which to $(\sigma(n-1), \varepsilon(n))$ associates $\sigma(n)$ may be not single-valued. In order to obtain well-defined Markovian dynamics, one should have a rule for *selecting* one specific Nash equilibrium; or, otherwise, one should content himself for having defined just a set of *possible* evolutions of the system. This point will be discussed in more details in a specific model, in Sect. 4.3.

4.2.3 Trend-Driven Dynamics

In many applications, the utility of an agent can be interpreted as the *return* of an investment. This return is determined by the *variation* of the value of an asset which, in turn, depends of the variation of *demand* for the asset. To model this situation, it is reasonable to assume that the utility U_i of agent i depends of the variation (*trend*) of a function $F_i(\sigma)$ of all actions. In the strategic case, which is the most meaningful in this interpretation, this amounts to define the following utility:

$$U_i(s_i, s^i; \sigma(n-1), \varepsilon_i(n)) = s_i \left[\mathrm{e}^i \left[F_i(s)\right] - F_i(\sigma(n-1)) + \mu_i \sigma_i(n-1) + \varepsilon_i(n) \right]. \tag{4.11}$$

We remark that in the absence of friction ($\mu_i = 0$ for all $i \in \Lambda$), the utility (4.8) does not depend on $\sigma(n-1)$, so that the resulting evolution $(\sigma(n))_{n\geq 0}$ is a possibly multiple-valued i.i.d. sequence. In the case of utility (4.11), this is not the case. Clearly, one could consider further generalizations in which a trend term is added to utility (4.8).

Note that this dynamics driven by the trend can be adapted to the non-strategic context, by letting

$$\begin{aligned} &U_i(s_i; \sigma(n-1), \sigma(n-2), \varepsilon_i(n)) \\ &\qquad := s_i \left[F_i(\sigma(n-1)) - F_i(\sigma(n-2)) + \mu_i \sigma_i(n-1) + \varepsilon_i(n)\right]. \end{aligned} \tag{4.12}$$

Rather than aiming at generality, we study some specific simple models, for which the thermodynamic limit ($|\Lambda| \to +\infty$) can be obtained explicitly, and compare the long-time behaviors of such models in the strategic and non-strategic case.

4.3 A Linear, Mean-Field Model

In this section, we consider an homogeneous, mean-field model, for which $\Lambda = \{1, 2, \ldots, N\}$, $\mu_i \equiv \mu \geq 0$, and

$$F_i(s) = k\, m_N(s), \tag{4.13}$$

where $k \geq 0$ and

$$m_N(s) := \frac{1}{N} \sum_{i=1}^{N} s_i.$$

Our aim is to analyze the $N \to +\infty$ limit of the random dynamics produced by the sequence of utility optimizations. We will consider all versions (4.4), (4.8), (4.11), (4.12) of the utility. We begin by briefly treating the case of non-strategic optimization, in order to better appreciate the effects of the game-theoretic setting.

4.3.1 The Non-strategic Case

We consider first the utility (4.4). In this case, we obtain the stochastic dynamics

$$\sigma_i(n) = \text{sign}\left[k\, m_N(\sigma(n-1)) + \mu \sigma_i(n-1) + \varepsilon_i(n)\right]. \tag{4.14}$$

Given a set A and $x \in A^N$, we introduce the empirical measure

$$\rho_N^x := \frac{1}{N} \sum_{i=1}^{N} \delta_{x_i}.$$

Thus, for instance,

$$\rho_N^{\sigma(n)} = \frac{1 + m_N(\sigma(n))}{2} \delta_1 + \frac{1 - m_N(\sigma(n))}{2} \delta_{-1},$$

while $\rho_N^{\sigma(n-1),\varepsilon(n)}$ is the *joint* empirical measure of $(\sigma(n-1), \varepsilon(n))$. Averaging over n Eq. (4.14), we obtain

$$m_N(\sigma(n)) = \int \mathrm{sign}\,[k\, m_N(\sigma(n-1)) + \mu s + e]\, \rho_N^{\sigma(n-1),\varepsilon(n)}(ds, de). \tag{4.15}$$

For each N, we fix a deterministic initial condition $\sigma(0)$ such that the following limit exists:

$$\lim_{N\to+\infty} m_N(\sigma(0)) =: m(0).$$

Then, one can prove by induction the following law of large numbers.

Proposition 2 *For every $h : \{-1, 1\} \times \mathbb{R}$ bounded and continuous*

$$\begin{aligned}\lim_{N\to+\infty} &\int h(s, e)\rho_N^{\sigma(n-1),\varepsilon(n)}(ds, de) \\ &= \int h(s, e)\left(\frac{1+m(n-1)}{2}\delta_1 + \frac{1-m(n-1)}{2}\delta_{-1}\right)(ds)d\eta(e),\end{aligned} \tag{4.16}$$

where $(m(n))_{n\geq 0}$ solves the recursion

$$\begin{aligned} m(n) &= (1+m(n-1))\eta(km(n-1)+\mu) \\ &\quad + (1-m(n-1))\eta(km(n-1)-\mu) - 1 =: G_1(m(n-1)), \end{aligned} \tag{4.17}$$

with initial condition $m(0)$. Moreover, for each $n \geq 1$, $m_N(\sigma(n))$ converges in probability to $m(n)$, as $N \to +\infty$.

Equation (4.17) describes the *macroscopic* dynamics of the system with a large number of agents. The long-time behavior of these dynamics is obtained by studying the steady-state solutions of (4.17).

Proposition 3 *1. Assume $\beta k \leq (1 + e^{-2\beta\mu})/2$. Then, $\overline{m} = 0$ is the unique fixed point for* (4.17). *Moreover, it is a global attractor, i.e., for every $m(0) \in [-1, 1]$*

$$\lim_{n\to+\infty} m(n) = 0.$$

2. For $\beta k > (1 + e^{-2\beta\mu})/2$, $\overline{m} = 0$ is an unstable fixed point. Moreover, there is $m^* > 0$ such that $\pm m^*$ are locally stable fixed points and for every $m(0) \in [-1, 1] \setminus \{0\}$

$$\lim_{n\to+\infty} m(n) \in \{-m^*, m^*\}.$$

Thus, (4.17) has the familiar behavior of the Curie–Weiss model. In comparison with the standard Curie–Weiss model, a slight difficulty is due to the fact that the function $G_1(\cdot)$ in (4.17) is non-necessarily concave in $[0, 1]$; one, however, shows that $G_1(\cdot)$ is actually concave for $\beta k \leq \frac{1+e^{-2\beta\mu}}{2}$ and at most one change of concavity occurs otherwise. With this remark, the standard proof for the Curie–Weiss model is easily adapted. Details are omitted here.

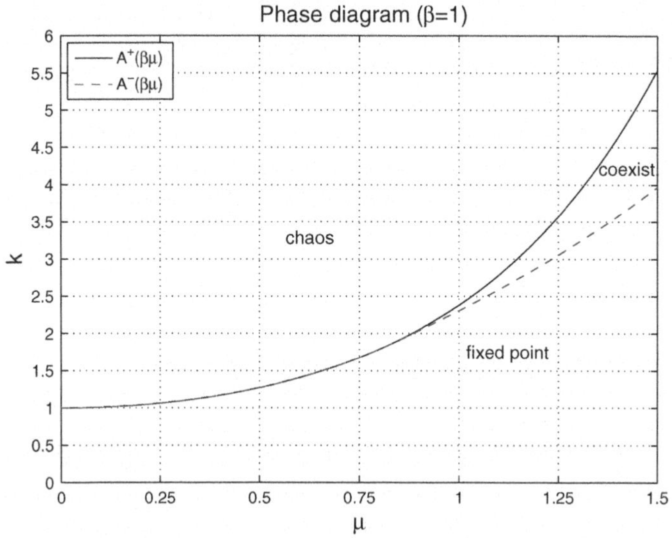

Fig. 4.1 Phase diagram of steady states

Let us now turn to the trend-dependent utility (4.11). The above argument for the derivation of the macroscopic dynamics can be repeated; Eq. (4.17) is now replaced by

$$\begin{aligned} m(n) &= (1 + m(n-1))\eta(k[m(n-1) - m(n-2)] + \mu) \\ &\quad + (1 - m(n-1))\eta(k[m(n-1) - m(n-2)] - \mu) - 1, \end{aligned} \tag{4.18}$$

where we need to assume $(m_N(0), m_N(1)) \to (m(0), m(1))$. The analysis of the steady states for (4.18) is considerably harder. Although not yet fully proved, the following picture is supported by strong numerical evidence (see Fig. 4.1).

Conjecture

Case $\mu = 0$. For $\beta k \leq 1$, $\overline{m} = 0$ is a globally attracting fixed point: For every $(m(0), m(1)) \in [-1, 1]^2$,

$$\lim_{n \to +\infty} m(n) = 0.$$

For $\beta k > 1$ all initial conditions, except $(m(0), m(1)) = (0, 0)$, converge to a periodic trajectory of period 6 of the form $(\ldots, x^*, x^*, 0, -x^*, -x^*, 0, \ldots)$ for a *unique* $x^* > 0$.

Case $\mu > 0$. There is a constant $A^-(\beta\mu) \leq A^+(\beta\mu) := [\cosh(2\beta\mu) + 1]/(2\beta)$ such that:

- for $\beta k \leq A^-(\beta\mu)$, all initial conditions are attracted to zero;
- for $\beta k \geq A^+(\beta\mu)$ for no initial condition, except $(m(0), m(1)) = (0, 0)$, the trajectory converges to 0;

- for $A^-(\beta\mu) < \beta k < A^+(\beta\mu)$, we have that $m(n) \to 0$ if and only if $(m(0), m(1)) \in \mathcal{N}$, where $\mathcal{N} \subsetneq [-1, 1]^2$ is a neighborhood of $(0, 0)$.

Finally, there is a constant $c > 0$ such that $A^-(\beta\mu) < A^+(\beta\mu)$ if and only if $\beta\mu > c$.

In the case $\mu > 0$, the trajectories that *do not* converge to zero appear to have a much more complex behavior. The long-time behavior is sensitive to the initial condition, and there is no evidence of locally stable periodic orbits. We observe that, for $\beta\mu > c$ and $A^-(\beta\mu) < \beta k < A^+(\beta\mu)$, the locally stable fixed point *coexists* with the "chaotic" phase.

4.3.2 The Strategic Case: The Utility with No Trend

In analogy with (4.14), we begin the analysis by the optimality condition for the utility (4.8):

$$\sigma_i(n) = \text{sign}\left[k\, \mathrm{e}^i[m_N(\sigma(n))] + \mu\sigma_i(n-1) + \varepsilon_i(n)\right]. \tag{4.19}$$

Equation (4.19) does not uniquely identify the dynamics, due to the possible multiplicity of Nash equilibria. Nevertheless, the following facts hold.

Proposition 4 *Assume that, for each $n \geq 1$, it is selected a Nash equilibrium, i.e., a solution of* (4.19). *Then, the resulting stochastic process* $(m_N(\sigma(n)))_{n\geq 0}$ *is* tight, *and each weak limit point satisfies a.s. the* implicit *equation*

$$\begin{aligned} m(n) &= (1 + m(n-1))\eta(k\, m(n) + \mu) + (1 - m(n-1))\eta(k\, m(n) - \mu) - 1 \\ &=: G_2(m(n), m(n-1)). \end{aligned} \tag{4.20}$$

For the proof of this result, see [3].

The dynamics are now described by the implicit recursion (4.20). By a standard fixed point argument, it is easily shown that, for every $x \in [-1, 1]$, the equation $y = G_2(y, x)$ has *at least* one solution in $[-1, 1]$. This is the macroscopic version of the existence of a Nash equilibrium (see Proposition 1). On the other hand, possible multiplicity of Nash equilibria translates to multiple solutions of the equation $y = G_2(y, x)$, producing a large set of *possible* dynamics, i.e., sequences $(m(n))_{n\geq 0}$ satisfying (4.20). For the study of these dynamics, it is useful to notice that, since $G_2(y, x)$ is linear in x, the equation $y = G_2(y, x)$ determines a function $x = \psi(y)$. Computing the derivative ψ', one shows that for $\beta k \leq 1$ the function ψ is strictly increasing on $\mathbb{R}$, and thus, it admits an inverse $\phi = \psi^{-1}$; moreover, ϕ maps $[-1, 1]$ into $[-1, 1]$. This simple computation leads to the following result, whose proof is omitted.

Proposition 5 *For $\beta k \leq 1$ and for any $\mu > 0$, Eq.* (4.20) *gives rise to a single-valued dynamics.*

1. *For $\beta k \leq (1 + e^{-2\beta\mu})/2$, $\overline{m} = 0$ is a globally attracting fixed point.*
2. *For $(1 + e^{-2\beta\mu})/2 < \beta k \leq 1$, $\overline{m} = 0$ is an unstable fixed point. Moreover, there is $m^* > 0$ such that $\pm m^*$ are locally stable fixed points and, for every $m(0) \in [-1, 1] \setminus \{0\}$*

$$\lim_{n \to +\infty} m(n) \in \{-m^*, m^*\}.$$

Note that the *high-temperature regime* $\beta k \leq (1 + e^{-2\beta\mu})/2$ coincides with that of the corresponding non-strategic model seen in Proposition 3. Moreover, as the threshold $(1 + e^{-2\beta\mu})/2$ is crossed, but $\beta k \leq 1$, the system enters a *polarized phase*, again as in the non-strategic model.

But, what happens for $\beta k > 1$? It is not hard to show that, as $\beta k > 1$, if μ is small enough, the function ψ is not monotonic, and, therefore, it cannot be globally inverted: Eq. (4.20) determines multi-valued dynamics. A partial description of these dynamics can be inferred by the following remarks (see also Fig. 4.2).

(a) For all values of the parameters, the equation $y = G_2(y, x)$ determines a function $y = \phi(x)$ mapping a neighborhood of the origin onto a neighborhood of the origin. This map is contractive, i.e., it has a Lipschitz constant $L < 1$ for $\beta k < (1 + e^{-2\beta\mu})/2$ and for $\beta k > (1 + e^{2\beta\mu})/2$. Thus, there is a *low-temperature regime*, in which if $m(0)$ is sufficiently close to 0, then $m(n) \to 0$ as $n \to +\infty$.
(b) For all $\beta k > (1 + e^{2\beta\mu})/2$, the equation $x = G_2(x, x)$ has exactly three solutions $-m^*, 0, m^*$, with $m^* > 0$. Moreover, the equation $y = G_2(y, x)$ determines a

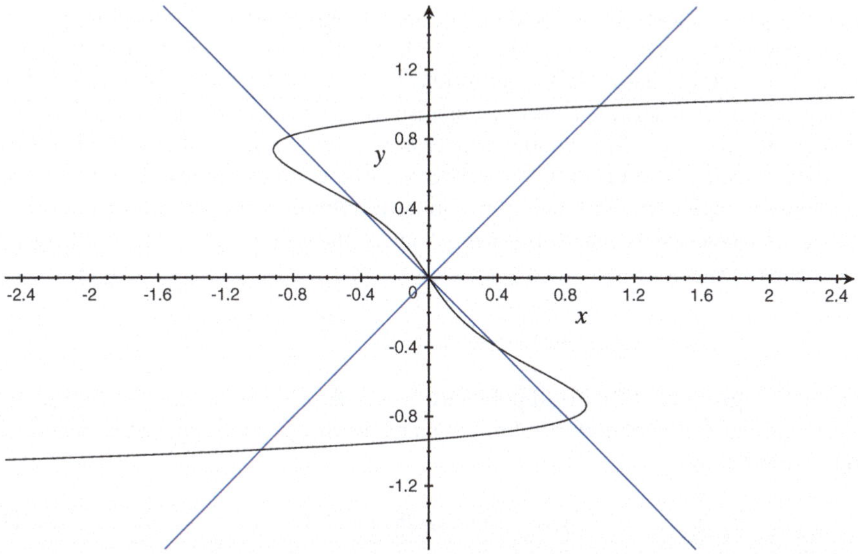

Fig. 4.2 Graph of the curve $y = G_2(y, x)$. Intersections with $y = x$ correspond to fixed points, intersection with $y = -x$ correspond to 2-cycles. Here the parameters are $\beta = 1, \mu = 0.7, k = 2.2$

contractive function $y = \phi(x)$ around $\pm m^*$. Thus, $\pm m^*$ can be seen as locally stable fixed points for the dynamics. Note that, however, for x close to $\pm m^*$, the equation $y = G_2(y, x)$ does not have necessarily a unique solution.
Unlike the non-strategic case, for large values of βk other steady-state solutions emerge.

(c) For βk sufficiently large, depending on the value of $\beta\mu$, the equation $x = G_2(x, -x)$ has five solutions $-x^{(2)}, -x^{(1)}, 0, x^{(1)}, x^{(2)}$, with $0 < x^{(1)} < x^{(2)}$. This means that the trajectories alternating $x^{(1)}$ and $-x^{(1)}$ (or $x^{(2)}$ and $-x^{(2)}$) solve (4.20), so they are possible dynamics for the system. Around each of these points, the equation $y = G_2(y, x)$ locally determines a function $y = \phi(x)$; the aforementioned 2-*cycle* (trajectory of period two) $(-x^{(1)}, x^{(1)})$ is locally stable (whenever it exists) for the dynamics determined by ϕ, while the 2-cycle $(-x^{(2)}, x^{(2)})$ is locally stable only for βk large.

The fact that the recursion (4.20) is multi-valued allows many other possible dynamics. For certain values of the parameters, the system could, for instance, shift from a neighborhood of a fixed point to a neighborhood of a 2-cycle. The selection of one specific dynamics depends on the selection of a particular Nash equilibrium, which corresponds, in the thermodynamic limit, to the choice of a particular solution of $y = G_2(y, x)$.

4.3.2.1 Selection of Nash Equilibria

Nash equilibria are fixed points for the best response map (4.10). In the limit of infinitely many agents, these fixed points become, given $m(n-1)$, fixed points for the map $y \mapsto G_2(y, m(n-1))$. One possible criterion for the selection of the fixed point is to assume that a Nash equilibrium is not instantaneously achieved, but it emerges as a result of a *learning* mechanism, known under the name of *Cournot adjustment* (see, e.g., [5]). Although there are several versions of this mechanism, we only consider it in the simplest version. Assume the dynamics (4.9), or (4.20) in the thermodynamic limit, take place in a discrete time that we call *macro-time*. Between two successive macro-times $n-1$ and n, the following learning mechanism evolves on a fast *micro-time*: for a given choice of $\xi(0) \in \{-1, 1\}^n$, we set recursively

$$\xi_i(k+1) := \operatorname{argmax}\left[U_i(\cdot, \xi^i(k); \sigma(n-1), \varepsilon_i(n))\right],$$

where U_i is given in (4.8). Fixed points for these micro-dynamics are, by definition, Nash equilibria. In the limit of infinitely many agents, $y_k := \frac{1}{N}\sum_i \xi_i(k)$ converges to the solution of

$$y_{k+1} = G_2(y_k, m(n-1)). \tag{4.21}$$

Since the map $y \mapsto G_2(y, m(n-1))$ is continuous and increasing, the limit $\lim_k y_k$ exists, and it is a fixed point. This allows to select a single solution to the equation

$y = G_2(y, m(n-1))$. The selected solution may depend on the choice of y_0. This virtually rules out those fixed points $y = G_2(y, m(n-1))$ that are *unstable* for the recursion (4.21); these points cannot indeed be obtained as limits in the above recursion, unless one chooses $y_0 = y$. To obtain a single-valued dynamics, one "natural" choice could be to set $y_0 := m(n-1)$.

We, now, briefly discuss the implications of this *learning* procedure. Consider, in particular, the 2-cycles discussed above. Suppose $-x^{(2)}, -x^{(1)}, 0, x^{(1)}, x^{(2)}$, with $0 < x^{(1)} < x^{(2)}$, are the solutions of $x = G_2(x, -x)$.

(a) Set, to begin with, $m(n-1) = -x^{(1)}$. It can be shown that $y = x^{(1)}$ is an *unstable* fixed point for the map $y \mapsto G_2(y, -x^{(1)})$: for no choice of y_0 the Cournot adjustment procedure can select the 2-cycle that alternates $-x^{(1)}$ and $x^{(1)}$.
(b) Set now $m(n-1) = -x^{(2)}$. It can be shown that $y = x^{(2)}$ is a locally stable fixed point for the map $y \mapsto G_2(y, -x^{(2)})$. However, $y_0 = -x^{(2)}$ is *not* in its basin of attraction.

Summing up, the following result can be proved.

Proposition 6 *Assume a single-valued dynamics is selected via the Cournot adjustment procedure, with* $y_0 = m(n-1)$. *Then, whenever* $\beta k > (1 + e^{-2\beta\mu})/2$ *and* $m(0) \neq 0$, *we have*

$$\lim_{n \to +\infty} m(n) \in \{-m^*, m^*\},$$

where m^* *is the unique strictly positive solution of* $x = G_2(x, x)$.

Thus, following the Cournot adjustment procedure with $y_0 = m(n-1)$, we obtain the same behavior as the corresponding non-strategic model. Stable 2-cycles may exist, indeed coexist with stable fixed points, but require different selection procedures.

4.3.3 The Strategic Case: The Utility with Trend

Finally, we are left with the trend-driven, strategic case, whose condition for optimality is given by

$$\sigma_i(n) = \text{sign}\left[k\, e^i[m_N(\sigma(n))] - k\, m_N(\sigma(n-1)) + \mu\sigma_i(n-1) + \varepsilon_i(n)\right]. \quad (4.22)$$

As for Proposition 4, we can obtain the following macroscopic description of the dynamics.

Proposition 7 *Assume that, for each* $n \geq 1$, *it is selected a Nash equilibrium, i.e., a solution of* (4.19). *Then, the resulting stochastic process* $(m_N(\sigma(n)))_{n \geq 0}$ *is* tight, *and each weak limit point satisfies a.s. the* implicit *equation*

$$\begin{aligned} m(n) &= (1+m(n-1))\eta(k\,m(n)-k\,m(n-1)+\mu) \\ &\quad +(1-m(n-1))\eta(k\,m(n)-k\,m(n-1)-\mu)-1 \\ &=: G_3(m(n), m(n-1)). \end{aligned} \tag{4.23}$$

The analysis of the steady-state solutions of (4.23) has been done in [3].

Proposition 8 *There is a constant $A(\beta\mu) \le \frac{1+e^{2\beta\mu}}{4}$ such that:*

1. *the equation $y = G_3(y, x)$ uniquely determines a function $y = \psi(x)$ around $(x, y) = (0, 0)$. Moreover, $\overline{m} = 0$ is stable for the resulting local dynamics if and only if $\beta k \le (1 + e^{2\beta\mu})/4$;*
2. *for $\beta k > A(\beta\mu)$ a 2-cycle exists, i.e., a strictly positive solution of $x = G_3(x, -x)$. For at least one 2-cycle, the dynamics can be locally made explicit, and the 2-cycle is stable for this local dynamics;*
3. *there is a constant $0 < c \le (1/2)\log 2$ such that, if $\beta\mu > c$, then $A(\beta\mu) < (1 + e^{2\beta\mu})/4$. In this case, stable fixed points coexist with a stable 2-cycle.*

Unlike the model corresponding to utility (4.8), in this case the analysis of the best response map reveals that 2-cycles survive the Cournot adjustment selection. Partial proofs and numerical simulations strongly support the following conjecture.

Conjecture Assume the Cournot adjustment selection with $y_0 = m(n-1)$ is used to obtain a single-valued dynamics $(m(n))_{n\ge 0}$. There is a constant $A(\beta\mu) \le (1 + e^{2\beta\mu})/4$ with the properties stated in Proposition 8 such that:

1. if $\beta k \le A(\beta\mu)$, then
$$\lim_{n\to+\infty} m(n) = 0$$
for every choice of $m(0)$;
2. if $\beta k > (1 + e^{2\beta\mu})/4$, then the trajectory $m(n)$ converges, as $n \to +\infty$, to a unique stable 2-cycle, for every $m(0) \ne 0$;
3. if $A(\beta\mu) < \beta k < (1 + e^{2\beta\mu})/4$, then there are $0 < \xi^{(1)} < \xi^{(2)}$ such that
   (i) the trajectory alternating $-\xi^{(1)}$ and $\xi^{(1)}$ is an unstable 2-cycle;
   (ii) the trajectory alternating $-\xi^{(2)}$ and $\xi^{(2)}$ is a locally stable 2-cycle;
   (iii) if $|m(0)| < \xi^{(1)}$, then $m(n)$ converges to 0 as $n \to +\infty$, while, if $|m(0)| > \xi^{(1)}$, then $m(n)$ converges to the stable 2-cycle.

Note that, even with the Cournot adjustment selection, the steady-state behavior deeply differs from the one of the corresponding non-strategic models.

4.4 Conclusion

We have introduced a modeling framework to include forecasting in the dynamics of interacting systems and discussed the problem of selecting a Nash equilibrium, making the resulting dynamics single-valued. In a simple mean-field spin model driven by the trend, the introduction of such forecasting dramatically changes the low-temperature dynamics, producing organized stable periodic behavior.

Acknowledgements We thank Marco LiCalzi and Paolo Pellizzari for inspiring discussions.

References

1. Bouchaud, J.P.: Crises and collective socio-economic phenomena: simple models and challenges. J. Stat. Phys. **151**, 567–606 (2013)
2. Brock, W.A., Durlauf, S.N.: Discrete choice with social interactions. Rev. Econ. Stud. **68**, 235–260 (2001)
3. Dai Pra, P., Sartori, E., Tolotti, M.: Strategic interaction in trend-driven dynamics. J. Stat. Phys. **152**, 724–741 (2013)
4. Dai Pra, P., Scoppola, B., Scoppola, E.: Sampling from a Gibbs measure with pair interaction by means of PCA. J. Stat. Phys. **149**, 722–737 (2012)
5. Fudenberg, D.: The Theory of Learning in Games, vol. 2. MIT Press, Cambridge (1998)
6. Landman, K.A., Binder, B.J., Newgreen, D.F.: Modeling development and disease in our "Second" brain. Cellular Automata, pp. 405–414. Springer, Berlin (2012)
7. Lämmer, S., Helbing, D.: Self-control of traffic lights and vehicle flows in urban road networks. J. Stat. Mech. Theory Exp. **2008**, P04019 (2012)
8. Myerson, R.B.: Game Theory. Harvard University Press, Cambridge (2013)
9. Piccoli, B., Tosin, A.: Time-evolving measures and macroscopic modeling of pedestrian flow. Arch. Ration. Mech. Anal. **199**, 707–738 (2011)

Chapter 5
Scaling and Inverse Scaling in Anisotropic Bootstrap Percolation

Aernout C.D. van Enter

Abstract In bootstrap percolation, it is known that the critical percolation threshold tends to converge slowly to zero with increasing system size, or, inversely, the critical size diverges fast when the percolation probability goes to zero. To obtain higher-order terms (i.e. sharp and sharper thresholds) for the percolation threshold in general is a hard question. In the case of two-dimensional anisotropic models, sometimes such correction terms can be obtained from inversion in a relatively simple manner.

5.1 Bootstrap Percolation Models

Bootstrap percolation models (also known in the literature as k-core percolation [1], neuropercolation [2, 3], jamming percolation [4], quorum percolation [5] or diffusion percolation [6]) are Cellular Automata, with a deterministic discrete-time dynamics. Often, however, probability is brought in, as one considers probabilistic initial conditions. Although bootstrap percolation models are not PCA in the proper sense, as CA combined with probability, they are close relations of PCA.

Bootstrap percolation models describe the growth of sets of occupied vertices (or sites) of a graph. At all vertices of a graph (whether finite or infinite), one places at an initial time with probability p a particle. The bootstrap percolation rule then requires each occupied vertex to stay occupied, and each empty vertex to become occupied whenever sufficiently many vertices in its neighbourhood are occupied. The choice of graph, the choice of "sufficiently many" and the choice of the neighbourhood determine the model. One is interested whether after sufficiently many iterations each vertex gets occupied or not, and how this depends on the value of p. In particular, one wants to know what happens for infinite graphs, or for sequences of increasing graphs. One also can consider more general rules, where an empty site gets occupied

A.C.D. van Enter (✉)
Johann Bernoulli Institute for Mathematics and Computer Science,
University of Groningen, P.O. Box 407, 9700 AK Groningen, The Netherlands
e-mail: a.c.d.van.enter@rug.nl

P.-Y. Louis and F.R. Nardi (eds.), *Probabilistic Cellular Automata*, Emergence, Complexity and Computation 27, https://doi.org/10.1007/978-3-319-65558-1_5

once a particular configuration, or one out of a particular set of configurations, in some neighbourhood is occupied. For example, the "modified" bootstrap percolation model requires that one neighbouring site along each lattice axis is occupied. The bootstrap percolation models have some obvious monotonicity properties, in particular the number of occupied vertices is growing in time, and there is stochastic monotonicity, in the sense that the occupation number of each vertex of one evolving configuration is always larger or equal than that of a second evolving configuration once this holds at the beginning.

Bootstrap percolation models have been applied in a variety of contexts, e.g. for the study of metastability [7] and for magnetic models [8], for the glass transition [4] and for capillary fluid flow [9], for study of neural networks [10, 11], rigidity [12], contagion [13], and they have also been studied for purely mathematical interest, including recreational mathematics [14–17].

Most interest is in the so-called critical models, in which the growth rule is such that a finite set of occupied sites (= vertices) cannot fill the infinite lattice, and at the same time all finite empty sets in an infinite occupied environment will be filled.

The simplest such models on (hyper-)cubic lattices are those where one considers the nearest neighbours of each site and requires half of them to be occupied to get occupied at the next step, or where one requires at least one occupied site among the neighbours along every axis (modified bootstrap percolation). In these models the most detailed results are known. In particular, it is known that on infinite lattices the percolation threshold is trivial ($p_c = 0$), that is, for every positive p every vertex of the infinite lattice will in the end be occupied with probability one [18, 19].

Moreover, on finite volumes the percolation threshold (now defined as the smallest value of p above which the volume will be occupied with probability above one half) scales as a $d - 1$-repeated logarithm of the size of the volume (i.e. $p_c = O(\frac{1}{\ln V})$ in $d = 2$, $p_c = O(\frac{1}{\ln \ln V})$ in $d = 3$, etc.). Such behaviour, with different constants for lower and upper bounds, was proven in [7, 20, 21], and with coinciding lower and upper bounds in [22–25]. These last types of results (i.e. $p_c = C\frac{1}{\ln V} + o(\frac{1}{\ln V})$ in $d = 2$, or $p_c = C\frac{1}{\ln \ln V} + o(\frac{1}{\ln \ln V})$ in $d = 3$, and similarly in general d with $d - 1$ times repeated logarithms for higher dimensions) have been called "sharp thresholds".

As for lower-order corrections, (estimates on those are also called "sharper" thresholds in the literature), in $d = 2$ the $o(\frac{1}{\ln V})$ terms were shown to be of order $O(\frac{1}{(\ln V)^{\frac{3}{2}}})$, see [26]. This strengthened earlier results of [27, 28]. For higher-dimensional results about "sharper" thresholds, see [29]. These sharper thresholds describe the systematic error which computational physicists in the past have run into, as is discussed e.g. in [30, 31].

However, another notion of sharp thresholds, based on a sharp-threshold theorem of Friedgut and Kalai, was presented in [32]. This ε-window—the window within which with large probability one will find the answer—provides an estimate for the statistical error, which is of order $O(\frac{\ln \ln V}{\ln^2 V})$ and hence much smaller than the systematic error. The statistical error being small with respect to the systematic error has been a source for various erroneous numerical estimates of percolation thresholds and their numerical precision in the past, as errors tended to be substantially underestimated.

In this contribution, I plan to describe to what extent the behaviour of bootstrap models is modified once the model becomes anisotropic, and in particular "unbalanced" (compare [33]). In particular, I will concentrate on the $(1, 2)-$model, introduced in [34], in which one considers an anisotropic neighbourhood consisting of the nearest neighbours along one axis, and the nearest and next-nearest neighbours along the other axis. The distinction between balanced and unbalanced rules is that in balanced cases the growth occurs with the same speed in different directions, whereas in unbalanced cases there are easy and hard directions for growth. It appears to be the case that in $d = 2$ a wide class of growth models is either balanced or unbalanced and that both classes display a characteristic scaling behaviour [35].

In higher dimensions, it turns out that the leading behaviour is ruled by the two "easiest" growth directions [36].

5.2 A Tractable Example: The (1, 2)-Model

In the $(1, 2)$-model, the neighbourhood of each site in $\mathbb{Z}^2$ consists of 2 sites in the east and west directions, and one site in the north and south directions. In picture form:

$$\mathcal{N} = \begin{matrix} & & \bullet & & \\ \bullet & \bullet & 0 & \bullet & \bullet \\ & & \bullet & & \end{matrix}$$

At every step, each empty site which has 3 of its neighbours (out of the 6 possible ones) occupied, becomes itself occupied, and every occupied site stays occupied forever. As an initial condition, we take a percolation configuration with initial occupation probability p. This model, which is critical, was introduced by Gravner and Griffeath [34], and they looked at its finite-size behaviour. The model is similar to, but somewhat easier to analyse than, the north-east-south model of Duarte [37], for which related but somewhat weaker results are known [38, 39].

The fact that $p_c = 0$ in the infinite lattice follows from an argument due to Schonmann, first given for Duarte's model [38, 40].

Indeed, let a $2 \times n$ rectangle be occupied, then the probability that this rectangle grows both eastwards and westwards is larger than the probability that at least 1 site in the columns east and west of this rectangle is occupied, which is $[1 - (1 - p)^n]^2$. The probability that this occurs in each column in a rectangle of size $l \times n$ we bound from below by $[1 - (1 - p)^n]^l$. Choose $n = \frac{C}{p} \ln \frac{1}{p}$, then this probability can be bounded by $(1 - p^C)^l$; once $C \geq 2$ and $l \geq \frac{1}{p^C}$, such a rectangle keeps growing in both directions with large probability; the fact that such an occupied and growing rectangle can occur with positive probability implies that somewhere in an infinite lattice such a nucleation centre will occur, and it will then fill up the whole lattice.

The question after this is how big a square volume should be for such a nucleation centre to occur with large probability (e.g. probability a half). The argument given above predicts that a $2 \times n$ rectangle occurs at some fixed location with probability at least $p^{2n} = e^{-O(\frac{1}{p} \ln^2 \frac{1}{p})}$ and that therefore the size of the square volume $V = N \times N$ should be the inverse of that probability, that is, if N (or V) $\geq e^{+O(\frac{1}{p} \ln^2 \frac{1}{p})}$, it can be filled with large probability. Inverting the argument implies an upper bound for the rate at which the percolation threshold decreases as a function of V, of the form

$$p_c \leq C_1 \frac{\ln^2 \ln V}{\ln V}. \tag{5.1}$$

An argument providing a lower bound for p_c of the same order, that is,

$$p_c \geq C_2 \frac{\ln^2 \ln V}{\ln V} \tag{5.2}$$

was developed in [41], using and correcting the analysis of [34].

In fact, one can improve the on above strategy, as follows (see [42], following [43]). One starts with a $2 \times \frac{2}{p} \ln \ln \frac{1}{p}$ rectangle, which has all its even (or odd) sites occupied, then at the next step, the whole rectangle is filled. After that, one grows with vertical steps of size 1 and horizontal steps of increasing size, through a sequence of rectangles R_n, which in the y-direction have size n and in the x-direction have size $\frac{1}{3p} \exp 3np$. This goes on until we reach the size $n = \frac{1}{3p} \ln \frac{1}{p}$. With this choice, the probability for a rectangle R_n to grow a step in the x-direction equals the probability to grow a step in the y-direction.

The probability of growing a step in the vertical direction from a rectangle R_n is approximately $8p^2 x_n$ (one needs two occupied sites close enough, the factor 8 here is of combinatorial origin) which equals $\frac{8p}{3} \exp 3np$. The probability of growing in the horizontal direction, over a distance $x_{n+1} - x_n$ equals a constant term $\frac{1}{e}$, for every n.

One thus needs to compute the product from $n_0 = \frac{2}{p} \ln \ln \frac{1}{p}$ to $n_f = \frac{1}{3p} \ln \frac{1}{p}$ over these probabilities. The result is

$$\begin{aligned}
\prod_{n=n_0}^{n=n_f} \frac{8p}{3e} \exp(3np) &= \frac{8p}{3e}^{(n_f - n_0)} \exp\left(3p \sum_{n=n_0}^{n=n_f} n\right) \\
&= \frac{8p}{3e}^{(n_f - n_0)} \exp\left(3p \left[\frac{1}{2} n_f(n_f - 1) - \frac{1}{2} n_0(n_0 - 1)\right]\right) \\
&= \exp\left(-\frac{1}{6p} \ln^2 \frac{1}{p} + \frac{1}{3} \ln \frac{8}{3p} \ln \frac{1}{p} + o\left(\frac{1}{p} \ln \frac{1}{p}\right)\right).
\end{aligned} \tag{5.3}$$

This is our main result, for the detailed proof that this strategy indeed provides the best estimate, see [42].

5.3 Inversion

If the probability for a nucleation centre to occur at a fixed location is given by an expression of the form $P = \exp(-\frac{C}{p})$, the necessary volume size to see such a nucleation centre with substantial probability in that volume, that is the "critical volume", will be $V_c = \exp(+\frac{C}{p})$, which is easily inverted, resulting in an expression of the form $p_c = C\frac{1}{\ln V}$ for the critical percolation threshold as a function of the volume.

However, if there are logarithmic corrections and subdominant terms as above, that is

$$V_c = \exp\left(\frac{C}{p}\ln^2\frac{1}{p} + \frac{C'}{p}\ln\frac{1}{p}\right), \tag{5.4}$$

to invert such expressions we need to perform some extra steps. We observe the following:

$$p_c = \frac{1}{\ln V}\left(C\ln^2\frac{1}{p_c} + C'\ln\frac{1}{p_c}\right). \tag{5.5}$$

We also notice that in the limit of V large and hence p_c small it holds that

$$\frac{1}{p_c} \leq \ln V \leq \frac{1}{p_c^{1+\varepsilon}} \tag{5.6}$$

and (by taking logarithms)

$$\ln\frac{1}{p_c} \leq \ln\ln V \leq (1+\varepsilon)\ln\frac{1}{p_c} \tag{5.7}$$

and

$$\ln\ln\frac{1}{p_c} \leq \ln\ln\ln V \leq \ln\ln\frac{1}{p_c} + \varepsilon. \tag{5.8}$$

Thus asymptotically, by substitution plus using the above estimates

$$\begin{aligned} p_c &= \frac{1}{\ln V}\left(C\ln^2\frac{1}{p_c} + C'\ln\frac{1}{p_c}\right) \\ &= \frac{1}{\ln V}\left(C(\ln\ln V - 2\ln\ln\ln V - \ln C + O(\varepsilon))^2 + C'\ln\ln V + O(\varepsilon)\right) \\ &= \frac{1}{\ln V}\left(C(\ln^2\ln V - 4\ln\ln\ln V\ln\ln V - 2\ln C\ln\ln V) + +C'\ln\ln V + O(\ln^2\ln\ln V)\right). \end{aligned} \tag{5.9}$$

Hence knowing the second term in the critical volume provides a third term in the critical probability, and we also notice that the second term in the critical probability does not depend on the constant C' of this second critical-volume term.

A related argument was used in [44] to estimate the ε-window. This analysis extended the analysis of [32], applying the sharp-threshold theorem of Friedgut and Kalai, and the ε-window turns out to have width $O(\frac{\ln^3 \ln V}{\ln^2 V})$.

Numerically, that is for computational physicists, e.g., these results are totally discouraging. Whereas in standard bootstrap percolation to obtain a 99% accuracy in p_c the order of magnitude of a square already needs to be of order $O(10^{3000})$ [27], in the $(1, 2)$-model one needs to go even higher, namely to a doubly exponential size of order $O(10^{10^{1400}})$. These findings illustrate the point made in [45] that Cellular Automata, despite being discrete in state, space and time, may still be ill-suited for computer simulations.

5.4 Generalisations: Related Models, Higher Dimensions and Other Graphs

In ordinary and modified bootstrap percolation, we have quite precise results. There is a variety of related models with similar behaviour, e.g. [46–49]. In particular, it is remarkable that the model of [47] is anisotropic, but nonetheless scales in the same way as ordinary bootstrap percolation; in the terms of [33], it is "balanced". A much wider class of models was recently considered in [50], in which some general order-of-magnitude results were derived for critical models. More recently [35], it was shown that this class consists of two subclasses, either the balanced ones, such as ordinary bootstrap percolation, which display similar asymptotic behaviour, or the unbalanced ones, in which logarithmic corrections of the type displayed in the $(1, 2)$-model occur. The essential distinction is that balanced models grow at the same rate in two different directions, whereas unbalanced models have an easy and a hard growth direction.

There exist also some results on bootstrap percolation in higher dimensions. In the anisotropic case, for the time being we only know order-of-magnitude results for (a, b, c)-models, in which neighbourhoods are considered which consist of neighbours at distances a, b and c $(a \leq b \leq c)$ along the different axes [36] (of which again half the sites need to be occupied to occupy an empty site). The result is that the scaling becomes doubly exponential, or, for the inverse quantities, doubly logarithmic, similarly to the isotropic models [20, 21], but with the behaviour controlled by the two-dimensional (a, b)-model. One bound is based on a variation of Schonmann's [19] induction-on-dimension argument, the other direction follows a similar strategy as [20]. To establish any form of a sharp threshold, however, is open for the time being.

In two dimensions, the $(1, b)$-models can be analysed along similar lines as the $(1, 2)$-model, which results in the same asymptotics, but with the (sharp and computable) constant $C = \frac{(b-1)^2}{2(b+1)}$, rather than $C = \frac{1}{6}$, as the leading term. For the sharp constant in the Duarte model, see [35].

A quite different family of results, in which there is a transition at a finite threshold p, occurs for bootstrap percolation models on either trees [51–54], random graphs [55, 56], or hyperbolic lattices [57]. Such transitions have a "hybrid" (mixed first-second order) character, in the sense that on the one hand, one finds that at the threshold the infinite cluster has a minimum density (so it jumps from zero, just as one expects at a first-order phase transition), while at the same time there are divergent correlation lengths and non-trivial critical exponents, which are characteristic for second-order (critical) phase transitions. Such hybrid "random first-order" transitions have been proposed to be characteristic for glass transitions. See, e.g. [58]. On regular lattices, models with this kind of behaviour cannot be constructed via the type of bootstrap percolation rules discussed above, but more complicated Cellular Automaton growth rules with this type of behaviour have been studied in [59, 60].

Acknowledgements I thank the organisers for their invitation to talk at the 2013 EURANDOM meeting on Probabilistic Cellular Automata, and I thank my colleagues and co-workers, Joan Adler, Jose Duarte, Hugo Duminil-Copin, Anne Fey-den Boer, Tim Hulshof and Rob Morris, as well as Susan Boerma-Klooster and Roberto Schonmann, for all they taught me. I thank Rob Morris for correcting me on the $(1, b)$-constant. Moreover, I thank Tim Hulshof for helpful advice on the manuscript.

References

1. Harris, A.B., Schwarz, J.M.: $\frac{1}{d}$ expansion for k-core percolation. Phys. Rev. E **72**, 046123 (2005)
2. Kozma, R.: Neuropercolation. http://www.scholarpedia.org/article/Neuropercolation
3. Kozma, R., Puljic, M., Balister, P., Bollobas, B., Freeman, W.J.: Phase transitions in the neuropercolation model of neural populations with mixed local and non-local interactions. Biol. Cybern. **92**, 367–379 (2005)
4. Toninelli, C.: Bootstrap and jamming percolation. In: Bouchaud, J.P., Mézard, M., Dalibard, J. (eds.) Les Houches school on Complex Systems, Session LXXXV, pp. 289–308 (2006)
5. Eckman, J.P., Moses, E., Stetter, E., Tlusty, T., Zbinden, C.: Leaders of neural cultures in a quorum percolation model. Front. Comput. Neurosci. **4**, 132 (2010)
6. Adler, J., Aharony, A.: Diffusion percolation: infinite time limit and bootstrap percolation. J. Phys. A **21**, 1387–1404 (1988)
7. Aizenman, M., Lebowitz, J.L.: Metastability effects in bootstrap percolation. J. Phys. A **21**, 3801–3813 (1988)
8. Adler, J.: Bootstrap percolation. Phys. A **171**, 452–470 (1991)
9. Lenormand, R., Zarcone, C.: Growth of clusters during imbibition in a network of capillaries. In: Family, F., Landau, D.P. (eds.) Kinetics of Aggregation and Gelation, pp. 177–180. Elsevier, Amsterdam (1984)
10. Amini, H.: Bootstrap percolation in living neural networks. J. Stat. Phys. **141**, 459–475 (2010)
11. Eckman, J.P., Tlusty, T.: Remarks on bootstrap percolation in metric networks. J. Phys. A Math. Theor. **42**, 205004 (2009)
12. Connelly, R., Rybnikov, K., Volkov, S.: Percolation of the loss of tension in an infinite triangular lattice. J. Stat. Phys. **105**, 143–171 (2001)
13. Lee, I.H., Valentiniy, A.: Noisy contagion without mutation. Rev. Econ. Stud. **67**, 47–67 (2000)

14. Bollobas, B.: The Art of mathematics: Coffee Time in Memphis, Problems 34 and 35. Cambridge University Press, Cambridge (2006)
15. New York Times Wordplay Blog (8 July 2013). http://wordplay.blogs.nytimes.com/2013/07/08/bollobas/
16. Winkler, P.: Mathematical Puzzles. A.K. Peters Ltd, p. 79 (2004)
17. Winkler, P.: Mathematical Mindbenders, A.K. Peters Ltd, p. 91 (2007)
18. van Enter, A.C.D.: Proof of Straley's argument for bootstrap percolation. J. Stat. Phys. **48**, 943–945 (1987)
19. Schonmann, R.H.: On the behavior of some cellular automata related to bootstrap percolation. Ann. Probab. **20**, 174–193 (1992)
20. Cerf, R., Cirillo, E.: Finite size scaling in three-dimensional bootstrap percolation. Ann. Prob. **27**(4), 1837–1850 (1999)
21. Cerf, R., Manzo, F.: The threshold regime of finite volume bootstrap percolation. Stoch. Process. Appl. **101**, 69–82 (2002)
22. Balogh, J., Bollobas, B., Morris, R.: Bootstrap percolation in three dimensions. Ann. Probab. **37**, 1329–1380 (2009)
23. Balogh, J., Bollobas, B., Duminil-Copin, H., Morris, R.: The sharp threshold for bootstrap percolation in all dimensions. Trans. Am. Math. Soc. **364**, 2667–2701 (2012)
24. Holroyd, A.E.: Sharp metastability threshold for two-dimensional bootstrap percolation. Probab. Theory Relat. Fields **125**, 195–224 (2003)
25. Holroyd, A.E.: The metastability threshold for modified bootstrap percolation in d dimensions. Electron. J. Probab. **11**(17), 418–433 (2006)
26. Morris, R.: The second term for bootstrap percolation in two dimensions. Manuscript in preparation. http://w3.impa.br/~rob/index.html
27. Gravner, J., Holroyd, A.E.: Slow convergence in bootstrap percolation. Ann. Appl. Probab. **18**, 909–928 (2008)
28. Gravner, J., Holroyd, A.E., Morris, R.: A sharper threshold for bootstrap percolation in two dimensions. Probab. Theory Relat. Fields **153**, 1–23 (2012)
29. Uzzell, A.E.: An improved upper bound for bootstrap percolation in all dimensions (2012). arXiv:1204.3190
30. Adler, J., Lev, U.: Bootstrap percolation: visualisations and applications. Braz. J. Phys. **33**, 641–644 (2003)
31. de Gregorio, P., Lawlor, A., Bradley, P., Dawson, K.A.: Clarification of the bootstrap percolation paradox. Phys. Rev. Lett. **93**, 025501 (2004)
32. Balogh, J., Bollobas, B.: Sharp thresholds in bootstrap percolation. Phys. A **326**, 305–312 (2003)
33. Duminil-Copin, H., Holroyd, A.E.: Finite volume bootstrap percolation with balanced threshold rules on $\mathbb{Z}^2$. Preprint (2012). http://www.unige.ch/duminil/
34. Gravner, J., Griffeath, D.: First passage times for threshold growth dynamics on $\mathbb{Z}^2$. Ann. Probab. **24**, 1752–1778 (1996)
35. Bollobás, B., Duminil-Copin, H., Morris, R., Smith, P.: The sharp threshold for the Duarte model. Ann. Probab. To appear (2016). arXiv:1603.05237
36. van Enter, A.C.D., Fey, A.: Metastability thresholds for anisotropic bootstrap percolation in three dimensions. J. Stat. Phys. **147**, 97–112 (2012)
37. Duarte, J.A.M.S.: Simulation of a cellular automaton with an oriented bootstrap rule. Phys. A **157**, 1075–1079 (1989)
38. Adler, J., Duarte, J.A.M.S., van Enter, A.C.D.: Finite-size effects for some bootstrap percolation models. J. Stat. Phys. **60**, 323–332 (1990); Addendum. J. Stat. Phys. **62**, 505–506 (1991)
39. Mountford, T.S.: Critical lengths for semi-oriented bootstrap percolation. Stoch. Proc. Appl. **95**, 185–205 (1995)
40. Schonmann, R.H.: Critical points of 2-dimensional bootstrap percolation-like cellular automata. J. Stat. Phys. **58**, 1239–1244 (1990)
41. van Enter, A.C.D., Hulshof, W.J.T.: Finite-size effects for anisotropic bootstrap percolation: logarithmic corrections. J. Stat. Phys. **128**, 1383–1389 (2007)

42. Duminil-Copin, H., van Enter, A.C.D., Hulshof, W.J.T.: Higher order corrections for anisotropic bootstrap percolation. Prob. Th. Rel. Fields. arXiv:1611.03294 (2016)
43. Duminil-Copin, H., van Enter, A.C.D.: Sharp metastability threshold for an anisotropic bootstrap percolation model. Ann. Prob. **41**, 1218–1242 (2013)
44. Boerma-Klooster, S.: A sharp threshold for an anisotropic bootstrap percolation model. Groningen bachelor thesis (2011)
45. Gray, L.: A mathematician looks at Wolfram's new kind of science. Not. Am. Math. Soc. **50**, 200–211 (2003)
46. Bringmann, K., Mahlburg, K.: Improved bounds on metastability thresholds and probabilities for generalized bootstrap percolation. Trans. Am. Math. Soc. **364**, 3829–2859 (2012)
47. Bringmann, K., Mahlburg, K., Mellit, A.: Convolution bootstrap percolation models, Markov-type stochastic processes, and mock theta functions. Int. Math. Res. Not. **2013**, 971–1013 (2013)
48. Froböse, K.: Finite-size effects in a cellular automaton for diffusion. J. Stat. Phys. **55**, 1285–1292 (1989)
49. Holroyd, A.E., Liggett, T.M., Romik, D.: Integrals, partitions and cellular automata. Trans. Am. Math. Soc. **356**, 3349–3368 (2004)
50. Bollobás, B., Smith, P., Uzzell, A.: Monotone cellular automata in a random environment. Comb. Probab. Comput. **24**(4), 687–722 (2015)
51. Balogh, J., Bollobas, B., Pete, G.: Bootstrap percolation on infinite trees and non-amenable groups. Comb. Probab. Comput. **15**, 715–730 (2006)
52. Bollobás, B., Gunderson, K., Holmgren, C., Janson, S., Przykucki, M.: Bootstrap percolation on Galton-Watson trees El. J. Prob. **19**, 13 (2014)
53. Chalupa, J., Leath, P.L., Reich, G.R.: Bootstrap percolation on a Bethe lattice. J. Phys. C **12**, L31 (1979)
54. Schwarz, J.M., Liu, A.J., Chayes, L.Q.: The onset of jamming as the sudden emergence of an infinite k-core cluster. Europhys. Lett. **73**, 560–566 (2006)
55. Balogh, J., Pittel, B.G.: Bootstrap percolation on the random regular graph. Random Struct. Algorithms **30**, 257–286 (2007)
56. Pittel, B., Spencer, J., Wormald, N.: Sudden emergence of a giant k-core. J. Comb. Theory B **67**, 111–151 (1996)
57. Sausset, F., Toninelli, C., Biroli, G., Tarjus, G.: Bootstrap percolation and kinetically constrained models on hyperbolic lattices. J. Stat. Phys. **138**, 411–430 (2010)
58. Bouchaud, J.P., Cipelletti, L., van Saarloos, W.: In: Berthier, L. (ed.) Dynamic Heterogeneities in glasses, Colloids, and Granular Media. Oxford University Press, Oxford (2011)
59. Jeng, M., Schwarz, J.M.: On the study of jamming percolation. J. Stat. Phys. **131**, 575–595 (2008)
60. Toninelli, C., Biroli, G.: A new class of cellular automata with a discontinuous glass transition. J. Stat. Phys. **130**, 83–112 (2008)

Chapter 6
The Sandpile Cellular Automaton

Antal A. Járai

Abstract This is a short introduction to the sandpile cellular automaton. It is aimed at non-specialist, who may not be familiar with statistical physics models. Technicalities are kept to a minimum and the emphasis is on motivation, clear definitions, key properties and some of the challenges.

6.1 Introduction

The Abelian sandpile model is a probabilistic cellular automaton. It was introduced by physicists Bak, Tang and Wiesenfeld [1] and Dhar [3] as an example for the concept of self-organized criticality (SOC). The aim of SOC was to provide a general mechanism by which spatially and temporally correlated structures with fractal geometry appear. Some of the motivating examples were: ecological systems, intensity of sunspots, current through resistors, sand flow and stock exchange indices. It is typical for SOC systems that small inputs accumulate over a long period of time, resulting in occasional sudden bursts of activity; for example, slow tectonic movements resulting in earthquakes.

The Abelian sandpile became one of the prime theoretical models of SOC. Dhar [3] discovered that it has special features (e.g. an Abelian group structure) that allow to compute various quantities, and he used the model as a test-ground for ideas about SOC. It is interesting to note that the model has been discovered in several equivalent forms, independently, in other disciplines. For example, in combinatorics it is known as the "chip-firing game"; see the references in [8]. In this note, we will give the definition of the model and state some of its main properties. Some of the key challenges in its analysis are outlined.

A.A. Járai (✉)
Department of Mathematical Sciences, University of Bath, Claverton Down, Bath BA2 7AY, UK
e-mail: A.Jarai@bath.ac.uk

P.-Y. Louis and F.R. Nardi (eds.), *Probabilistic Cellular Automata*, Emergence, Complexity and Computation 27, https://doi.org/10.1007/978-3-319-65558-1_6

There is a large body of work on Abelian sandpiles and SOC. We mention a few pointers for the reader interested in the more technical aspects. The book [15] describes physical and biological examples where the concept of SOC is thought to apply. The survey [4] gives the theoretical physics perspective on SOC with a good deal of it devoted to properties of the Abelian sandpile. Mathematical aspects of sandpiles are covered by [8, 9, 28].

6.2 Definition, Examples and Some Remarks

An example of the sandpile model in $d = 2$ dimensions is shown in Fig. 6.1. Each cell in a 4×4 grid is occupied by a number of indistinguishable particles (grains of sand, chips, units of energy, etc.), and this collection is called a *sandpile*. When the number of particles is 0, 1, 2 or 3, the cell is called *stable*, otherwise it is called *unstable*. By a *stable sandpile* we mean one where all the cells are stable.

We now define a stochastic dynamics on the collection of all stable sandpiles. Given a stable sandpile, select one of the cells at random, each cell having equal probability, and add a particle there. If after addition the cell is still stable, as in Fig. 6.1, this completes the definition of a single step of the dynamics. Suppose that after an addition the cell becomes unstable, as in the first square of Fig. 6.2. In this case the system undergoes a so-called "avalanche", defined by a cellular automaton (CA) rule. The rule is that if there is any unstable cell (where the number of particles is at least 4), that cell "topples", and sends one particle to each of its neighbours. We continue to topple unstable cells until each cell becomes stable again. That this must happen eventually follows from the requirement that when a toppling occurs on the boundary, one or more particles leave the system. Once the avalanche is over, the new stable sandpile that has just been reached defines the result of a single step of the dynamics. The dynamics is continued by selecting a new cell at random for addition, independently of the past history of the process.

One may ask whether it matters in what order topplings are carried out (in Fig. 6.2 they were carried out in parallel). The answer is that it does not; this is called the *Abelian property*, that gave the model its name [3]. Similarly to the example given, we can define the dynamics on larger grids, in higher dimensions, or on more general finite graphs. In general, the threshold at which a cell becomes unstable is defined as the number of neighbouring cells. We require that there exist a *sink*, that is a

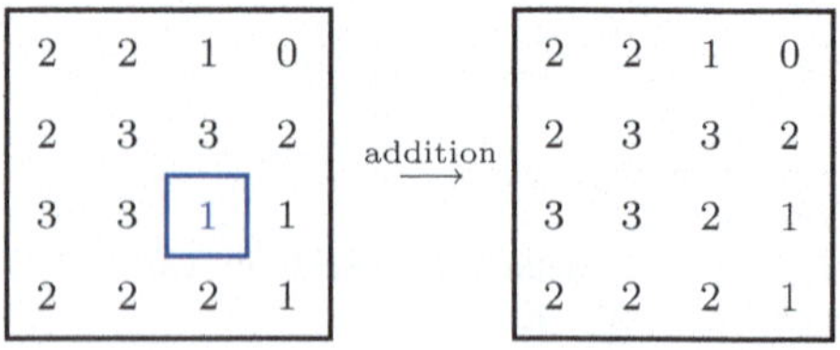

Fig. 6.1 Example of the Abelian sandpile on a 4×4 grid. A particle has been added at one of the cells

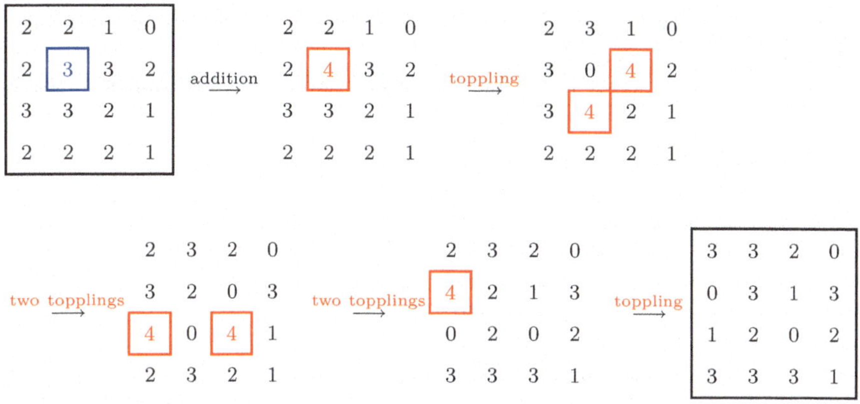

Fig. 6.2 A particle addition resulting in a sequence of topplings. The initial configuration and the final configuration reached after the avalanche are the framed squares. The intermediate configurations during the avalanche are unframed

distinguished vertex of the graph where particles are lost (in the example above, all boundary cells can be viewed as being connected to a sink). Oriented graphs can also be considered [8]. The more general setting of graphs, and in particular sandpiles on heterogeneous networks may be of interest for applications.

Mathematically, the Abelian sandpile model is a Markov chain, where the state-space is the collection of all stable sandpiles (where no cell can topple), and transitions are triggered by random additions. One is interested in the long-run behaviour, that is, the stationary state of the Markov chain.

The sandpile dynamics is not intended as a realistic model of sand particles toppling down a slope. In a more realistic model, topplings should occur if the discrete gradient of the height function in some direction exceeds a critical value, and in this case particles should move in that direction. Such models will typically not be Abelian (the order of topplings does matter), and this can make their analysis more difficult. The Abelian model, while not realistic, is a useful abstraction that already contains some essential features of avalanches observed in experiments [15]. One such feature is that in the stationary state, the size of avalanches spans all length scales allowed by system size.

In order to see heuristically that avalanches will span all length scales, consider a large grid and suppose we start with an empty initial condition, that is no particles. Initially, there will be very few avalanches, and the ones that occur will be small (only few topplings). As more and more particles get added, cells will have sufficient number of particles to allow avalanches to spread out more. The only limit to the growth of the extent of avalanches is that at the boundary of the grid particles are lost. Numerical simulations show that on a large grid the probability distribution of the size of avalanches (at stationarity) follows a power law. The probability of seeing an avalanche with s topplings is approximately $cs^{-\tau}$, for all $1 \leq s \leq s_{\max}$, where the cut-off value $s_{\max}$ depends on the system size. The value of the exponent τ depends on the spatial dimension. The power law distribution signifies that there is no typical length scale for the avalanches.

At stationarity, there will be a delicate balance between two mechanisms: particle additions act to increase the number of particles, avalanches on the other hand act to spread them out and dissipate "excess" through the boundary. There will be a critical density of particles that is just enough to allow avalanches to propagate. Fey, Levine and Wilson [7] and Levine [18] analyze this "approach to criticality" rigorously. The existence of a critical density is in analogy with the behaviour of real sand: if sand is poured very slowly onto a table, the conical pile will have a characteristic slope that is just large enough to allow particles to topple down.

Since in the sandpile model there is no characteristic spatial scale, the correlation between the number of particles at different spatial locations can also be expected to decay as a power law. It is indeed so; see Sect. 6.3.4. The way SOC manifests itself in the sandpile model is that the critical density of particles, the power law avalanche size and power law correlations are reached and maintained due to the dynamics. Therefore the reason to call it "self-organized".

So far our discussion was focused on a model where particles can only leave the system through the boundary. It is also natural to consider models where particles can be dissipated from the bulk; this could arise naturally in applications. If a positive fraction of cells in the bulk are connected to the sink, correlations will decay exponentially, and there will be a characteristic length scale. The same will be true if we pass to a continuous height model where a small amount of dissipation takes place at every cell; see [12].

While historically the main interest has been in the stationary state of the sandpile Markov chain, there are some very natural questions to ask about the deterministic CA itself. For example, place a large number n of particles at the origin of the two-dimensional grid $\mathbb{Z}^2$, and no particles elsewhere. Let the pile stabilize via the topplings. What can we say about the final configuration, as $n \to \infty$? Pegden and Smart have recently shown [24] that the intriguing fractal patters seen in simulations of this problem have a scaling limit.

6.3 Key Properties

This section summarizes some of the main properties of Abelian sandpiles that can be derived in a mathematically rigorous way. The reader is referred to [3, 8, 9, 28] for proofs.

6.3.1 The Least Action Principle

We noted earlier that when an arbitrary sandpile is stabilized via topplings, the order of topplings does not matter: each cell will topple the same number of times regardless of what order is chosen. Stabilization in fact satisfies a stronger property, called the *least action principle* that has been introduced in [6]. Suppose that we allow *illegal topplings*, that is, ones that make the number of particles negative at the toppled cell. Then in any sequence of (possibly illegal) topplings whose final result is stable, each cell is toppled at least as many times as in a legal stabilizing sequence. That is, during legal stabilization each cell carries out the minimum amount of work necessary to reach a stable configuration, even among possibly illegal sequences.

The least action principle was used by Fey, Levine and Peres [6], to prove bounds on the asymptotic shape of the stabilization of a single column of particles. One way the principle can be used is if we have a good a priori guess of the number of times each vertex will topple, also called the *odometer function*. The odometer function was used by Levine and Peres [19] to prove shape theorems for aggregation models closely related to the sandpile cellular automaton. The least action principle also plays an important role in the scaling limit result of Pegden and Smart [24].

6.3.2 Recurrent Sandpiles, Group Structure, Dhar's Formula

A sandpile is called *recurrent*, if it is a recurrent state of the associated Markov chain. Equivalently, the sandpile is visited infinitely often by the Markov chain, with probability 1. A further equivalent formulation is the following property: η is a recurrent sandpile, if given any sandpile σ, it is possible to add particles to σ and then stabilize via topplings to get η. Let us write $\mathcal{R}$ for the collection of recurrent sandpiles.

Typically, only a small fraction of the stable sandpiles are recurrent. In fact, the number of recurrent sandpiles can be expressed in terms of the combinatorial Laplacian. Suppose $\Lambda \subset \mathbb{Z}^d$ is a finite set, and consider the Abelian sandpile in Λ. Define the entries of a $|\Lambda| \times |\Lambda|$ matrix Δ as follows

$$\Delta_{x,y} = \begin{cases} 2d & \text{if } x = y; \\ -1 & \text{if } |x - y| = 1; \\ 0 & \text{otherwise.} \end{cases}$$

The number of recurrent sandpiles in Λ is $\det(\Delta)$; see below.

The following binary operation $\oplus$ turns $\mathcal{R}$ into an Abelian group [3]. Given $\eta, \zeta \in \mathcal{R}$, add them elementwise: $\eta + \zeta$, and then stabilize via topplings to obtain $\eta \oplus \zeta \in \mathcal{R}$. It can be shown that the group obtained, called the *sandpile group* is isomorphic to $\mathbb{Z}^\Lambda / \mathbb{Z}^\Lambda \Delta$, where the subgroup we factor out with in this formula is the integer row span of the matrix Δ. The index of this subgroup is easily seen to be $\det(\Delta)$.

Once the sandpile Markov chain reaches the set $\mathcal{R}$, it never leaves it. The stationary distribution of the Markov chain is seen to be the uniform distribution on $\mathcal{R}$. Writing $n_{x,z}(\eta)$ for the number of topplings at z caused by adding a particle at x to the sandpile

η, we have the relation:

$$(\eta \oplus \mathbf{1}_x)(y) = \eta(y) + \mathbf{1}_x(y) - \sum_z n_{x,z}(\eta)\Delta_{z,y},$$

where $\mathbf{1}_x$ is the sandpile with a single particle at x and no particles elsewhere. Averaging over η, we get

$$0 = \mathbf{1}_x(y) - \sum_z \mathbb{E}[n_{x,z}]\Delta_{z,y},$$

and hence

$$\mathbb{E}[n_{x,z}] = \Delta^{-1}_{x,z}.$$

This extremely useful formula for the average number of topplings caused by adding a single particle is known as *Dhar's formula*.

6.3.3 Bijection with Spanning Trees

The Matrix-Tree Theorem in combinatorics [2] states that $\det(\Delta)$ equals the number of spanning trees of the graph obtained from Λ by adding the sink vertex. Based on Dhar's burning algorithm [3], Majumdar and Dhar [23] gave an explicit bijection between the collection of recurrent sandpiles $\mathcal{R}$ and the collection of spanning trees. This bijection has been very fruitful in proving things about the sandpile model and is outlined here.

Start with an arbitrary stable sandpile, as in the top-left corner of Fig. 6.3. Any cell that has at least as many particles as neighbours is declared burnt and is removed. These cells are represented as red squares in Fig. 6.3. Now this step is iterated: any cell that has at least as many particles as unburnt neighbours is declared burnt and is removed. It can be shown that the sandpile we started with is recurrent if and only if all cells burn eventually [3, 9, 23]. The bijection is constructed by linking each cell to a cell burnt one step before, using the sink vertex as the root of the tree; see [9, 23] for details.

As noted earlier, the stationary distribution is uniform on $\mathcal{R}$, so the bijection carries this over to the uniform distribution on spanning trees. The latter object is called the Uniform Spanning Tree (UST), and has been studied extensively; see the book [21]. Facts about the uniform spanning tree can be used to construct a sandpile measure on a large class of infinite graphs [13]. If $G = (V, E)$ is a connected infinite graph, let $V_1 \subset V_2 \subset \cdots \subset V$ be a sequence of finite vertex sets such that $\cup_{n=1}^{\infty} V_n = V$. Let ν_n be the stationary distribution of the sandpile Markov chain on V_n (where a sink vertex has been added with the boundary vertices of V_n connected to it). Then

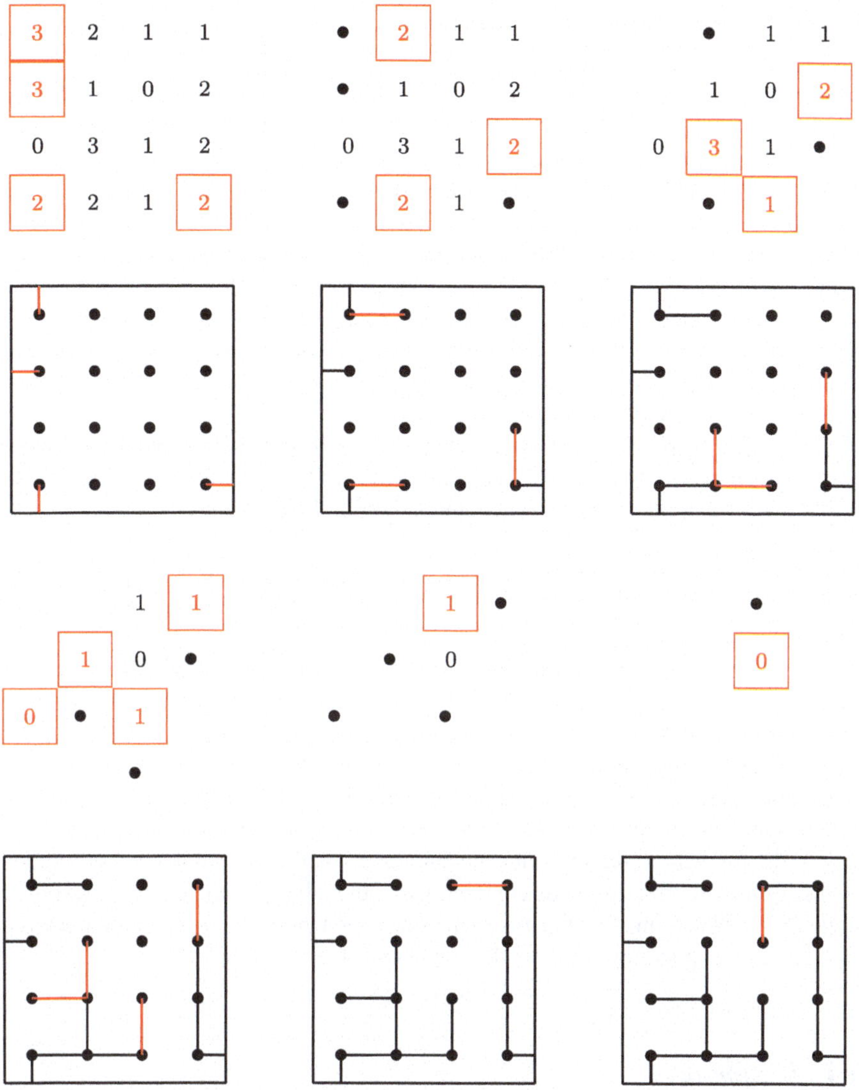

Fig. 6.3 The burning algorithm of Dhar and the bijection arising from it. The *red squares* are cells being burnt. *Black dots* indicate cells that were burnt one step before. The thick frame represents the sink vertex

provided certain conditions on G, the distributions ν_n converge weakly to a measure ν on stable sandpiles on G that is independent of the chosen sequence $V_1, V_2, \ldots$.

6.3.4 Some Computable Quantities

Majumdar and Dhar [22] computed the probability that for the sandpile on $\mathbb{Z}^2$ the origin has no particles. Based on the idea that $|\mathcal{R}| = \det(\Delta)$, the answer can be written in terms of a finite determinant with entries involving the simple random walk potential kernel [29]. The answer is: $p(0) = 2/\pi^2 - 4/\pi^3$.

Using the same method, Majumdar and Dhar also computed the covariance between the events that the origin and cell x, respectively, have no particles, and found that in all dimensions $d \geq 2$ this behaves as $c|x|^{-2d}$ for large x, with some $c < 0$. The 2D model is especially interesting here, due to conjectured conformal covariance properties of the scaling limit; see e.g. [14]. Dürre showed that in any smooth domain in $\mathbb{C}$, the k-point correlations, appropriately rescaled, have a scaling limit that is conformally covariant.

Computing the probabilities of 1, 2, etc. particles, respectively, proved much more difficult even in two dimensions. The combined result of the works [14, 17, 25, 26] is that in $d = 2$ dimensions the remaining height probabilities are:

$$p(1) = \frac{1}{4} - \frac{1}{2\pi} - \frac{3}{\pi^2} + \frac{12}{\pi^3} \qquad p(2) = \frac{3}{8} + \frac{1}{\pi} - \frac{12}{\pi^3} \qquad p(3) = \frac{3}{8} - \frac{1}{2\pi} + \frac{1}{\pi^2} + \frac{4}{\pi^3}.$$

It has been a mystery for a long time why the *particle density* $\sum_{i=0}^{3} ip(i)$ comes out to be the simple rational number $17/8$. Recently, Kassel and Wilson [16] gave a beautiful simple proof of this fact that only relies on symmetry properties of the lattice. Their method applies to other two-dimensional lattices as well.

Detailed computations are possible for sandpiles on a D-regular tree [5], where various types of sandpile configurations can be counted using generating functions. Due to the tree structure, an avalanche started at a cell o is completely determined by the connected cluster of cells containing o with all heights equal to the maximum value $D - 1$. It is found that the avalanche size exponent is $\tau = 3/2$. It is conjectured that the same exponent is valid in all dimensions $d \geq 5$; see [27].

6.4 Challenges

We close this note by outlining some challenges that are not covered by the discussions in the previous sections.

Dynamical properties. A lot is known about the stationary distribution of the Abelian sandpile model, for example via the bijection with spanning trees. We know much less about dynamical properties. Dhar's formula gives control of the first moment of topplings in avalanches, but we have no workable formula for the second moment. Similarly, we have no good understanding of how two successive avalanches are correlated. Any progress on such characteristics would be very valuable.

General networks. Sandpiles on general networks can be useful from the point of view of applications. The basic theory is not dependent on graph structure. However, in order to make quantitative statements about avalanches, some assumptions have to be made about the underlying graph. There is work in progress on sandpiles on Galton–Watson trees [11]. It would be important to have work on graphs that are not trees nor close to trees.

Critical exponents. There is no fully rigorous argument establishing the value of the avalanche exponent τ, even in dimensions $d \geq 5$, where it is expected that the value is the same as on a 3-regular tree: $\tau = 3/2$. Lyons, Morris and Schramm [20] proved bounds for the *radius exponent*, that is, the probability that adding a particle at the origin, say, causes a toppling at distance R away. The bound they get in $d \geq 5$ is believed to be optimal up to a logarithmic factor.

Avalanches in two dimensions. Járai and Redig [10] showed that in dimensions $d \geq 3$ the sandpile measure in $\mathbb{Z}^d$ has the finite avalanche property: adding a single particle to the configuration produces a finite number of topplings almost surely. We believe the same is true in dimension $d = 2$, but there is no proof so far.

Scaling limit in two dimensions. As mentioned earlier, computations made by physicists, e.g. [14], point to conformal covariance properties in the scaling limit in 2D. It would be important to have rigorous proofs of such statements.

References

1. Bak, P., Tang, C., Wiesenfeld, K.: Self-organized criticality. Phys. Rev. A **38**, 364–374 (1988)
2. Bollobás, B.: Modern graph theory. Graduate Texts in Mathematics, vol. 184. Springer, New York (1998)
3. Dhar, D.: Self-organized critical state of sandpile automaton models. Phys. Rev. Lett. **64**, 1613–1616 (1990)
4. Dhar, D.: Theoretical studies of self-organized criticality. Phys. A **369**, 29–70 (2006)
5. Dhar, D., Majumdar, S.N.: Abelian sandpile model on the Bethe lattice. J. Phys. A **23**, 4333–4350 (1990)
6. Fey, A., Levine, L., Peres, Y.: Growth rates and explosions in sandpiles. J. Stat. Phys. **138**, 143–159 (2010)
7. Fey, A., Levine, L., Wilson, D.B.: Approach to criticality in sandpiles. Phys. Rev. E **82**, 031121 (2010). https://doi.org/10.1103/PhysRevE.82.031121
8. Holroyd, A.E., Levine, L., Mészáros, K., Peres, Y., Propp, J., Wilson, D.B.: Chip-firing and rotor-routing on directed graphs. In and Out of Equilibrium 2. Progress in Probability, vol. 60. Birkhäuser, Basel (2008)
9. Járai, A.A.: Sandpile Models. Lecture Notes for the 9th Cornell Probability Summer School (2014). arXiv:1401.0354
10. Járai, A.A., Redig, F.: Infinite volume limit of the abelian sandpile model in dimensions $d \geq 3$. Probab. Theory Relat. Fields **141**, 181–212 (2008)
11. Járai, A.A., Werning, N.: Minimal configurations and sandpile measures. J. Theor. Probab. **27**, 153–167 (2016)
12. Járai, A.A., Ruszel, W., Saada, E.: Sandpiles on Galton-Watson trees. In preparation (2014)
13. Járai, A.A., Redig, F., Saada, E.: Approaching criticality via the zero dissipation limit in the abelian avalanche model. J. Stat. Phys. **159**(6), 1369–1407 (2015)

14. Jeng, M., Piroux, G., Ruelle, P.: Height variables in the Abelian sandpile model: scaling fields and correlations. J. Stat. Mech. Theory Exp. P10015+63 (2006). https://doi.org/10.1088/1742-5468/2006/10/P10015
15. Jensen, H.J.: Self-Organized Criticality: Emergent Complex Behavior in Physical and Biological Systems. Cambridge Lecture Notes in Physics, vol. 10. Cambridge University Press, Cambridge (1998)
16. Kassel, A., Wilson, D.B.: Looping rate and sandpile density of planar graphs. Am. Math. Mon. **123**(1), 19–39 (2016)
17. Kenyon, R., Wilson, D.B.: Spanning trees of graphs on surfaces and the intensity of loop-erased random walk on planar graphs. J. Am. Math. Soc. **28**(4), 985–1030 (2015)
18. Levine, L.: Threshold state and a conjecture of Poghosyan, Poghosyan, Priezzhev and Ruelle. Comm. Math. Phys. **335**(2), 1003–1017 (2015)
19. Levine, L., Peres, Y.: Strong spherical asymptotics for rotor-router aggregation and the divisible sandpile. Potential Anal. **30**, 1–27 (2009)
20. Lyons, R., Morris, B.J., Schramm, O.: Ends in uniform spanning forests. Electron. J. Probab. **13**, 1702–1725 (2008). https://doi.org/10.1214/EJP.v13-566
21. Lyons, R., Peres, Y.: Probability on Trees and Networks. Cambridge University Press, New York (2016)
22. Majumdar, S.N., Dhar, D.: Height correlations in the Abelian sandpile model. J. Phys. A **24**, L357–L362 (1991)
23. Majumdar, S.N., Dhar, D.: Equivalence between the Abelian sandpile model and the $q \to 0$ limit of the Potts model. Phys. A **185**, 129–145 (1992)
24. Pegden, W., Smart, C.K.: Convergence of the Abelian sandpile. Duke Math. J. **162**, 627–642 (2013)
25. Poghosyan, V.S., Priezzhev, V.B., Ruelle, P.: Return probability for the loop-erased random walk and mean height in the Abelian sandpile model: a proof. J. Stat. Mech. Theory Exp. P10004+12 (2011). https://doi.org/10.1088/1742-5468/2011/10/P10004
26. Priezzhev, V.B.: Structure of two-dimensional sandpile I. height probabilities. J. Stat. Phys. **74**, 955–979 (1994)
27. Priezzhev, V.B.: The upper critical dimension of the abelian sandpile model. J. Stat. Phys. **98**, 667–684 (2000)
28. Redig, F.: Mathematical aspects of the abelian sandpile model. Mathematical Statistical Physics. Elsevier, Amsterdam (2006)
29. Spitzer, F.: Principles of Random Walk. Graduate Texts in Mathematics, vol. 34, 2nd edn. Springer, New York (1976)

Chapter 7
Ising Model on the Torus and PCA Dynamics: Reversibility, Irreversibility, and Fast Tunneling

Carlo Lancia and Benedetto Scoppola

Abstract In this chapter, we present a class of PCA (Probabilistic Cellular Automata) that can be used for approximate sampling the Gibbs measure. We list a series of results about them, restricting the discussion to the nearest-neighbor Ising model. For both the weakly and strongly coupled spins, we show how it is possible to explicitly evaluate the accuracy of our approximation scheme. Moreover, in the strong coupling regime (low temperature), we show how our procedure may drastically improve the known results about the convergence of the system to the stationary distribution. An important ingredient in this context is the use of an irreversible dynamics, which let new interesting states (the so-called Ising waves) arise.

7.1 Introduction

A very interesting model in statistical mechanics, whose introduction dates back to the '50s of the past century, is the nearest-neighbor Ising model. Despite its simplicity, it has many interesting features, and it has been used as an intellectual laboratory in order to understand many physical phenomena.

The model is defined as follows. Consider a square subset $\Lambda \subset \mathbb{Z}^d$ of side n. In what follows we will impose periodic boundary conditions, i.e., we will consider $\Lambda = \mathbb{Z}^d \mod n$. A configuration $\sigma \in \{1, -1\}^\Lambda$ on Λ is a set of spin σ_x, $x \in \Lambda$. The Ising model is defined by the following Hamiltonian:

$$H(\sigma) = -\sum_{<x,y>} J\sigma_x\sigma_y\,. \tag{7.1}$$

C. Lancia
Department of Medical Statistics and Bioinformatics, Leiden University Medical Center, Leiden, The Netherlands
e-mail: lancia@mat.uniroma2.it; c.lancia@math.leidenuniv.nl

B. Scoppola (✉)
Dipartimento di Matematica, Università di Roma: Tor Vergata, via Della Ricerca Scientifica 1, 00133 Roma, Italy
e-mail: scoppola@mat.uniroma2.it

P.-Y. Louis and F.R. Nardi (eds.), *Probabilistic Cellular Automata*, Emergence, Complexity and Computation 27, https://doi.org/10.1007/978-3-319-65558-1_7

where the sum runs on pairs of nearest-neighbor sites. For simplicity, we will consider only the ferromagnetic case, i.e., $J > 0$. At equilibrium, the system is described by the Gibbs measure

$$\pi_G(\sigma) = \frac{e^{-H(\sigma)}}{Z_G}, \tag{7.2}$$

where the partition function Z_G is defined by

$$Z_G = \sum_{\sigma} e^{-H(\sigma)}. \tag{7.3}$$

The Ising model can be studied using either a static or a dynamic approach. The former focuses on the expected value of interesting quantities with respect to the Gibbs measure. It is possible to show (see for instance [7] and references therein) that for $d \geq 2$ and J small,

$$\pi_G(\sigma_x) = 0, \qquad \lim_{|y|\to\infty} \lim_{L\to\infty} \pi_G(\sigma_x \sigma_{x+y}) = 0, \tag{7.4}$$

where y is a vector oriented as a coordinate direction of Λ.

On the other hand, if J is sufficiently large, then $\pi_G(\sigma_x) = 0$ for obvious symmetry reasons, but

$$\lim_{|y|\to\infty} \lim_{L\to\infty} \pi_G(\sigma_x \sigma_{x+y}) = m_J^2, \tag{7.5}$$

with $m_J \neq 0$. According to (7.5), the typical configurations have an excess of spins oriented in one direction, and the spontaneous magnetization m_J measures this phenomenon. It turns out that the spontaneous magnetization m_J abruptly appears for $J > J_c$ (phase transition). When $d = 2$, it is possible to explicitly compute the partition function, and hence the critical coupling J_c; for $d > 3$, no analytic solution of the system has been found so far.

A different way to study the Ising system is the dynamical approach based on the theory of MCs (Markov Chains) — see [14] and references therein. It is possible to define MCs, i.e., matrices of transition probabilities $P_{\sigma,\sigma'}$, that have the Gibbs measure $\pi_G(\sigma)$ as their unique stationary distribution.

A simple yet important example is the so-called Metropolis dynamics. At each step of the chain, a site $x \in \Lambda$ is chosen uniformly at random and the spin in x is flipped with probability $e^{-\Delta H_+}$, where the quantity ΔH_+ is the positive determination[1] of

$$\Delta H = H\left(\sigma^{(x)}\right) - H(\sigma), \tag{7.6}$$

[1]In other words, $\Delta H_+ = \Delta H$ if ΔH is nonnegative and $\Delta H_+ = 0$ otherwise.

and $\sigma^{(x)}$ is the configuration that is identical to σ in all sites $y \neq x$ and has a flipped spin at site x.

It is easy to see that the measure π_G is reversible for this dynamics, and hence it is the unique stationary distribution of the chain. The numerical computation of the MC evolution is quite easy; thus the Metropolis dynamics gives in principle a fast an easy way to sample the Gibbs measure. However, it is not difficult to prove that the Metropolis dynamics converges fast to the stationary distribution only when J is sufficiently small, while the convergence is very slow when $J > J_c$. That behavior is due to the so-called *metastability phenomenon*.

Let us imagine that the chain is started from the configuration $\sigma_x = -1$ for each $x \in \Lambda$. Since $J > J_c$, the system has spontaneous magnetization, and the chain will rapidly tend to a configuration in which (7.5) is satisfied due to the presence of an excess of negative spins. Then, the system will pass to the set of the configurations with an excess of positive spins only by virtue of extremely rare fluctuations. Typically, the time needed to see this phenomenon, known as *tunneling*, is exponential in the volume $|\Lambda|$.

There exists a vast literature on metastability and tunneling time, see [15] and references therein. In particular, there is an almost complete control on the typical path followed by the system in order to tunnel from the states with negative magnetization to the ones with positive magnetization. However, it is clear that the dynamical approach applied to the Ising system has a very slow time of convergence to the stationary state when $J > J_c$ because, by the symmetry of the Hamiltonian, the two configurations σ and $-\sigma$ have the same probability at equilibrium.

In this chapter, we discuss the possibility to study a dynamical approach to the Ising model by means of a class of PCA (Probabilistic Cellular Automata). We will define a class of MC (see [3, 4, 11]) in which at each step all the spins have, in principle, the chance to flip simultaneously. That class is interesting for many reasons.

1. This dynamics is simple enough to be very suitable for parallel computing, giving a concrete possibility to simulate the statistical mechanics in a much faster way with respect to the standard single spin-flip Metropolis dynamics, see Sect. 7.2.1 below.
2. The possibility to sample the Gibbs measure using PCA is in general a nontrivial problem, which has a long history. The study of PCA in the context of Equilibrium Statistical Mechanics dates back to [8, 12], where various features of the infinite-volume limit have been investigated, in particular its space-time Gibbsian nature. However, it is not always possible to construct a PCA having a Gibbsian measure as its stationary distribution, see for instance [5]. In [10], explicit conditions are given on μ for the existence of a PCA that is *reversible* with respect to μ. As a consequence of [10], no PCA can be reversible with respect to the two-dimensional Ising model. In [2, 12], a PCA is introduced whose invariant (reversible) measure π is related to the Ising model as follows: The projection of π on the *even* sites, i.e., those $(i, j) \in \mathbb{Z}^2$ with $i + j$ even, coincides with the same projection of the Ising model, and the same holds for the odd sites; however,

opposite to the Ising model, spins at even sites are independent under π from spins at odd sites. When the nearest-neighbor interaction of the Ising model is generalized to a general pair interaction, this simple structure is lost. The class of PCA presented in this chapter is a concrete solution of the problem of sampling the Gibbs measure of the Ising model by means of PCA. In particular, we will show that in some regions of the parameters the Gibbs measure is as close as desired to the stationary measure of the PCA.
3. Even when the PCA stationary measure is not close to the Gibbs measure, the dynamics shows interesting features.
4. With this technique is possible to define PCA that are not reversible, having nonetheless the property that their stationary measure is very close to the Gibbs one. It is possible to prove that the irreversible dynamics for $J > J_c$ is much faster than the reversible one, even when both have a stationary measure close to the Gibbs measure.

The remainder is organized as follows: Sect. 7.2 presents the class of PCA that will be studied throughout the chapter, Sect. 7.3 reviews the recent literature on this class of PCA, and Sect. 7.4 collects some numerical results and some open problems.

7.2 Reversible and Irreversible PCA on 2D Torus

Let us consider the discrete two-dimensional torus $\Lambda = \mathbb{Z}^2 \mod n$. In the following, we will consider a spin system whose configurations are the elements of the set $\mathcal{X} = \{-1, +1\}^\Lambda$. Each site of the torus is uniquely determined by the couple (i, j), where $0 \le i, j \le n-1$. Given a configuration $\sigma \in \mathcal{X}$, the value of the spin located at the site (i, j) is $\sigma_{i,j}$. The spin $\sigma_{i,j}$ has four nearest neighbors, namely $\sigma_{i,j+1}, \sigma_{i+1,j}, \sigma_{i,j-1}$, and $\sigma_{i-1,j}$. For each pair $(\sigma, \tau) \in \mathcal{X} \times \mathcal{X}$, let us consider the following Hamiltonian:

$$H(\sigma, \tau) = -\sum_{i,j=0}^{n-1} \sigma_{i,j}(J^\uparrow \tau_{i,j+1} + J^\rightarrow \tau_{i+1,j} + J^\downarrow \tau_{i,j-1} + J^\leftarrow \tau_{i-1,j} + q\, \tau_{i,j}) \, . \tag{7.7}$$

Equation (7.7) describes the interaction between two possible system configurations in terms of the coupling constants $J^\uparrow$, $J^\rightarrow$, $J^\downarrow$, $J^\leftarrow$, and q. Having in mind the definition of a random dynamics, that interaction can be used for weighing the potential transition from σ to τ. In the resulting dynamics, the spin $\tau_{i,j}$ *feels* a local field

$$h_{ij}(\sigma) = J^\uparrow \sigma_{i,j-1} + J^\rightarrow \sigma_{i-1,j} + J^\downarrow \sigma_{i,j+1} + J^\leftarrow \sigma_{i+1,j} + q\, \sigma_{i,j} \tag{7.8}$$

that is produced by the spins of the old configuration that are located at the four nearest-neighbor sites $(i, j+1)$ (N/North), $(i+1, j)$ (E/East), $(i, j-1)$ (S/South), and $(i-1, j)$ (W/West), and by the spin which is located in the very same site (i, j).

For $\sigma, \tau \in \mathcal{X} \times \mathcal{X}$, let us consider the MC defined by the following transition matrix:

$$P(\sigma, \tau) = \frac{e^{-H(\sigma,\tau)}}{Z_\sigma}, \tag{7.9}$$

where $Z_\sigma = \sum_{\sigma' \in \mathcal{X}} e^{-H(\sigma,\sigma')}$ is a normalization constant. If we now assume that $J^\uparrow, J^\rightarrow, J^\downarrow, J^\leftarrow, q > 0$, each couple of neighboring sites $((i, j), (i \pm 1, j \pm 1))$ will give a negative contribution to the overall energy if the spins agree, i.e., $\tau_{i,j} = \sigma_{i\pm1,j\pm1}$; conversely, that couple will give a positive contribution if the spins disagree, i.e., $\tau_{i,j} = -\sigma_{i\pm1,j\pm1}$. The same argument holds for the interaction between $\tau_{i,j}$ and $\sigma_{i,j}$; the term $q\,\sigma_{i,j}\tau_{i,j}$ will contribute in decreasing the system energy if the spins agree, while it will contribute in increasing the system energy if they disagree.

As usual with ferromagnetic Ising systems, the update rule (7.9) favors transitions to new configurations where spins are aligned to their neighbors. However, the presence of the term $q\,\sigma_{i,j}\tau_{i,j}$ contrasts with the natural tendency of a ferromagnetic Ising system to get its spins aligned to each other, and in fact it represents a sort of inertial contribution to the system Hamiltonian. An example may help to understand this very important point.

Let us imagine that $\sigma_{i,j}$ agrees with its N and E neighbors, and it disagrees with its S and W neighbors; further, let $J^\uparrow = J^\rightarrow = J^\downarrow = J^\leftarrow > 0$. In a *classical* Ising system, the inertial contribution to the Hamiltonian would not be present, and the value of the spin at site (i, j) would be updated with uniform probability. Under the action of the inertial term $q\,\sigma_{i,j}\tau_{i,j}$, the updated spin $\tau_{i,j}$ is more likely of staying in accordance with the old spin $\sigma_{i,j}$ than flipping to $-\sigma_{i,j}$.

7.2.1 Parallel Implementation of the Dynamics

In Sect. 7.1, we have mentioned that the dynamics (7.9) is "very suitable for parallel computing". MCs can be easily implemented and simulated on any computer using the so-called *random mapping representation*, an algorithmic procedure that uses independent uniform random variables $\{u_k\}_k \in [0, 1]$ to determine the next value of the chain (see, e.g., [9]). Using (7.7) and (7.8),

$$H(\sigma, \tau) = -\sum_{i,j=0}^{n-1} h_{i,j}(\sigma)\tau_{i,j}\,.$$

Easy computations[2] yield

[2]See, e.g., [4].

$$Z_\sigma = \sum_{\sigma' \in \mathcal{X}} e^{-H(\sigma,\sigma')} = \prod_{i,j=0}^{n-1} 2\cosh h_{i,j}(\sigma)\,,$$

so that

$$P(\sigma,\tau) = \prod_{i,j=0}^{n-1} \frac{e^{\tau_{i,j} h_{i,j}(\sigma)}}{2\cosh h_{i,j}(\sigma)}\,. \tag{7.10}$$

The update rule (7.10) can be recast as follows

$$P(\sigma,\tau) = \prod_{i,j=0}^{n-1} P\left(\tau_{i,j} \,|\, \sigma\right)\,, \tag{7.11}$$

where

$$P\left(\tau_{i,j} \,|\, \sigma\right) = \frac{e^{\tau_{i,j} h_{i,j}(\sigma)}}{2\cosh h_{i,j}(\sigma)}\,.$$

The probability of updating the spin at site (i, j) to the value $+1$ is thus $e^{h_{i,j}(\sigma)}/[2\cosh h_{i,j}(\sigma)]$, whereas the probability of updating it to the value -1 is $e^{-h_{i,j}(\sigma)}/[2\cosh h_{i,j}(\sigma)]$. Since the update rule depends on σ only, each site can be updated simultaneously.

The update procedure for the spin at vertex (i, j) is sketched in Listing 7.1. The number of floating point operations required for the update of a single spin is very low. Indeed, the problem of simulating the evolution of such a MC is not particularly complex, involving only simple operations on many data elements. This is particularly suited for a GPU (graphic processing unit) because it closely resembles the operations involved in graphic applications. As no dependency is present in the probabilistic update rule for the evolution of the spins, they can be updated simultaneously, making the problem embarrassingly parallel. Therefore, a natural choice to efficiently implement such a MC seems to be CUDA (Compute Unified Device Architecture).

CUDA maps geometry onto blocks of threads executing simultaneously. It is then natural to map a configuration σ, which can be easily stored as a square matrix of size $n \times n$, onto a square grid of blocks. Each block is able to run multiple threads, simultaneously updating a portion of the configuration matrix σ. The update rule (7.11) ensures that no inter-thread dependencies will arise, because the transition probabilities depend only on the previous time-step configuration.

The only aspect that must be carefully regarded is the memory access, because the memory is the main potential performance inhibitor in GPUs. Dividing the two-dimensional matrix into a set of square tiles is a good strategy to efficiently use the GPU-shared memory and minimize the number of global memory reads. However, the presence of periodic boundary conditions causes a nonuniform access pattern on

Listing 7.1 Update procedure for the spin in site *(i,j)*

```
for each neighbor (k,l) of (i,j)
    p -= J((k,l)) * sigma((k,l))
end for
p *= -J
p -= q * sigma((i,j))
p = 1 / (1 + exp(2p))

if rand() < p then
    tau((i,j)) = 1
else
    tau((i,j)) = -1
end if
```

the boundary, that is, the impossibility to achieve coalesced memory and high memory bandwidth. Nevertheless, the use of a GPU infrastructure dramatically reduces the computational time needed to simulate a step of the chain. An experiment presented in [11] shows, for example, that less than a minute is required to simulate 500 steps of the chain (7.9) when the side of the torus is of the order of 10^5.

7.2.2 *Stationary Measure of the Dynamics*

Equation (7.9) defines the transition kernel — a $2^{n^2} \times 2^{n^2}$ matrix — of a MC (X_t) that takes value on $\mathcal{X}$, the set of all possible configurations. The chain is naturally described by the following probability distribution over $\mathcal{X}$:

$$\mu^t(\tau) = \mathbb{P}(X_t = \tau) \; .$$

By the *Total Probability Theorem*,

$$\mu^t(\tau) = \sum_{\sigma \in \mathcal{X}} \mu^{t-1}(\sigma)\, P(\sigma, \tau) \, .$$

If a probability distribution π may be found such that

$$\pi(\tau) = \sum_{\sigma \in \mathcal{X}} \pi(\sigma)\, P(\sigma, \tau) \, , \tag{7.12}$$

then π is called the *stationary* distribution, because it is invariant under a step of the chain. Under very mild conditions,[3] a stationary measure exists and its uniqueness

[3] Irreducibility and aperiodicity, see, e.g., [9].

is guaranteed. Moreover, the system will asymptotically tend to the stationary state, in the sense that

$$\lim_{t\to\infty} d_{\mathrm{TV}}\left(\mu^t, \pi\right) = 0\,, \tag{7.13}$$

where $d_{\mathrm{TV}}\left(\cdot,\cdot\right)$ is the usual *total variation distance*, see, e.g., [9]; whenever that happens, the chain is called *ergodic*. We will not consider other-than-ergodic MCs so both existence and uniqueness of π are guaranteed, and the limit (7.13) is satisfied.

Suppose now that $J^{\uparrow} = J^{\downarrow}$ and $J^{\leftarrow} = J^{\rightarrow}$; then, the Hamiltonian (7.7) is a symmetric function of σ and τ. Using the *detailed balance condition*

$$\pi(\sigma)P(\sigma,\tau) = \pi(\tau)P(\tau,\sigma)$$

it is immediate to verify that the MC (7.9) is reversible with respect to

$$\pi(\sigma) = \frac{Z_\sigma}{Z}\,, \tag{7.14}$$

where $Z = \sum_{\sigma\in\mathcal{X}} Z_\sigma$ is a normalization constant. Reversibility infers the existence of a stationary measure and the property of *time reversal*, that is to say, the chain evolves regardless of the *time arrow* if started with initial distribution $\mu^0 = \pi$, see, e.g., [9].

Let us now consider a *completely asymmetric* model, where the interaction happens in two spatial directions only, e.g., only in the N and E directions because $J^{\downarrow} = J^{\leftarrow} = 0$. The Hamiltonian is no longer symmetric and, it is impossible to find a reversible measure.[4] Let us consider the stationary distribution π given by (7.14) and see if it satisfies (7.12), for example. Then,

$$\sum_{\sigma\in\mathcal{X}} \pi(\sigma)\,P(\sigma,\tau) = \sum_{\sigma\in\mathcal{X}} \frac{Z_\sigma}{Z}\,\frac{e^{-H(\sigma,\tau)}}{Z_\sigma} = \frac{\sum_{\sigma\in\mathcal{X}} e^{-H(\sigma,\tau)}}{Z}\,. \tag{7.15}$$

Equation (7.15) is satisfied whenever the system is *dynamically reversible* [1], i.e.,

$$\sum_{\sigma\in\mathcal{X}} e^{-H(\sigma,\tau)} = \sum_{\sigma\in\mathcal{X}} e^{-H(\tau,\sigma)}\,. \tag{7.16}$$

Therefore, we have found that if the system is dynamically reversible, then the stationary measure is again given by (7.15). The same result is given in [11], where the dynamical reversibility (7.16) is called *Weak Balance Condition* instead. Using Peierls contours, the very same paper also proves that the completely asymmetric model on the torus is actually dynamically reversible.

[4]The chain has in fact a unique stationary measure that does not satisfy the detailed balance condition.

7.3 Gibbsian Regime of the Stationary Measure and Fast Mixing

The relations between the stationary measure (7.14) and the Gibbs measure (7.2) have been studied in two papers, namely [3, 4]. The former focuses on a spin system with an interaction more general than (7.1). The starting Hamiltonian is

$$H(\sigma) = -\sum_{x,y} J_{xy}\sigma_x\sigma_y\,. \tag{7.17}$$

where the sum runs on all the pairs of sites x, y, and the interaction J_{xy} is summable. The resulting PCA Hamiltonian is, therefore,

$$H(\sigma,\tau) = -\sum_{x,y} J_{xy}\sigma_x\tau_y + q\sum_{x}\sigma_x\tau_x\,. \tag{7.18}$$

The nearest-neighbor Ising model is obviously a particular case of the more general class of spin systems (7.18). Restricted to the case discussed in this chapter, the results presented in [4] can be summarized as follows:

Theorem 1 *Consider the isotropic* PCA *Ising dynamics for $d = 2$, i.e., the* MC *whose transition probabilities are defined by*

$$P(\sigma,\tau) = \frac{e^{-H(\sigma,\tau)}}{Z_\sigma}\,, \tag{7.19}$$

where

$$H(\sigma,\tau) = -\sum_{<x,y>} J\sigma_x\tau_y + q\sum_{x}\sigma_x\tau_x\,, \tag{7.20}$$

$$Z_\sigma = \sum_{\tau} e^{-H(\sigma,\tau)}\,,$$

and the first sum runs of nearest-neighbors sites on a two-dimensional torus Λ. Consider the stationary (reversible) distribution of this dynamics

$$\pi^{PCA}(\sigma) = \frac{Z_\sigma}{Z}\,. \tag{7.21}$$

For each $q \geq 0$, let $\delta := e^{-2q}$. Suppose:

(a) $\delta = \delta(|\Lambda|)$ *is such that* $\lim_{|\Lambda|\to\infty} \delta^2|\Lambda| = 0$;
(b) $J < J_c$,

where J_c is the critical coupling of the Ising model. Then,

$$\lim_{|\Lambda|\to\infty} d_{TV}\left(\pi^{PCA}, \pi^{G}\right) = 0. \tag{7.22}$$

Remark 1 In [4], Theorem 1 is written in terms of the so-called *Dobrushin condition* (Eq. (20) in [4]). The Dobrushin condition is well known (see for instance [7]) to hold up to the critical value of the coupling for the two-dimensional nearest-neighbor Ising model.

Remark 2 The proof of Theorem 1 is based on the fact that in the regime $J < J_c$ the correlations tend exponentially to vanish at large distance. This ingredient is carefully applied (see Eq. (18) in [4]) using a general result due to Föllmer [6].

Remark 3 One might think that the requested smallness of the parameter δ brings to a dynamics that is not really parallel. As a matter of fact, this theorem implies that at each step a number of spins that is of the order of n^{α} is updated, where $\alpha < 1$ and n is the side of the square box Λ. In order to fully parallelize the procedure, one has to choose at each step a set of candidate sites and then try to update them. This can be written in such a way that the parallel algorithm runs really in a time inversely proportional to the number of available processors if the memory is shared among them.

Remark 4 We expect that analogously to the case of Metropolis dynamics, the PCA (7.19) should be fast in converging to equilibrium. A simple coupling argument allows to prove (see [13, Chap. 15]) that for $J < J_c$ the mixing time of the Metropolis dynamics is proportional to $|\Lambda| \log |\Lambda|$. Due to the slightly parallel nature of the dynamics, it is reasonable to expect that the mixing time of (7.19) is even shorter. This will be the subject of further work.

As discussed in previous section, an interesting possibility to sample the Gibbs measure is the use of *irreversible* PCA. A result on this subject has been recently proved in [3]. Let us briefly summarize the two main theorems proved therein. The first is analogous to Theorem 1 above in the context of large J.

Theorem 2 *Consider the completely asymmetric PCA Ising dynamics for $d = 2$, i.e., the* MC *whose probability transition is defined by*

$$P(\sigma, \tau) = \frac{e^{-H(\sigma,\tau)}}{Z_\sigma}, \tag{7.23}$$

where

$$H(\sigma, \tau) = -\sum_{i,j=0}^{L-1} J\sigma_{i,j}(\tau_{i,j+1} + \tau_{i+1,j} + q\,\tau_{i,j}), \tag{7.24}$$

$$Z_\sigma = \sum_\tau e^{-H(\sigma,\tau)},$$

and i, j *are the coordinates of a two-dimensional torus* Λ *of side* L*. Consider the stationary (irreversible) distribution of this dynamics*

$$\pi^{PCA}(\sigma) = \frac{Z_\sigma}{Z}, \tag{7.25}$$

and define the low temperature regime as the following choice of parameters:

$$J = J(L) = k \log L, \qquad q = q(L) = c\frac{\log L}{L}. \tag{7.26}$$

Then, there exists a constant $C > 0$ *such that*

$$d_{TV} \pi_{PCA} \pi_G \leq C\left(\frac{1}{L^{\frac{c}{2}-1}} + \frac{1}{L^{2k-2}}\right). \tag{7.27}$$

Remark 5 In this case, the dynamics is really parallel. The value of the parameter q is now very small and tends to vanish as the volume $|\Lambda|$ increases. Clearly, it can not be $q = 0$ since in this case (see discussion above) the stationary distribution is known to be very different from the Gibbs measure. Note also that in this regime the measure π_{PCA} tends to be completely polarized, i.e., it is mainly concentrated on the two configurations $\sigma_{i,j} = 1$ and $\sigma_{i,j} = -1$ for all the sites (i, j).

One expects that the low temperature regime defined in Theorem 2 gives rise to very slow convergence to the stationary distribution, or equivalently that the system exhibits very long tunneling times. A careful evaluation of the dynamics, however, allows to prove the following result:

Theorem 3 *In the low temperature regime with parameter* k *and* c *such that* $c > \frac{1}{2}$ *and* $k - 4c > 4$,

$$\lim_{L\to\infty} d_{PCA}(L^{8k}) = 0,$$

where

$$d_{PCA}(t) = \sup_\sigma d_{TV} P^t(\sigma, \cdot)\pi_{PCA}(\cdot).$$

Remark 6 Theorem 3 provides an example of a polynomial mixing time in the context of low temperature spin system. The following three essential ingredients are exploited to obtain this result: first, the system has to be defined on a torus, in order to give a complete control of the stationary distribution; second, the dynamics has to be irreversible in order to ensure the existence of suitable exit trajectories from the metastable states $\sigma \pm 1$; third, the dynamics has to be massively parallel, i.e., q has to be small.

Remark 7 Proving Theorem 3, one sees that there are states in the system that tend to be relatively stable, in the sense that in the limit $J \to \infty$ (zero temperature), they tend to be reproduced indefinitely. This states are the so-called *diagonal configuration*, in which a single NW-SE diagonal of spin with orientation different with respect to their nearest neighbors moves in the NE direction with unit velocity. When, by a fluctuation, a single spin close to some spin of the diagonal becomes parallel to the orientation of the diagonal itself, it can change all the signs of its own diagonal with finite probability. Hence, the number of diagonal with opposite orientation can increase with finite probability. This gives a polynomial estimate of the tunneling time. As we shall see in the next chapter, when the system is defined with slightly smaller J, although the stationary measure π_{PCA} is quite different from the Gibbs measure, the typical configurations are qualitatively similar to the diagonal configurations discussed above, giving rise to the phenomenon of the so-called *Ising waves*.

7.4 Numerical Experiments in the Non-Gibbsian Regime

In this section, we study through a simulative approach the totally asymmetric PCA model on the two-dimensional torus. As we have discussed in Sect. 7.3 above, it is possible to fully characterize the stable configurations of the model in the regime $J^{\uparrow}, J^{\rightarrow} \gg J_c$ and $q \ll 1$. However, the typical time to see the creation of a diagonal configuration is exponential in J and polynomial in the volume n^2 (with a rather large degree, though). In other words, if the number of sites n^2 is sufficiently large, then the system started from $\sigma_{i,j} = -1$ for each $i, j = 0, \ldots, n-1$ will not exhibit any of those stable configuration before an unacceptably long time has passed. Nevertheless, the intuition behind Theorem 3 and Remark 7 should remain valid if we consider $J^{\uparrow}, J^{\rightarrow}$ large and q small enough.

Figure 7.1 shows the configuration of the system across nine successive steps of the dynamics on a 50×50 two-dimensional torus. The system, which was started from maximum negative magnetization, has already formed a few diagonal lines of plus spins after a very short time. Those lines are quite stable and propagate from one step of the dynamics to the next under a mechanism very similar to the one discussed in Sect. 7.3. As a result, the lines of plus spins steadily drift in the NE direction; in [11], this phenomenon is referred to as *Ising waves*.

We conclude the section by addressing the tunneling phenomenon, i.e., the abrupt passage from a configuration with negative magnetization (resp. positive) to a state with positive (resp. negative) magnetization. The idea is to confront the tunneling time (the time to switch from the configuration $\sigma_{i,j} = -1$ to $\sigma_{i,j}$) in the symmetric and in the totally asymmetric models.

If the parameters $J^{\uparrow}$, $J^{\rightarrow}$, and q are chosen in the regime studied in [3], then the totally asymmetric model has a tunneling time that is polynomial in the volume. However, in that regime the exponent is very large, and again the phenomenon happens over such a long timescale that it is impossible to experience it in any practical situation. Because of the waves-formation mechanism highlighted in Remark 7, we

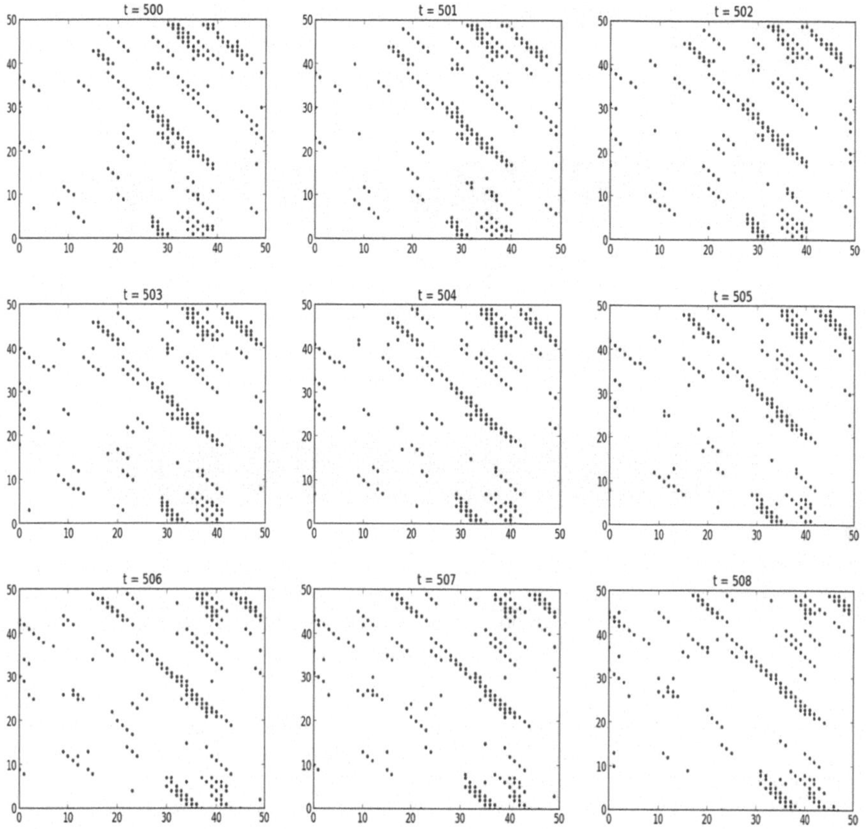

Fig. 7.1 Nine consecutive system configurations (plus spins in a sea of minuses) for the totally asymmetric model on a 50×50 two-dimensional torus. The diagonal lines of plus spins steadily move in the NE direction, giving rise to the so-called *Ising waves*

expect that, with $J^{\uparrow}$, $J^{\rightarrow}$ large and q small, the totally asymmetric model will exhibit tunneling in a much shorter time than the symmetric counterpart, even in a regime where the stationary distribution is not close to the Gibbs measure.

To demonstrate the soundness of our intuition, we have simulated the evolution of a symmetric PCA model ($J^{\uparrow} = 0.42$, $J^{\rightarrow} = 0.42$, $J^{\downarrow} = 0.42$, $J^{\leftarrow} = 0.42, q = 0.1$) and confronted it with the totally asymmetric counterpart ($J^{\uparrow} = 1.57$, $J^{\rightarrow} = 1.57$, $J^{\downarrow} = 0.0$, $J^{\leftarrow} = 0.0$, $q = 0.1$). The reader must not be puzzled by the sum of the $J^{\cdot}$ being not equal in the two Hamiltonian. We are in fact in a regime where the stationary distributions are not close to the Gibbs measure, and the role of q is crucial in establishing equilibrium. As a result, no intuition is available for setting up the parameter so to have comparable models. A reasonable way to set up the models is to calibrate them by trial and error until they show the same spontaneous

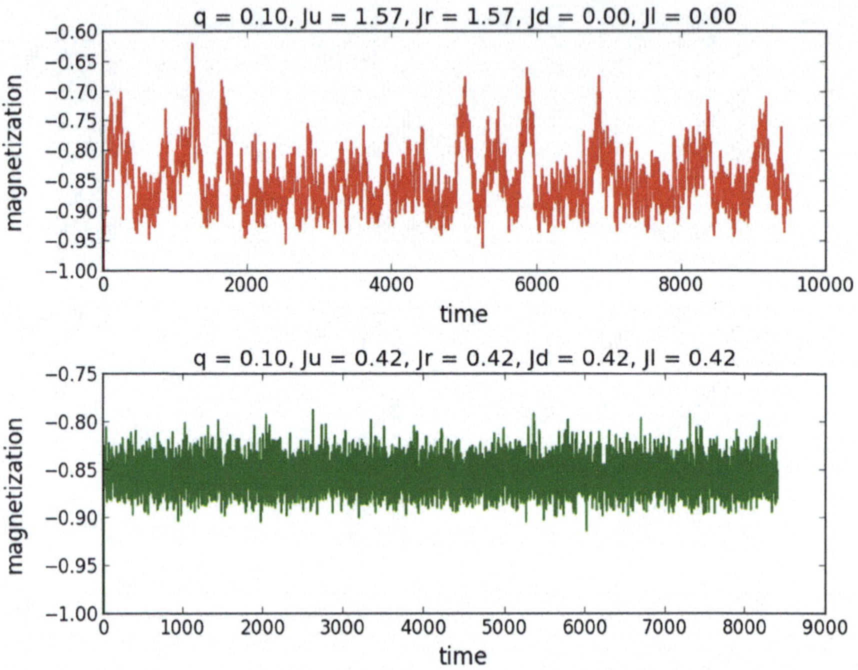

Fig. 7.2 Symmetric versus totally asymmetric model, calibration of the parameters. The choice of the parameters guarantees that both the symmetric and the totally asymmetric model exhibit spontaneous magnetization, and that the value of the magnetization is the same. Note that `Ju` correspond to $J^{\uparrow}$, `Jr` to $J^{\rightarrow}$, `Jd` to $J^{\downarrow}$, and `Jl` to $J^{\leftarrow}$

magnetization. Figure 7.2 demonstrates through a short simulation that the chosen settings satisfy the spontaneous-magnetization criterium.

Figure 7.3 shows the evolution of the magnetization in both models through a long simulation. The behavior of the totally asymmetric case is displayed in the top sub-figure, whereas the behavior of the symmetric model is shown in the bottom sub-figure. While the magnetization of the symmetric—and hence reversible— model remains very close to the initial value, the totally asymmetric PCA appears to be much more mobile, exhibiting several times the tunneling phenomenon during the same time lapse.

The experiment strongly suggests that the totally asymmetric PCA is fast mixing even in a regime where the hypotheses of Theorems 2 and 3 do not hold, but the spins are still strongly coupled and the inertial term is small enough. Most likely, this is due to the irreversibility of the model and the presence of a mechanism similar to the one that leads in [3] to the formation of diagonal configurations.

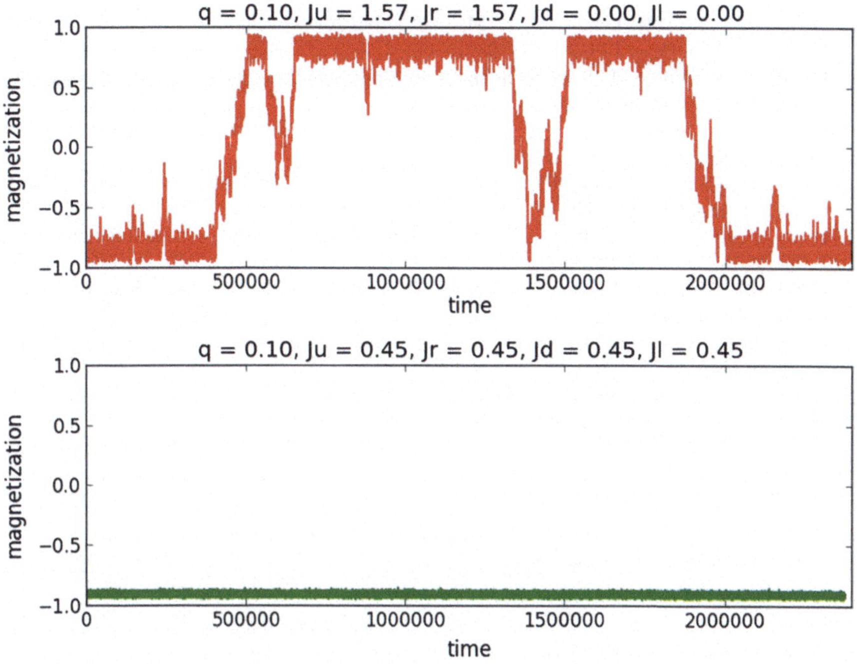

Fig. 7.3 Symmetric versus totally asymmetric model, tunneling. The totally asymmetric model (top sub-figure) is much more mobile than the symmetric counterpart (bottom sub-figure) and explores the state space $\mathcal{X}$ faster and more efficiently. Note that `Ju` correspond to $J^{\uparrow}$, `Jr` to $J^{\rightarrow}$, `Jd` to $J^{\downarrow}$, and `Jl` to $J^{\leftarrow}$

Acknowledgements We thank the organizers of the workshop "Probabilistic Cellular Automata: Theory, Applications and Future Perspectives" (Eurandom 2013, TU Eindhoven) for the possibility to share ideas on this new and promising subject.

References

1. Anderson, B.D.O., Kailath, T.: Forwards, backwards, and dynamically reversible Markovian models of second-order processes. IEEE Trans. Circuits Syst. **26**, 956–965 (1979)
2. Cirillo, E.N.M., Nardi, F.R.: Metastability for a stochastic dynamics with a parallel heat bath updating rule. J. Stat. Phys. **110**, 183–217 (2003)
3. Dai Pra, P., Scoppola, B., Scoppola, E.: Fast mixing for the low temperature 2D Ising model through reversible parallel dynamics. J. Stat. Phys. **159**, 1–20 (2015)
4. Dai Pra, P., Scoppola, B., Scoppola, E.: Sampling from a Gibbs measure with pair interaction by means of PCA. J. Stat. Phys. **149**, 722–737 (2012)
5. Dawson, D.A.: Synchronous and asynchronous reversible Markov systems. Canad. Math. Bull **17**, 633–649 (1974)
6. Föllmer, H.: A covariance estimate for Gibbs measures. J. Funct. Anal. **46**, 387–395 (1982)
7. Gallavotti, G.: Statistical Mechanics: A Short Treatise. Springer, Berlin (1999)

8. Goldstein, S., Kuik, R., Lebowitz, J.L., Maes, C.: From PCA's to equilibrium systems and back. Commun. Math. Phys. **125**, 71–79 (1989)
9. Häggström, O.: Finite Markov Chains and Algorithmic Applications, vol. 52. Cambridge University Press, Cambridge (2002)
10. Kozlov, O., Vasilyev, N.: Reversible Markov chains with local interaction. Multicomponent random systems **6**, 451–469 (1980)
11. Lancia, C., Scoppola, B.: Equilibrium and non-equilibrium Ising models by means of PCA. J. Stat. Phys. **153**, 641–653 (2013)
12. Lebowitz, J.L., Maes, C., Speer, E.R.: Statistical mechanics of probabilistic cellular automata. J. Stat. Phys. **59**, 117–170 (1990)
13. Levin, D.A., Peres, Y., Wilmer, L.E.: Markov Chains and Mixing Times. American Mathematical Society, Providence (2009)
14. Martinelli, F.: Dynamical analysis of low-temperature Monte Carlo cluster algorithms. J. Stat. Phys. **66**, 1245–1276 (1992)
15. Martinelli, F.: Lectures on Glauber dynamics for discrete spin models. Lectures on Probability Theory and Statistics, pp. 93–191. Springer, Berlin (1999)

Chapter 8
Synchronization in Interacting Reinforced Stochastic Processes

Pierre-Yves Louis and Ida G. Minelli

Abstract We present a family of interacting stochastic processes introduced in [13] whose individual dynamics follow a reinforcement updating rule. This is a natural generalization of PCA dynamics on a continuous spin space. The interaction changes the long-time behavior of each process and the speed of evolution, producing a phenomenon of synchronization.

8.1 A Natural Generalization of PCA Dynamics

Most of the probabilistic cellular automata (PCA) have a finite local set of values associated to each constituting site (also known as *spin space*). This is the original Von Neumann's definition. Since then, a great variety of generalizations were considered. For application purposes, the usual $S = \{0, 1\}$ or $S = \{-1, +1\}$ may be replaced by a more elaborated one: Drossel and Schwabl forest fire PCA model [15] uses $S = \{0, 1, 2\}$ for empty/occupied by a tree/burning; the cellular Potts model (CPM) presented in this book uses a finite space [9], whose cardinality is as large as the number of biological cells to be modeled. Finally, see [8] for a PCA with a countable spin space S and an ecological modeling context. Many probabilistic models inspired by statistical mechanics are defined with a continuous spin space, where $S = [0, 1]$ or $S = S^1$. They can be used to build *continuous spin*-PCA dynamics (see for instance Chap. 12 in [4]). *Coupled maps lattices* may be seen as generalization of deterministic CA with continuous spin space [19]. Some models like *discrete q-valued rotators* use instead a regular discrete version: $\{0, 1/N, \dots, 1 - 1/N, 1\}$ where N is a parameter.

P.-Y. Louis
Laboratoire de Mathématiques et Applications UMR 7348, Université de Poitiers, CNRS, 11 Boulevard Marie et Pierre Curie, Téléport 2 - BP 30179, Poitiers, France
e-mail: pierre-yves.louis@math.cnrs.fr

I.G. Minelli (✉)
Dipartimento di Ingegneria e scienze dell'informazione e matematica, Universitá dell'Aquila, Via Vetoio (Coppito 1), 67100 L'Aquila, Italy
e-mail: ida.minelli@dm.univaq.it

P.-Y. Louis and F.R. Nardi (eds.), *Probabilistic Cellular Automata*, Emergence, Complexity and Computation 27, https://doi.org/10.1007/978-3-319-65558-1_8

We consider here discrete-time stochastic processes which share the main aspects of the traditional PCA: Multicomponent stochastic dynamics, time and space are discrete, *synchronous* updating scheme. The main model considered is a reinforced process with a *mean field* interaction based on Pólya-type urns. Pólya urns models are used in computer science (Chap. 8 in [22]) and well-known schemes in bio-sciences (Chap. 9 in [22]).

8.2 Introduction to the Main Concepts

8.2.1 What Is Synchronization?

Synchronization occurs in many natural contexts and is a common topic of different scientific fields. This is a general concept for a phenomenon observed in multicomponent dynamical evolutions. The following are constituting aspects:

- Notion of unit (cell, component, individual) with a proper dynamics,
- Finite (possibly large) number of units,
- Interaction among units which influences their dynamics,
- The units after some time adopt the same kind of behavior, each individual behavior being coordinated to a global common characteristic.

Among the different perspectives on this large and multidisciplinary topic, we cite [1, 5, 26, 28] and the recent mathematical works by [6, 7, 10, 18]

8.2.2 A Basic Model of Self-reinforcement: Classical Pólya urn

In social science or biology, reinforcement is defined as an action which increases the frequency of a certain behavior. We may define a reinforced process as a stochastic process where an event which has occurred many times in the past has a bigger probability to occur in the future. For a survey on this kind of processes, see [25]. There is a big variety of reinforced processes which is described in terms of urn models. The simplest of them is the Pólya urn model which we briefly describe below.

- At time 0, a urn contains a red balls and b blue balls.
- At each discrete time t > 0, a ball is drawn out and it is replaced in the urn together with c balls of the same color.

We denote by Z_t the proportion of red balls in the urn at time t, and we are interested in the distribution of Z_t when t is large. An easy calculation shows that $(Z_t)_{t\geq 0}$ is a bounded *martingale*; thus, it converges almost surely and in L^p to a random variable Z_∞. Moreover, it can be proved that Z_∞ has beta distribution of parameters a/c and b/c.

8.2.3 *Two Examples of Generalized Pólya urns*

An interesting feature of urn models is that they may exhibit very different behaviors, even when we make seemingly slight changes in the reinforcement scheme. An example is given by the so-called Friedman's urn (see, e.g., [17]): At each step, the ball selected is replaced by α balls of the same color and β balls of the color not drawn, where $\alpha > \beta > 0$. The following result holds:

Theorem 1 (stated in [16]) *The proportion Z_t of red balls converges a.s. to* $1/2$.

Other generalizations of Pólya urn have a *weighted reinforcement*, i.e., the probability of drawing a given color depends in a nonlinear way, on the frequency with which that color was drawn in the past, as can be seen in the following model, which has been introduced by B. Davis [14]. The nth time a red (respectively, blue) ball is drawn, a_n (respectively b_n) red balls are added in the urn. The model has a stronger reinforcement than the previous ones: The number of balls which are added in the urn when a color is drawn increases with the number of times that color appeared in the past. Let $S_n = \sum_{k=0}^{n} a_k$, $S'_n = \sum_{k=0}^{n} b_k$ and suppose $\sum_{n=0}^{\infty} 1/S_n$, $\sum_{n=0}^{\infty} 1/S'_n < \infty$. Then, the asymptotic behavior of Z_t is given by the following result which has been proved using an exponential embedding argument devised by Rubin.

Theorem 2 (stated in [14]) *Denote by G (respectively, G') the event "the total number of blue (respectively, red) draws is finite" then $P(G)$, $P(G') > 0$ and $P(G) + P(G') = 1$.*

8.3 Interacting Urns and Synchronization

A Pólya urn is a basic model which describes a process with a property of self-reinforcement. However, the process of reinforcement is often influenced by the environment, so we are led to consider systems of urns in which we introduce some kind of interaction among them.

Recently, systems of interacting urns have been studied, among which we cite [20, 21, 24]. The models in [20, 21] have a strong reinforcement mechanism, and the conditional probability of drawing a color depends on the frequency with which that color was drawn in the past both in the given urn and in the whole system. Under certain conditions on these probabilities, the authors show that there is a phenomenon of *fixation*, i.e., depending on the strength of interaction, all or part of the urns draw eventually the same color. So, if the interaction among urns is sufficiently strong, the urns *synchronize*, which means that the proportion of a given color in the urns converges a.s. to the same random variable (which takes values in the set $\{0, 1\}$).

In the next section, we propose a new model of interacting urns with a weaker reinforcement mechanism and we show that in this case we have *synchronization* whatever the strength of interaction is.

8.3.1 Mean Field Interacting Pólya urns

We consider a system of N Pólya-type urns in which we introduce a *group interaction*. Let us fix a parameter $\alpha \in [0, 1]$, which represents the strength of interaction. We denote by $Z_t(i)$ the proportion of red balls in urn i for $i = 1, \ldots, N$ and by $Z_t = (1/N)\sum_{i=1}^N Z_t(i)$ the proportion of red balls in the whole system at time t.

- At time 0, there are a red and b blue ball in each urn.
- At each time $t > 0$, in each urn, *independently* between the urns, a red ball is added with a probability $\alpha Z_t + (1-\alpha)Z_t(i)$.

More precisely, consider a family $\{U(t, i);\ t, i \in \mathbb{N}\}$ of i.i.d. random variables with uniform distribution on $[0, 1]$. Let $m = a + b$ be the number of balls in each urn at time 0, $\mathcal{F}_t = \sigma\big(U(s, i);\ 0 \le s \le t, i \in \mathbb{N}\big)$ and denote by $Y_t(i) \in \{0, 1\}$ the color of the ball added in urn i at time t (with the convention that 1 means "red" and 0 means "blue"). The following recursion holds:

$$\begin{cases} Z_0^N(i) = \dfrac{a}{m} \\[2mm] Z_{t+1}^N(i) = \dfrac{t+m}{t+m+1} Z_t^N(i) + \dfrac{1}{t+m+1} Y_t(i), \end{cases}$$

where $Y_t(i) = I_{\{U(t+1,i) \le \alpha Z_t^N + (1-\alpha) Z_t^N(i)\}}$. Conditionally on $\mathcal{F}_t$, the random variables $Y_t(i)$ for $i = 1, \ldots, N$ are independent with Bernoulli distribution of parameter $\alpha Z_t + (1-\alpha)Z_t(i)$.

In what follows, we will suppose $a = b = 1$ to simplify notations, but the results hold for any values of a and b.

8.3.2 First Remarks

Our object of interest is the process $\{Z_t(i)\}_t$. By the definition, it follows that

$$\mathbb{E}(Z_{t+1}(i) - Z_t(i)|\mathcal{F}_t) = \frac{\alpha}{t+3}(Z_t - Z_t(i)). \tag{8.1}$$

Note that the case $\alpha = 0$ corresponds to N independent Pólya urns, each converging a.s. to its own random limit $Z_\infty(i)$.

When $\alpha > 0$, $\{Z_t(i)\}_t$ is not a martingale, but if we sum over i in (8.1) we obtain that $\{Z_t\}_t$ is still a bounded martingale, thus it converges a.s. and in L^p to a random limit Z_∞. In particular, $\mathbb{E}(Z_t) = 1/2$ for all $t \in \mathbb{N}$.

Moreover, since, for each fixed t, $(Z_t(1), \ldots, Z_t(N))$ is exchangeable, it holds for all i, $\mathbb{E}(Z_t(i)) = \frac{1}{2}$. Note also that $\{(Z_t(1), \ldots, Z_t(N))\}_{t \geq 0}$ is a Markov process with values in $(0, 1)^N$. At each time $t + 1$, all the random variables $Z_t(i)$ change their values according to the following rule:

- $Z_t(i)$ change in $[(t+2)/(t+3)]Z_t(i) + 1/(t+3)$ with conditional probability p,
- $Z_t(i)$ change in $[(t+2)/(t+3)]Z_t(i)$ with conditional probability $1 - p$,

where $p = (\alpha/N) \sum_j Z_t(j) + (1-\alpha) Z_t(i)$.

8.3.3 Simulations

Simulations ($N = 10$) show that when α is sufficiently large, the urns have a "conformist" behavior, i.e., they tend to synchronize on the same proportion of red balls,

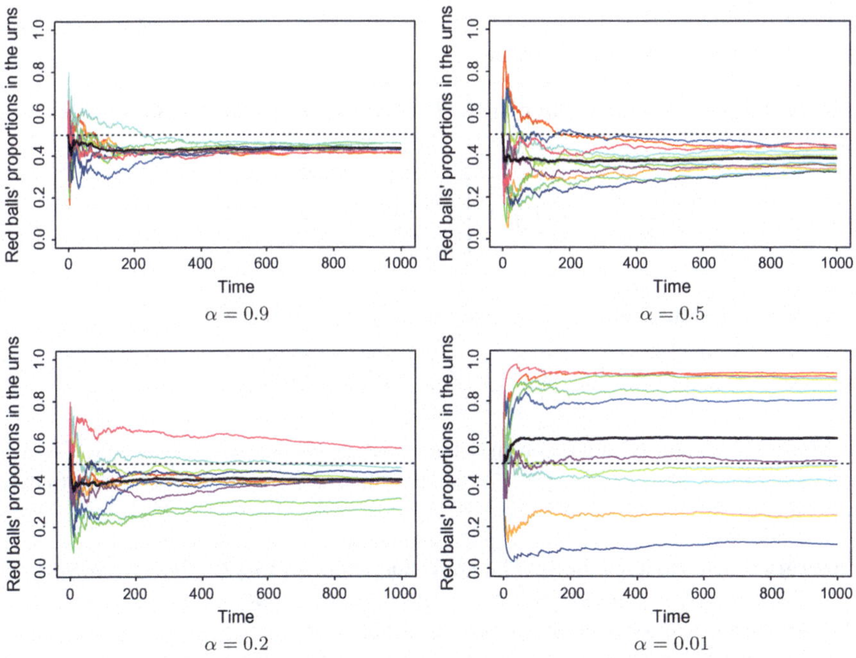

Fig. 8.1 Through mean field interacting urns. $N = 10$ urns are considered for different values of α. Each urn starts with 1 *red*/1 *blue ball*. The time is the x-axis. Each step of time represents the synchronous updating of N urns. Represented along the y-axis are the trajectories of the proportion of *red balls* in each urn. The *black curve* is associated with the mean field $\{Z_t\}_t$

which coincides with the global one. For small values of α, the picture seems to be less clear, and one may suppose that there exists some critical value of α below which the urns do not synchronize.

Really, it turns out that synchronization occurs in any case but still there is a critical value of α for which there is a change in the speed of convergence to the asymptotic global proportion of red balls Z_∞ (Fig. 8.1).

8.4 Main Results

What do we observe in such a model?

- Even when the influence α of the environment is small, when time t goes to $+\infty$, the urns *synchronize* almost surely, i.e., the $\{Z_t(i)\}_t$'s converge all to *the same* random variable.
- The limiting random variable Z_∞ has expectation 1/2 and a variance which depends on the number of urns.
- The variance of Z_∞ is of order 1/N. This means that as the number of urns increases, the limiting distribution of $Z_t(i)$ becomes more and more concentrated around its mean value.

Our main results are the following:

Theorem 3 (L^2 synchronization [13]) *The following asymptotic estimates hold:*

$$\mathbb{E}\left[(Z_t(i) - Z_t)^2\right] \sim \begin{cases} t^{-2\alpha} & \text{for } 0 < \alpha < 1/2 \\ t^{-1}\log(t) & \text{for } \alpha = 1/2 \\ t^{-1} & \text{for } 1/2 < \alpha \leq 1. \end{cases}$$

Theorem 4 (Almost sure synchronization [13]) *For each $i = 1, 2, \ldots, N$*

$$\lim_{t\to+\infty} Z_t(i) = Z_\infty \qquad \text{almost surely.}$$

Moreover, a central limit theorem holds for $\{Z_t\}_{t\geq 0}$ when $N \to +\infty$ which shows that the variance of Z_t is of order $1/N$ and that fluctuations around the mean of $\{Z_t\}_t$ are Gaussian.

Theorem 5 (Space asymptotic: Gaussian fluctuations [13]) *Let $W_t^N := \sqrt{N}(Z_t - 1/2)$. The stochastic process $\{W_t^N : t \geq 0\}$ converges weakly, as $N \to +\infty$, to the Gauss–Markov process solution of the recursion*

$$\begin{cases} W_{t+1} = W_t + \dfrac{1}{(t+3)^2}\left(\dfrac{1}{4} - (1-\alpha)^2 x_t^\infty\right) B_{t+1} \\ W_0 = \delta_0 \end{cases} \tag{8.2}$$

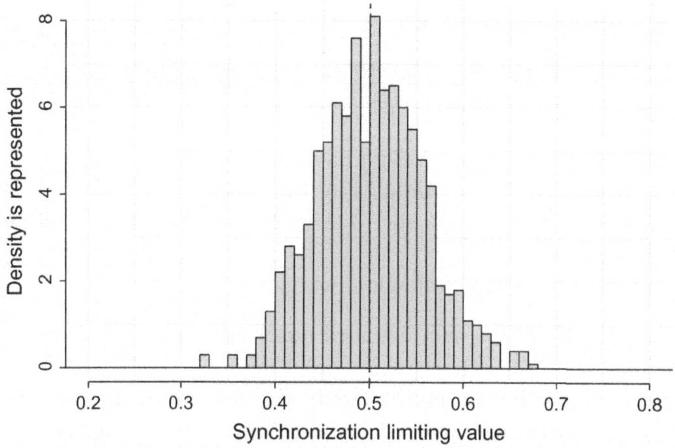

Fig. 8.2 For $N = 30$ urns interacting through mean field, with $\alpha = 0.8$, sample of 1000 independent realizations of Z_t for t large

where $\{B_t : t \geq 1\}$ *is a sequence of i.i.d.* $\mathcal{N}(0, 1)$ *and* x_∞ *is the solution of a deterministic recursive equation in* t.

Note that we have synchronization but there is *no fixation*, i.e., it does *not* hold $Z_\infty \in \{0, 1\}$ a.s. The Fig. 8.2 shows an approximation of Z_t for t large through an histogram.

8.5 Sketch of Synchronization's Proof

Recall that a discrete-time process $\{X_t\}_{t \geq 0}$ is defined to be a *quasi-martingale* if

$$\sum_{t=0}^{+\infty} \mathbb{E}\big[\big|\mathbb{E}(X_{t+1} - X_t|\mathcal{F}_t)\big|\big] < +\infty \text{ holds.}$$

If $\{X_t\}_{t \geq 0}$ is bounded, it can be proved (see [23]) that there exists $X_\infty \in L^p$ such that $\lim_{t \to +\infty} X_t = X_\infty$ almost surely and in L^p.

The idea of the synchronization's proof is the following

1. Show that $\{Z_t\}_{t \geq 0}$ is a bounded martingale which converges a.s. and in L^p to a random variable Z_∞ which has mean $1/2$ and variance lower than $1/(4N)$.
2. Write a recursive equation for $x_t = \mathbb{E}[(Z_t(i) - Z_t)^2]$:

$$x_{t+1} = f(t)\, x_t + g(t)$$

where

$$f(t) := 1 - \frac{A}{t+3} + \frac{B}{(t+3)^2}, \quad g(t) := \frac{\frac{N-1}{N}\left(\frac{1}{2} - \mathbb{E}[Z_t^2]\right)}{(t+3)^2},$$

and A and B are constant depending on α and N and obtain that the limit of x_t as t goes to $+\infty$ is 0 (L^2 synchronization).
Note that g depends on the quantity $\frac{1}{2} - \mathbb{E}[Z_t^2]$, which is controlled through the estimate in step 1.

3. Derive from Theorem 3 the following two facts:
   - For each i, $\{Z_t(i)\}_{t\geq 0}$ is a bounded *quasi-martingale*, and hence it has an almost sure limit $Z_\infty(i)$.
   - The limit $Z_\infty(i)$ has to be necessarily Z_∞. More precisely, all the $\{Z_t(i)\}_t$ converge almost surely and in L^p to the same random variable Z_∞ as t goes to $+\infty$.

Remark 1 The case $\alpha = 0$ (no interaction) corresponds to a system of N independent Pólya urns. In this case, we have convergence as well, but there is no synchronization among the urns, i.e., the correlation coefficient of two urns is zero in the independent case, while it converges to 1 in the interacting case.

Moreover, in the independent case, each sequence $\{Z_t(i)\}_{t\geq 0}$ for $i = 1, \ldots, N$ converges to a uniform random variable, which means that when t is very large, any fraction of red balls can be observed with equal probability, while in the interacting case the presence of the other urns makes some values of $Z_t(i)$ more likely to occur. In particular, when the number of urns increases, the variance of Z_∞ is smaller so the probability that a single urn has a proportion of red balls different from the expected one becomes smaller too.

8.6 Related Models of Interacting Urns

A remarkable fact in the model described above is that urns synchronize on the limiting global proportion of red balls no matter how strong the interaction is. We may guess that this depends on the fact that Z_t has less fluctuations than the proportion $Z_t(i)$ of red balls in each single urn. In the first two examples below, we modify the model by introducing some more stable objects in the interaction term. In the third model, we introduce more fluctuations in the interaction term. The model can be seen as a first attempt to consider a short-range interaction (which in general means that for each urn i, Z_t is replaced by the local proportion of red balls in a set of neighboring urns).

8.6.1 Forced Pólya urn Model

Consider first the following simple model where the updating scheme is perturbed by a fixed driving proportion f. One urn whose composition evolves through the proportion $\{Z_t\}_t$ solution of:

$$Z_{t+1} = \frac{t+2}{t+3} Z_t + \frac{1}{t+3} \mathbb{1}_{\{U_t < \alpha f + (1-\alpha) Z_t\}}$$

where $f \in [0, 1]$ is fixed, $\{U_t\}_t$ are i.i.d. uniform on $[0, 1]$ and $\alpha \in [0, 1]$ is fixed. It is straightforward using the previous method [13] to state $\lim_{t \to +\infty} Z_t = f$ a.s.. The case of N (noninteracting) urns follows in a straightforward way. It exhibits a trivial synchronization phenomenon toward the common deterministic driving proportion. See Fig. 8.3 as illustration.

Remark the following fact. Let Z_n be the proportion of red balls in a Friedman urn where the reinforcement matrix is symmetric and at each step we put in the urn a balls of the color drawn and b balls of the color not drawn. Let Z_n be the conditional probability to draw a red ball, given $\mathcal{F}_n$. Moreover, denote by $\bar{Z}_n$ the frequency of red balls in the first n steps. Note that for a classical Pólya urn interacting with the fixed proportion $\frac{1}{2}$, we have $Z_n = \frac{n}{n+2} \bar{Z}_n + \frac{1}{n+2} \simeq \bar{Z}_n$, where $a_n \simeq b_n$ means $\lim_{n \to \infty} \frac{a_n}{b_n} = 1$. Now, for the Friedman urn we have $X_n = a \sum_{k=1}^{n} Y_k + b \sum_{k=1}^{n} (1 - Y_k) + 1$, where X_n is the number of red balls after n steps, and

$$Z_n = \frac{X_n}{2 + (a+b)n} = \frac{(a-b)n}{2 + (a+b)n} \bar{Z}_n + \frac{bn+1}{2+(a+b)n} \simeq \frac{a-b}{a+b} \bar{Z}_n + \frac{b}{a+b} \text{ a.s.}$$

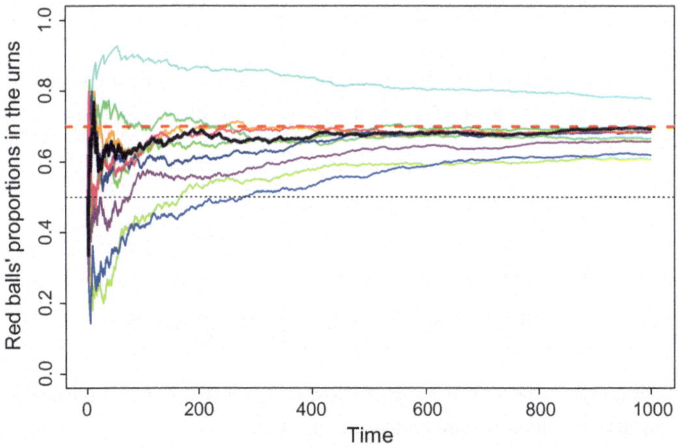

Fig. 8.3 $N = 10$ independent urns are considered, each driven by a fixed proportion 0.7. $\alpha = 0.5$. Each urn starts with 1 *red*/1 *blue ball*. Represented are the trajectories of the proportion of *red balls* in each urn. The *red horizontal dotted line* is representing the driving proportion

and the last quantity can be written as $\frac{a-b}{a+b}\bar{Z}_n + \left(1 - \frac{a-b}{a+b}\right)\frac{1}{2}$. So, if we pose $1 - \alpha = \frac{a-b}{a+b}$, the conditional probability of drawing a red ball behaves like the one of the classical Pólya urn interacting with the fixed proportion $\frac{1}{2}$.

8.6.2 Urns with a "Preferred" Color

In the following model, we introduce some spatial inhomogeneity by considering mean field interacting *biased* urns. In the dynamics of each urn, there is an interplay among self-reinforcement, a mean field interaction and a fixed proportion.

First, we consider the case of two interacting biased Pólya urns. Each urn starts with one red/one blue ball. One urn is indexed with R (resp. B) denoting that it is biased in favor of red (resp. blue) balls. It means a fixed driving proportion 1 (resp. 0). The updating scheme is as before: add one ball at a time, same color of the randomly chosen one. Nevertheless, with probability ρ ($\rho \in [0, 1]$), use the usual $\alpha/1 - \alpha$ scheme, with probability $1 - \rho$, add one red (resp. blue) ball. X_t^R (resp. X_t^B) denotes the number of red balls in the red urn (resp. in the blue urn). The proportion of red balls is $Z_t^R := X_t^R/(t + 2)$. The following recursive equations hold:

$$\begin{cases} X_{t+1}^R = X_t^R + \mathbb{1}_{\{U<\rho(\alpha Z_t+(1-\alpha)Z_t^R)+(1-\rho)\cdot 1\}} \\ X_0^R = 1 \\ \\ X_{t+1}^B = X_t^B + \mathbb{1}_{\{U<\rho(\alpha Z_t+(1-\alpha)Z_t^B)+(1-\rho)\cdot 0\}} \\ X_0^B = 1 \\ Z_t := \dfrac{Z_t^R + Z_t^B}{2} \end{cases}$$

The case $\rho = 1$ means nonbiased urns and a.s. synchronization toward the random limit of the mean field process $\{Z_t\}_t$. When $\rho < 1$, we have a.s. convergence toward deterministic limits

$$\lim_{t\to+\infty} Z_t^R = \frac{1}{2}\left(1 + \frac{1-\rho}{1-(1-\alpha)\rho}\right) \quad \text{a.s.}$$

$$\lim_{t\to+\infty} Z_t^B = \frac{1}{2}\left(1 - \frac{1-\rho}{1-(1-\alpha)\rho}\right) \quad \text{a.s.}$$

The natural generalization of this result follows straightforward and holds for N interacting biased urns. This is illustrated with Fig. 8.4.

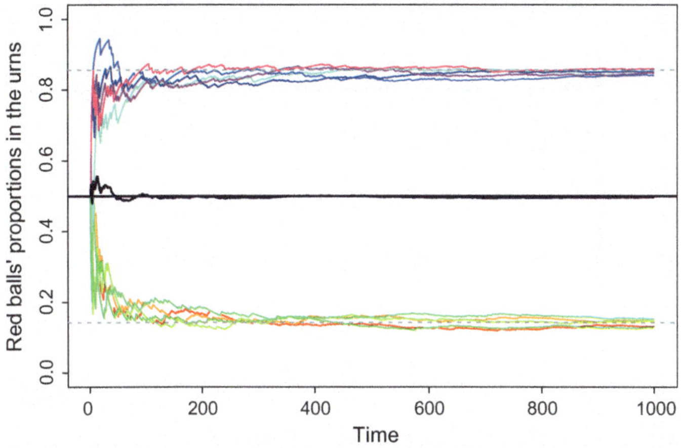

Fig. 8.4 $N = 10$ urns interacting through mean field ($\alpha = 0.6$) are considered. This first (resp. last) five urns are biased toward *red* (resp. *blue*) *balls*. The biased is applied with $\rho = 0.4$. Each urn starts with 1 *red*/1 *blue ball*. Represented are the trajectories of the proportion of *red balls* in each urn. The *gray horizontal dotted lines* are the computed deterministic limiting proportions. The *black curve* is the mean field's trajectory

8.6.3 A Simple Finite Range Interacting Case

We consider urns organized in a tree structure. So, we take a finite rooted tree T where the root urn 0 is a free evolving Polya urn and the updating rule of urn $i \in T$ with $i \neq 0$ uses the parameter $\alpha Z_t(\wp(i)) + (1-\alpha)Z_t(i)$ where $\wp(i)$ denotes the unique mother urn of the urn i in the tree structure.

The number $X_t(i)$ of red balls in the urn system is defined through

$$\begin{cases} X_{t+1}(i) = X_t(i) + \mathbb{1}_{\{U(t+1,i)<\alpha Z_t(\wp(i))+(1-\alpha)Z_t(i)\}} \text{ when } i \neq 0 \\ X_{t+1}(i) = X_t(i) + \mathbb{1}_{\{U(t+1,i)<Z_t(i)\}} \text{ when } i = 0 \\ X_0(i) = 1 \text{ for all } i \in T \end{cases}$$

With an argument similar to the one shown in Sect. 8.5, it can be proved that for this model we have a.s. synchronization toward the random limit of the root which is a free Pólya urn.

Figure 8.5 shows one sample of trajectories for a binary tree. Each curve denotes the time evolution (represented on the x-axis) of the proportion $\{Z_t(i)\}_t$, and the line's width is larger when the urn is higher in the generations. The color is related to filiation. The main two families which started at generation 1 share a dominating red (resp. yellow) tonality.

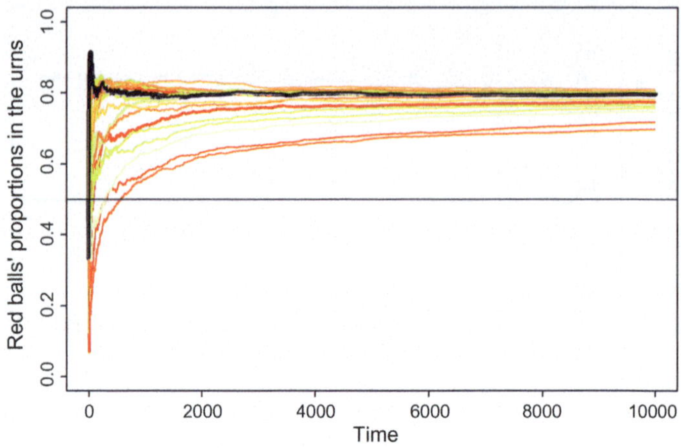

Fig. 8.5 Interacting urns through a binary tree structure. $N = 15$ urns are considered (root and three generations), $\alpha = 0.55$. Each urn starts with 1 *red*/1 *blue ball*. Represented are the trajectories of the proportion of *red balls* in each urn. The *black curve* is associated to the free Pólya urn at the root

8.7 Open Questions

Some open questions are still unsolved. We ask if for some variants of the model there is a phenomenon of phase transition, i.e., synchronization occurs only for some values of the strength of interaction. For instance, this may happen when the interaction among urns is local or when there is a stronger mechanism of self-reinforcement. This last behavior could be observed by changing the replacement scheme in the original model. Recall that for the models considered above we have a linear self-reinforcement, which means that the probability of adding a ball of a given color is linear in the frequency with which that color appeared in the past. We may consider mean field interacting urns with a nonlinear self-reinforcement, for instance, a "Rubin-type" self-reinforcement, that is a model with

$$\begin{cases} X_{t+1} = X_t + \mathbb{1}_{\{U < \alpha w(X_t(i))/[w(X_t(i))+w(Y_t(i))]+(1-\alpha)Z_t\}} \\ X_0 = 1 \end{cases}$$

where $\{Z_t\}_t$ is the mean field process $Z_t = (1/N)\sum_i Z_t(i)$, $X_t(i)$ (resp. $Y_t(i)$) is the number of red (resp. blue) balls in urn i at time t, and $\{w(k)\}_k$ is a sequence of weights having suitable properties and defining the reinforcement procedure (see for example [14, 20, 21]).

Generalizations

Since this chapter was written, several research papers were published and other models of interacting urns or reinforced processes have been considered. In [27], synchronization of interacting Friedman urns has been proved, obtaining for the variance of $Z_n(i) - Z_n$ the same scaling as in [13]. In [12], a general class of interacting reinforced processes is analyzed. These processes are characterized by two parameters, describing the size of reinforcement and the strength of interaction, and include as particular cases models which are equivalent to the ones considered here and in [27]. In particular, it is shown that synchronization is a general property and that scaling of the above variance changes with the reinforcement's size.

Moreover, for all such models it is natural to study fluctuations around the limits. In [11], central limit theorems for $\{Z_n\}_{n\geq 1}$ and $\{Z_n(i)\}_{n\geq 1}$ in the case of interacting Pólya urns have been proved. In [12], these results are extended to processes of the class mentioned above by proving functional central limit theorems (see references therein for general literature about fluctuations theorems). See for further developments [2, 3].

References

1. Acebrón, J.A., Bonilla, L.L., Vicente, C.J.P., Ritort, F., Spigler, R.: The Kuramoto model: a simple paradigm for synchronization phenomena. Rev. Mod. Phys. **77**(1), 137 (2005)
2. Aletti, G., Ghiglietti, A.: Interacting generalized Friedman's urn systems. Stoch. Process. Appl. (2016)
3. Aletti, G., Crimaldi, I., Ghiglietti, A.: Synchronization of reinforced stochastic processes with a network-based interaction. arXiv:1607.08514
4. Alonso-Sanz, R.: Discrete Systems with Memory. World Scientific Series on Nonlinear Science Series A, vol. 75. World Scientific, Singapore (2011)
5. Arenas, A., Díaz-guilera, A., Kurths, J., Moreno, Y., Zhou, C.: Synchronization in complex networks **469**(3), 93–153 (2008)
6. Berglund, N., Fernandez, B., Gentz, B.: Metastability in interacting nonlinear stochastic differential equations: I. From weak coupling to synchronization. Nonlinearity **20**(11), 2551–2581 (2007)
7. Bertini, L., Giacomin, G., Poquet, C.: Synchronization and random long time dynamics for mean-field plane rotators. Probab. Theory Relat. Fields 1–61 (2013)
8. Birkner, M., Depperschmidt, A.: Survival and complete convergence for a spatial branching system with local regulation. Ann. Appl. Probab. **17**(5/6), 1777–1807 (2007)
9. Chen, N., Glazier, J.A., Alber, M.S.: A parallel implementation of the cellular potts model for simulation of cell-based morphogenesis, pp. 58–67
10. Collet, F., Dai Pra, P., Sartori, E.: A simple mean field model for social interactions: dynamics, fluctuations, criticality, pp. 1–37 (2010)
11. Crimaldi, I., Dai Pra, P., Minelli, I.G.: Fluctuation theorems for synchronization of interacting Polya urns. Stoch. Process. Appl. **126**(3), 930–947 (2016)
12. Crimaldi, I., Dai Pra, P., Louis, P.Y., Minelli, I.G.: Syncronization and functional central limit theorems for interacting reinforced random walks (2016). arXiv:1602.06217
13. Dai Pra, P., Louis, P.Y., Minelli, I.G.: Synchronization via interacting reinforcement. J. Appl. Probab. **51**(2), 556–568 (2014)
14. Davis, B.: Reinforced random walk. Probab. Theory Relat. Fields **84**(2), 203–229 (1990)
15. Drossel, B., Schwabl, F.: Self-organized critical forest-fire model. Phys. Rev. Lett. **69**(11), 1629–1632 (1992)
16. Freedman, D.A.: Bernard Friedman's urn. Ann. Math. Stat. **36**(3), 956–970 (1965)
17. Friedman, B.: A simple urn model. Commun. Pure Appl. Math. **2**(1), 59–70 (1949)
18. Jahnel, B., Kuelske, C.: Synchronization for discrete mean-field rotators. Electron. J. Probab. **19**(14), 1–26 (2014)

19. Kaneko, K.: Simulating Spatiotemporal Chaos with Coupled Map Lattices. Springer Proceedings in Physics, pp. 260–271. Springer, Berlin (1992)
20. Launay, M.: Interacting urn models (2012). arXiv:1101.1410
21. Launay, M., Limic, V.: Generalized interacting urn models (2012). arXiv:1207.5635
22. Mahmoud, H.M.: Pólya Urn Models. CRC Press, Boca Raton (2008)
23. Métivier, M.: Semimartingales, de Gruyter Studies in Mathematics, vol. 2. Walter de Gruyter & Co., Berlin (1982)
24. Paganoni, A.M., Secchi, P.: Interacting reinforced-urn systems. Adv. Appl. Probab. **36**(3), 791–804 (2004)
25. Pemantle, R.: A survey of random processes with reinforcement. Probab. Surv. **4**(1–79), 25 (2007)
26. Pikovsky, A., Rosenblum, M., Kurths, J., Hilborn, R.C.: Synchronization: a universal concept in nonlinear science. Am. J. Phys. **70**(6), 655–655 (2002)
27. Sahasrabudhe, N.: Synchronization and fluctuation theorems for interacting Friedman urns. J. Appl. Probab. **53**(4), 1221–1239 (2016)
28. Strogatz, S.: Review of sync: the emerging science of spontaneous order (2003)

Chapter 9
Nonequilibrium Physics Aspects of Probabilistic Cellular Automata

Christian Maes

Abstract Probabilistic cellular automata (PCA) are used to model a variety of discrete spatially extended systems undergoing parallel-updating. We propose an embedding of a number of classical nonequilibrium concepts in the PCA-world. We start from time-symmetric PCA, satisfying detailed balance, and we give their Kubo formula for linear response. Close-to-detailed balance we investigate the form of the McLennan distribution and the minimum entropy production principle. More generally, when time-symmetry is broken in the stationary process, there is a fluctuation symmetry for a corresponding entropy flux. For linear response around nonequilibria we also give the linear response which is now not only entropic in nature.

9.1 PCA and Physics

Despite numerous programmes, ambitions and studies there is no derivation of the dynamics of probabilistic cellular automata (PCA) from more microscopic physical rules or from more fundamental physics as generally understood. There is of course always the possibility to look at discrete time steps for a sequential-(continuous)-time interacting particle system, but that will not yield PCA as the latter are always *non-strategic* in the sense that their conditional probability given the previous configuration is a product distribution. Moreover, from the point of view of continuous time, the discrete time-step appears to introduce another important time-scale into the physical problem, which would need to be accounted for. Alternatively there is of course always the possibility, repeatedly entertained, that it is PCA that are *more* fundamental, and that the logic should in fact be reversed: the more standard physical descriptions must then be derived from PCA rules, [1]. In particular, thinking of PCA as physics on the Planck scale, classical space-time would emerge as a coarse-grained feature of quantum gravity, [2].

C. Maes (✉)
Instituut voor Theoretische Fysica, KU Leuven, Celestijnenlaan 200D,
B-3001 Leuven, Belgium
e-mail: christian.maes@fys.kuleuven.be

P.-Y. Louis and F.R. Nardi (eds.), *Probabilistic Cellular Automata*, Emergence, Complexity and Computation 27, https://doi.org/10.1007/978-3-319-65558-1_9

Whatever point of view, it is not automatic to transfer continuous time physical notions to the domain of PCA. What has been done in the past is to connect d-dimensional PCA with $(d+1)$-dimensional equilibrium statistical mechanical models, and we will review the main relation in the next section. For example, the study of phase transitions in PCA which may be useful for the understanding of robustness of large parallel computations, will benefit from (e.g. renormalization group) techniques in equilibrium statistical mechanics of critical phenomena. The main motivation of the present paper is however to search for analogues of *non* equilibrium concepts, and to give PCA-versions of some recent results in stochastic kinetics.

Recently indeed much discussion was devoted to the application of a thermodynamic formalism to smooth dynamics [3], including a thermodynamic discussion of stochastic processes modeling systems in weak contact with different equilibrium reservoirs, [4, 5]. Nonequilibrium statistical physics is obviously expected to be incomplete when restricting it to such concepts as energy, work, heat and entropy(production) even when including the study of their fluctuations, but it is a good start to see how already these notions appear and play in physically motivated stochastic dynamics for open systems. The situation for PCA is then even worse. Our basic method will not start from detailed balance as expressed in terms of an energy function or potential, but rather begins with estimating time–reversal breaking. That is the content of Sects. 9.3–9.4, where we repeat the fluctuation symmetry for the source term of time–symmetry breaking. We then continue with that source term "entropy production" in the following sections where we discuss the minimum entropy production principle and the McLennan–Zubarev distribution. We end by giving the linear response formula for general PCA.

9.2 Notation

We only consider translation-invariant probabilistic cellular automata on the cubic lattice $\mathbb{Z}^d$, characterized by the one-site updating

$$p_i(a|\eta) = \text{Prob}[X_n(i) = a|X_{n-1} = \eta] \tag{9.1}$$

for state space $K = \{+1, -1\}^{\mathbb{Z}^d}$, $a = \pm 1, \eta \in K$. We refer to [6, 7] as general references. $(X_n, n = 0, 1, 2, \ldots)$ is a discrete time Markov process on K, with

$$\text{Prob}[X_n(i) = a_i, i \in V|X_{n-1} = \eta] = \prod_{i \in V} p_i(a_i|\eta) \tag{9.2}$$

for all finite $V \subset \mathbb{Z}^d$. We prefer of course to have $p_i(a_i|\eta)$ to depend locally on neighboring $\eta(j)$, $j \sim i$ only.

As a parameterization we choose to write

$$p_i(a|\eta) = \frac{1}{2}(1 + a\, h_i(\eta)) \tag{9.3}$$

where $|h_i| \leq 1$ on d-dimensional configurations. Again, $h_i(\eta)$ is a local and translation invariant function of $\eta \in K$. The formal $(d+1)$-dimensional Hamiltonian is

$$H(\sigma) = -\sum_{i,n} \log p_{i,n}(\sigma_n(i)|\sigma_{n-1}) \tag{9.4}$$

for $\sigma = (\sigma_n(i), i \in \mathbb{Z}^d, n \in \mathbb{N})$. For local PCA the relative Hamiltonian $H(\sigma) - H(\sigma')$ where $\sigma = \sigma'$ outside some finite volume $\Lambda \subset \mathbb{Z}^{d+1}$ makes mathematical sense. That is in fact the start of the connection between $(d+1)$-dimensional equilibrium statistical mechanics and PCA as dynamics on discrete configurations on $\mathbb{Z}^d$, [8, 9]. The present paper will emphasize the *non*equilibrium aspects, and these start from realizing that the Hamiltonian H does not need to be reflection–invariant in the temporal (or, $(d+1)$th)–direction.

9.3 Detailed Balance

In contrast to continuous time interacting particle systems, PCA as defined above cannot produce any given Gibbs distribution as stationary. In particular, detailed balance is not so naturally installed for PCA. Remember indeed that the updating is in parallel with each spin being updated independently given the previous configuration, so that is it is not immediate how to minimize an energy function, or how to install a Lyapunov function, especially with local interactions. The change of $X_n(i)$ can be determined by some cost function $\mathcal{L}(X_n(i), X_n(j), j \sim i)$ but while $X_n(i) \to X_{n+1}(i)$ changes also its neighbors $X_n(j) \to X_{n+1}(j)$ get updated similarly and simultaneously, which may prevent gradient flow. That is not to say that we cannot build invertible cellular automata, indeed we can, [10], but the very concept of (semi-bounded) energy appears deeply related to a continuous time process. Constructions involving the Hamiltonian formalism for integer-valued variables and integer time steps, are, to say the least, quite cumbersome.

Coming back to *probabilistic* cellular automata, a stationary process $(X_n, n \in \mathbb{Z})$, is time-reversible (statistically symmetric under $n \to -n$) when in (9.3)

$$h_i(\eta) = \tanh\left[\lambda_i + \sum_j J_{ij}\eta_j\right] \tag{9.5}$$

for some λ_i and symmetric $J_{ij} = J_{ji}$. The stationary distribution is then, formally,

$$\nu(\eta) = C \exp \sum_i \Big\{ \lambda_i \eta_i + \log 2 \cosh \Big[\lambda_i + \sum_j J_{ij} \eta_j \Big] \Big\}. \tag{9.6}$$

Note in fact that the corresponding interaction has at least three–body interaction; to obtain a simpler nearest neighbor-interaction appears impossible. It is then also true that, in contrast with continuous (sequential) time, not all equilibrium distribution can be reached as stationary distribution. For example, the standard Ising model cannot be obtained; see however [11, 12]. An alternative is working on bipartite lattices, with alternate updating in the way of [13].

Detailed balance can formally be written as the condition that, pretending first we have a Markov chain with transition probability $p(\eta|\eta')$,

$$p(\eta|\eta')\nu(\eta') = S(\eta, \eta') \tag{9.7}$$

is symmetric. As a consequence then,

$$\nu(\eta) = \nu(-1) \prod_i \frac{1 + \eta_i h_i(-1)}{1 - h_i(\eta)} \tag{9.8}$$

where "-1" stands for the configuration which is constant equal to -1, see [14]. That gives rise to a well–defined Hamiltonian on K. We call such ν equilibrium distributions even though there is no thermodynamic notion of equilibrium here. We can for example examine what happens to them under a small perturbation. We are then talking about linear response around equilibrium.

Suppose we start in equilibrium (with expectations $\langle\cdot\rangle_{\text{eq}}$) and we perturb ($\rightarrow \langle\cdot\rangle^h_{\text{eq}}$) by letting

$$p_i^h(\sigma_n(i)|\sigma_{n-1}) = \frac{p_i(\sigma_n(i)|\sigma_{n-1})}{z_i(\sigma_{n-1})}\, e^{\frac{h_n}{2}[V_i(\sigma_n) - V_i(\sigma_{n-1})]} \quad n = 1, 2 \ldots \tag{9.9}$$

where all V_i are local and only a finite number are non-zero, and the h_n are small amplitudes. The linear response on an observable O at time $n > m$ is found to be

$$\frac{\partial}{\partial h_m} \langle O(\sigma_n)\rangle^h_{\text{eq}} \,(h = 0) = \frac{1}{2} \sum_i \langle [V_i(\sigma_{m+1}) - V_i(\sigma_{m-1})]\, O(\sigma_n)\rangle_{\text{eq}} \tag{9.10}$$

where the subscript reminds us that the reference (unperturbed) process is equilibrium time-reversal symmetric. The right–hand side is an equilibrium time–correlation function. We recognize the analogue of the Kubo formula (or the fluctuation–dissipation theorem) around equilibrium, [15].

9.4 Breaking Detailed Balance

A measure for breaking detailed balance is given by

$$J_{i,n}(\sigma) := \log \frac{p_i(\sigma_n(i)|\sigma_{n-1})}{p_i(\sigma_{n-1}(i)|\sigma_n)} \tag{9.11}$$

which is a local function on $\mathbb{Z}^{d+1}$ (involving just two-time layers). The reason is that there is always $G_{L,N}(\sigma)$ with uniform bound $||G_{L,N}|| \leq c(d)NL^{d-1}$ so that

$$W_{N,L}(\sigma) := \sum_{n=-N+1}^{N-1} \sum_{|i|\leq L-1} J_{i,n}(\sigma) + G_{L,N}(\sigma) \tag{9.12}$$

is antisymmetric under time–reversal $(\theta_{L,N}\sigma)_n(i) := \sigma_{-n}(i)$ for $(i,n) \in \Lambda_{L,N}$ which is a rectangular shaped region centered at the origin with time-extension $2N+1$ and spatial volume $(2L+1)^d$. Under detailed balance, for the equilibrium process then $\langle W_{N,L}\rangle_{\text{eq}} = 0$.

There is actually a further symmetry, called fluctuation symmetry, in the following sense:
For $L = L(N) \leq N$ growing to infinity with time N, the limit

$$e(\lambda) := \lim_N \frac{1}{|\Lambda_{L,N}|} \log\langle e^{-\lambda \sum_{(i,n)\in\Lambda_{L-1,N-1}} J_{i,n}}\rangle \tag{9.13}$$

exists for all real λ and $e(\lambda) = e(1-\lambda)$. The expectation $\langle\cdot\rangle$ is for a general local PCA in the stationary regime. We refer to [16, 17] for a proof and extensions within the context of Gallavotti–Cohen symmetries, [18].

9.5 Entropy Production Rate Density

For a stationary distribution ν we consider its extension (the stationary Markov process) P_ν on $\mathbb{Z}^{d+1}$. In analogy with continuous time [17], we *define* the mean entropy production rate per unit volume as the space–time relative entropy density with respect to time-reversal

$$\text{MEP}_\nu := s(P_\nu|P_\nu\theta) = -s(P_\nu) + \langle \log p_0(\sigma_0(0)|\sigma_1)\rangle = \langle J_0\rangle \tag{9.14}$$

where J_0 is found from (9.11) with $i = 0 = n$ and $s(P_\nu)$ is the statistical mechanical equilibrium entropy of the $(d+1)$dimensional Gibbs measure, also called Kolmogorov-Sinai entropy,

$$s(P_\nu) = -\langle \sum_a p_0(a|\sigma_{-1}) \log\, p_0(a|\sigma_{-1}) \rangle. \tag{9.15}$$

Whether MEP_ν truly corresponds to an entropy production is unclear, as we have not obtained PCA as subsystem or as reduced description after contact with heat baths etc. It is rather to be seen here as the expected rate of time-reversal breaking. Clearly MEP_ν is non-negative, and equals zero at detailed balance. It is the first λ-derivative of $e(\lambda)$ in (9.13).

If the process is not stationary but has reached probability distribution μ, we define the expected entropy production rate in μ as

$$\text{EP}[\mu] := \langle J_0(\sigma)\rangle_\mu + S(\mu P) - S(\mu) \tag{9.16}$$

where the first term takes the expectation of (9.11) over the two-time layer (σ_0, σ_1) when σ_0 is averaged with probability distribution μ on K. The $S(\mu)$ and the $S(\mu P)$ are Shannon entropy densities for μ and its (single step) update μP (with stochastic matrix P). In fact we can also write $\text{EP}[\mu]$ itself as a relative entropy density of P_μ restricted to two-time layers with respect to $P_{\mu P}\theta$ on these two times and with θ exchanging the two times, or formally

$$\text{EP}[\mu] \propto \sum_{\sigma_0,\sigma_1} \mu(\sigma_0)p(\sigma_1|\sigma_0) \log \frac{\mu(\sigma_0)p(\sigma_1|\sigma_0)}{\mu P(\sigma_1)p(\sigma_0|\sigma_1)}. \tag{9.17}$$

The functional $\text{EP}[\mu]$ is non-negative, convex and vanishes under detailed balance when μ is the equilibrium distribution. There is in fact a unique minimizer, which we could call the Prigogine distribution.

9.6 Minimum Entropy Production Principle

It turns out that when operating close to detailed balance the stationary distribution can also be characterized as minimum of a functional which very much resembles the entropy production rate density. In other words the stationary ν equals a minimizer of an entropy production-like functional. We give here the argument for any fixed finite volume (perhaps with periodic boundary conditions) on which the PCA gets defined, which is the case of a (discrete time) Markov chain. We follow below the straightforward variational method of [19]. Whether the minimum entropy production principle or a close relative of it can also be derived as a consequence of dynamical large deviation theory in the way of [20], remains an open question.

Consider

$$\sigma[\mu] := \sum_{x,y} \mu(x)p(x,y) \log \frac{\mu(x)p(x,y)}{\mu(y)p(y,x)} \tag{9.18}$$

and take the variation with respect to $\mu(x)$ to find

$$\sum_y p(x,y)\log\frac{\mu(x)p(x,y)}{\mu(y)p(y,x)} - \frac{\mu P(x)}{\mu(x)} = \text{constant.} \tag{9.19}$$

We now like to show that (9.19) is indeed satisfied to first order around equilibrium. The latter is quantified via a dimensionless parameter $\varepsilon \ll 1$. We take $\mu = \nu(1+\varepsilon g)$ and $p(x,y) = t(x,y)(1+\varepsilon m(x,y))$ with detailed balance for $\nu(x)t(x,y) = t(y,x)\nu(y)$. Then the first term in the left-hand side of (9.19) becomes

$$\sum_y t(x,y)(1+\varepsilon m(x,y))\log\frac{(1+\varepsilon g(x))(1+\varepsilon m(x,y))}{(1+\varepsilon g(y))(1+\varepsilon m(y,x))} = 1+\varepsilon v(x) \tag{9.20}$$

(expanding to first order in ε) where

$$\begin{aligned} v(x) &= \sum_y t(x,y)[g(x)+m(x,y)-g(y)-m(y,x)] = g(x) - \sum_y t(x,y)g(y) \\ &\quad - \sum_y t(x,y)m(y,x) \end{aligned} \tag{9.21}$$

from using $\sum_y t(x,y) = 1$ and $\sum_y t(x,y)m(x,y) = 0$. The second term in (9.19) contains $\mu P(x) = \nu(x)(1+\varepsilon\tilde{g}(x))$, where

$$\begin{aligned} \nu(x)(1+\varepsilon\tilde{g}(x)) &= \sum_y t(y,x)(1+\varepsilon m(y,x))\nu(y)(1+\varepsilon g(y)) \\ &= \nu(x) + \varepsilon\sum_y t(y,x)m(y,x)\nu(y) + \varepsilon\sum_y t(y,x)\nu(y)g(y) \\ &= \nu(x) + \varepsilon\nu(x)\sum_y t(x,y)m(y,x) + \varepsilon\nu(x)\sum_y t(x,y)g(y) \\ &\implies \tilde{g}(x) = \sum_y t(x,y)[m(y,x)+g(y)] \end{aligned}$$

where we used detailed balance $t(y,x)\nu(y) = \nu(x)t(x,y)$. Therefore,

$$\frac{\mu P(x)}{\mu(x)} = 1+\varepsilon\tilde{g}(x) - \varepsilon g(x) = 1+\varepsilon\left[\sum_y t(x,y)[m(y,x)+g(y)] - g(x)\right] \tag{9.22}$$

which we must compare with (9.21) to see that indeed (9.19) is satisfied.

Remark that $\sigma[\mu]$ not quite equals (9.17) for $\sigma_0 \to x, \sigma_1 \to y, p(\sigma_1|\sigma_0) \to p(x, y)$. We really would have to consider instead of $\sigma[\mu]$ the entropy production functional

$$\mathrm{EP}[\mu] = \sum_{x,y} \mu(x)p(x, y) \log \frac{\mu(x)p(x, y)}{\mu P(y)p(y, x)} \tag{9.23}$$

However, taking the variation of that one, we find that the stationary distribution does *not* satisfy it even to first order around equilibrium. In other words we should not expect that the stationary distribution of a PCA equals the Prigogine distribution even in linear order.

9.7 McLennan–Zubarev Formula

Close-to-detailed balance we can give an expression for the stationary distribution. In [21] is explained a rigorous derivation for continuous time. Let us here look at a (discrete time, irreducible and aperiodic) Markov chain $X_n, n \geq 0$, for $X_n \in K$ finite.

From the previous section we know that the distribution μ coincides with the stationary distribution ν to linear order in ε when it satisfies (9.19). So we can get μ correct to first order by plugging it in (9.19): (using $\mu = \mu P$),

$$\sum_y [p(x, y) - \delta_{x,y}] \log \mu(y) = \sum_y p(x, y) \log \frac{p(x, y)}{p(y, x)} + \text{constant.} \tag{9.24}$$

We substitute again $\mu(x) = \nu(x)(1 + \varepsilon g(x))$ and $p(x, y) = t(x, y)(1 + \varepsilon m(x, y))$ and we must have that (9.21) is constant:

$$g(x) - \sum_y t(x, y)g(y) = \sum_y t(x, y)m(y, x) + \text{constant} \tag{9.25}$$

which we must solve for g. One should however be aware that the (detailed balance) matrix L with element $t(x, y) - \delta_{x,y}$ is singular. We can however use the constant to project on the subspace orthogonal to the constant functions. That is the so called pseudo-inverse L^{-1} for which we have

$$L^{-1} f(x) = -\sum_{n=0}^{\infty} P^n[f - \langle f \rangle_{\mathrm{eq}}] \tag{9.26}$$

to be used for the function $f(x) = \sum_y t(x, y)m(y, x)$. That gives the correction g to equilibrium, yielding the McLenna–Zubarev form. To work out the analogous McLennan–Zubarev form for PCA (in the thermodynamic limit) and to show it is a Gibbsian distribution at least in the high noise regime is left here as an open problem.

9.8 Linear Response

For the perturbation (9.9), but now starting from a general distribution ρ and not restricting ourselves to detailed balance, we have the nonequilibrium response formula

$$\langle O(\sigma_n)\rangle_\rho^h - \langle O(\sigma_n)\rangle_\rho = \sum_i \sum_{m=1}^{n-1} \frac{h_m}{2}\{\langle [V_i(\sigma_{m+1}) - V_i(\sigma_m)]\, O(\sigma_n)\rangle_\rho - \langle \langle V_i(\sigma_{m+1}) - V_i(\sigma_m)|\sigma_m\rangle\, O(\sigma_n)\rangle_\rho\} + O(h^2) \qquad (9.27)$$

It is the generalization of the Kubo-like formula (9.10) to nonequilibrium processes. It contains a frenetic contribution following the line of [22]. See [23] for an update on linear response around nonequilibria in continuous time.

References

1. Wolfram, S.: A new kind of Science. Wolfram Media (2002)
2. 't Hooft, G.: Hamiltonian formalism for integer-valued variables and integer time steps and a possible application in quantum physics, Report-no: ITP-UU-13/29, SPIN-13/21. arXiv:1312.1229
3. Ruelle, D.: Smooth dynamics and new theoretical ideas in nonequilibrium statistical mechanics. J. Stat. Phys. **95**, 393 (1999)
4. Sekimoto, K.: Stochastic Energetics. Lecture Notes in Physics. Springer, Berlin (2010)
5. Maes, C., van Wieren, M.H.: Thermoelectric phenomena via an interacting particle system. J. Phys. A: Math. Gen. **38**, 005–1020 (2005)
6. Ilachinski, A.: Cellular Automata: A Discrete Universe. World Scientific Pub Co Inc, Singapore (30 July 2001). Reprint edition
7. Dobrushin, R., Kryukov, V., Toom, A. (eds.): Stochastic cellular systems: ergodicity, memory, morphogenesis. Nonlinear Science: Theory and Applications. Manchester University Press, Manchester (1990)
8. Goldstein, S., Kuik, R., Lebowitz, J.L., Maes, C.: From PCA to equilibrium system and back. Commun. Math. Phys. **125**, 71–79 (1989)
9. Georges, A., Le Doussal, P.: From equilibrium spin models to probabilistic cellular automata. J. Stat. Phys. **54**, 1011–1064 (1989)
10. Toffoli, T., Margolus, N.H.: Invertible cellular automata: a review. Phys. D **45**, 229–253 (1990)
11. Dai Pra, P., Scoppola, B., Scoppola, E.: Sampling from a Gibbs measure with pair interaction by means of PCA. J. Stat. Phys. **149**(4), 722–737 (2012)
12. Lancia, C., Scoppola, B.: Equilibrium and non-equilibrium Ising models by means of PCA. J. Stat. Phys. **153**(4), 641–653 (2013)
13. Domany, E., Kinzel, W.: Equivalence of Cellular Automata to Ising Models and Directed Percolation. Phys. Rev. Lett. **53**, 311 (1984)
14. Stavskaja, O.N.: Gibbs invariant measures for Markov chains on finite lattices with local interaction. Math. USSR Sb. **21**, 395 (1973)
15. Kubo, R.: The fluctuation-dissipation theorem. Rep. Prog. Phys. **29**, 255 (1966)
16. Maes, C.: The fluctuation theorem as a Gibbs property. J. Stat. Phys. **95**, 367–392 (1999)
17. Maes, C., Redig, F., Van Moffaert, A.: On the definition of entropy production, via examples. J. Math. Phys. **41**, 1528–1554 (2000)

18. Gallavotti, G., Cohen, E.G.D.: Dynamical ensembles in nonequilibrium statistical mechanics. Phys. Rev. Lett. **74**, 2694 (1995)
19. Jiu-li, L., Van den Broeck, C., Nicolis, G.: Z. Phys. B Conden. Matter **56**, 165 (1984)
20. Maes, C., Netočný, K.: Minimum entropy production principle from a dynamical fluctuation law. J. Math. Phys. **48**, 053306 (2007)
21. Maes, C., Netočný, K.: Rigorous meaning of McLennan ensembles. J. Math. Phys. **51**, 015219 (2010)
22. Baiesi, M., Maes, C., Wynants, B.: Nonequilibrium linear response for Markov dynamics, I: jump processes and overdamped diffusions. J. Stat. Phys. **137**, 1094–1116 (2009)
23. Baiesi, M., Maes, C.: An update on nonequilibrium linear response. New J. Phys. **15**, 013004 (2013)

Part II
Computer Science and Discrete Dynamical Systems

Chapter 10
An Example of Computation of the Density of Ones in Probabilistic Cellular Automata by Direct Recursion

Henryk Fukś

Abstract We present a method for computing probability of occurrence of ones in a configuration obtained by iteration of a probabilistic cellular automata (PCA), starting from a random initial configuration. If the PCA is sufficiently simple, one can construct a set of words (or blocks of symbols) which is complete, meaning that probabilities of occurrence of words from this set can be expressed as linear combinations of probabilities of occurrence of these words at the previous time step. One can then set up and solve a recursion for block probabilities. We demonstrate an example of such PCA, which can be viewed as a simple model of diffusion of information or spread of rumours. Expressions for the density of ones are obtained for this rule using the proposed method.

10.1 Introduction

Binary probabilistic cellular automata (PCA) in one dimension are one of the most frequently studied types of cellular automata, and one of the most natural and most frequently encountered problems in PCA is what the author proposes to call *the density response problem*: If the proportion of ones in the initial configuration drawn from a Bernoulli distribution is ρ_0, what is the expected proportion of ones after t iterations of the PCA rule?

Of course, one could ask a similar question about the probability of occurrence of longer blocks of symbols after t iterations of the PCA rule. Due to the complexity of PCA dynamics, it is clear that questions of this type are rather hopeless if one wants to know the answer for an arbitrary rule. In spite of this, it may still be possible to provide the answer if the rule is sufficiently simple.

One of the methods which can be used to do this is studying the structure of preimages of short blocks and detecting patterns present in them. This approach has been successfully used by the author for a number of deterministic CA rules, such as elementary rules 172, 142, 130 (Refs. [3, 4, 7] respectively), and several others. It has

H. Fukś (✉)
Department of Mathematics and Statistics, Brock University, St. Catharines, ON, Canada
e-mail: hfuks@brocku.ca

P.-Y. Louis and F.R. Nardi (eds.), *Probabilistic Cellular Automata*, Emergence, Complexity and Computation 27, https://doi.org/10.1007/978-3-319-65558-1_10

also been used for a special class of PCA known as single-transition α-asynchronous rules [8].

In this chapter, however, we would like to describe yet another method of computing probabilities of blocks of symbols, by setting up a system of recursive equations which can then be explicitly solved. Such a recursive system can be easily constructed for any rule for probabilities of *all* blocks, but it is normally too big and too complex to be solved. In certain cases, however, one can find a smaller set of blocks for which the recursion is solvable. We will present one such example, using a PCA which can be viewed as a simple model of diffusion of innovations or spread of rumours.

To give the reader a flavour of what to expect, let us informally define the aforementioned PCA rule. Suppose we have an infinite one-dimensional lattice where each site is occupied by an individual who has already adopted the innovation (1) or who has not adopted it yet (0). Initially, the proportion of adopters is ρ_0. Once the individual adopts the innovation, he remains in state 1 forever. Individuals in state 0 can change their states to 1 (adopt the innovation) with probabilities depending on the state of nearest neighbours: If only the right (resp., left) neighbour already adopted, the probability is p (resp., q), and if both of them already adopted, the probability is r. What is the proportion of adopters ρ_t of after t iterations of the rule, assuming that the initial configuration is drawn from a Bernoulli distribution? We will show that the explicit formula for ρ_t can derived,

$$\rho_t = \begin{cases} 1 - E\left((\rho_0 q - 1)(\rho_0 p - 1)\right)^t - F(1-r)^t & \text{if } pq\rho_0^2 - (p+q)\rho_0 + r \neq 0, \\ 1 - (G + Ht)(1-r)^{t-1} & \text{if } pq\rho_0^2 - (p+q)\rho_0 + r = 0, \end{cases}$$

where E, F, G, H are constants depending on parameters p, q, r, and ρ_0.

In order to accomplish this, we will start from some general theoretical remarks, considering PCA as maps in the space of shift-invariant probability measures, similarly as done in [9–11], and other works. More precisely, we will look at orbits of uniform Bernoulli measures under the action of PCA.

10.2 Probabilistic Cellular Automata

Probabilistic CA are often defined as stochastic dynamical systems. In this article, we will concentrate on Boolean CA in one dimension. Let $s_i(t)$ denote the state of the lattice site i at time t, where $i \in \mathbb{Z}$, $t \in \mathbb{N}$. We will further assume that $s_i(t) \in \{0, 1\}$ and we will say that the site i is occupied (empty) at time t if $s_i(t) = 1$ (resp., $s_i(t) = 0$).

In a probabilistic cellular automaton, lattice sites simultaneously change states from 0 to 1 or from 1 to 0 with probabilities depending on states of local neighbours. A common method for defining PCA is to specify a set of local transition probabilities. For example, in order to define a nearest neighbour PCA one has to specify the probability $w(s_i(t+1))|s_{i-1}(t), s_i(t), s_{i+1}(t))$ that the site $s_i(t)$ with nearest neighbours $s_{i-1}(t), s_{i+1}(t)$ changes its state to $s_i(t+1)$ in a single time step.

A more formal definition of nearest neighbour PCA can be constructed as follows. Let r be a positive integer, called *radius of PCA*, and let us consider a set of independent Boolean random variables $X_{i,\mathbf{b}}$, where $i \in \mathbb{Z}$ and $\mathbf{b} \in \{0, 1\}^{2r+1}$. Probability that the random variable $X_{i,\mathbf{b}}$ takes the value $a \in \{0, 1\}$ will be assumed to be independent of i and denoted by $w(a|\mathbf{b})$,

$$Pr(X_{i,\mathbf{b}} = a) = w(a|\mathbf{b}). \tag{10.1}$$

Obviously, $w(1|\mathbf{b}) + w(0|\mathbf{b}) = 1$ for all $\mathbf{b} \in \{0, 1\}^{2r+1}$. The update rule for the PCA is then defined by

$$s_i(t + 1) = X_{i,\{s_{i-r}(t),\ldots,s_i(t),\ldots,s_{i+r}(t)\}}. \tag{10.2}$$

Note that new random variables X are used at each time step t, that is, random variables X used at the time step t are independent of those used at previous time steps.

Having the above definition in mind, we note that in order to fully define a nearest neighbour PCA rule (i.e., rule with $r = 1$), it is enough to specify eight transition probabilities $w(1|x_1x_2x_3)$ for all $x_1, x_2, x_3 \in \{0, 1\}$. Remaining eight probabilities, $w(0|x_1x_2x_3)$, can be obtained by $w(0|x_1x_2x_3) = 1 - w(1|x_1x_2x_3)$.

In any dynamical system, the main object of interest is the orbit of the system starting from a given initial point, and properties of this orbit. In the case of PCA, we often assume that the initial condition is "random" or "disordered", typically meaning that each $s_i(0)$ is set to 1 with a given probability ρ_0 and to 0 with probability $1 - \rho_o$, independently of each other. We then want to answer question of the type "After t iterations, what is the proportion of sites in state 1?" or "After t iterations, what is the probability of finding a pair of adjacent zeros"? In order to pose and answer questions of this kind rigorously, we will present an alternative definition of PCA, as maps in the space of probability measures.

10.2.1 Orbits of Probability Measures

Let $\mathcal{A} = \{0, 1\}$ and $X = \mathcal{A}^{\mathbb{Z}}$. A finite sequence of elements of $\mathcal{A}$, $\mathbf{b} = b_1b_2, \ldots, b_n$ will be called a *block* (or *word*) of length n. Set of all blocks of elements of $\mathcal{A}$ of all possible lengths will be denoted by $\mathcal{A}^{\star}$.

A *cylinder set* generated by the block $\mathbf{b} = b_1b_2, \ldots, b_n$ and anchored at i is defined as

$$[\mathbf{b}]_i = \{\mathbf{x} \in \mathcal{A}^{\mathbb{Z}} : \mathbf{x}_{[i,i+n)} = \mathbf{b}\}. \tag{10.3}$$

The set of probability measures on the σ-algebra generated by cylinder sets of X will be denoted by $\mathfrak{M}(X)$. Details of construction of such measures, using Hahn–Kolmogorov theorem, can be found in [5]. These details, however, are not essential for our subsequent considerations. Given a probability measure $\mu \in \mathfrak{M}(X)$, the measure of a cylinder set $[\mathbf{b}]_i$, denoted by $\mu([\mathbf{b}]_i)$, is often informally called a "probability of occurrence of block $\mathbf{b}$ at site i".

Let the function $w : \mathcal{A} \times \mathcal{A}^{2r+1} \to [0, 1]$, whose values are denoted by $w(a|\mathbf{b})$ for $a \in \mathcal{A}$, $\mathbf{b} \in \mathcal{A}^{2r+1}$, satisfying $\sum_{a\in\mathcal{A}} w(a|\mathbf{b}) = 1$, be called *local transition function* of *radius* r, and let its values be called *local transition probabilities*. A *probabilistic cellular automaton* with local transition function w is a map $F : \mathfrak{M}(X) \to \mathfrak{M}(X)$ defined as

$$(F\mu)([\mathbf{a}]_i) = \sum_{\mathbf{b}\in\mathcal{A}^{|\mathbf{a}|+2r}} w(\mathbf{a}|\mathbf{b})\mu([\mathbf{b}]_{i-r}) \text{ for all } i \in \mathbb{Z}, \mathbf{a} \in \mathcal{A}^\star, \tag{10.4}$$

where we define

$$w(\mathbf{a}|\mathbf{b}) = \prod_{j=1}^{|\mathbf{a}|} w(a_j|b_j b_{j+1} \dots b_{j+2r}). \tag{10.5}$$

When the function w takes values in the set $\{0, 1\}$, the corresponding cellular automaton is called a *deterministic CA*.

In this paper, we will exclusively deal with shift-invariant probability measures for which $\mu([\mathbf{b}]_i)$ is independent of i. We will, therefore, drop the index i and simply write $\mu([\mathbf{b}])$. Moreover, we will be interested in orbits of Bernoulli measures ν_λ defined for $\lambda \in [0, 1]$ by

$$\nu_\lambda([\mathbf{b}]) = \lambda^{\#1(\mathbf{b})}(1-\lambda)^{\#0(\mathbf{b})} \text{ for any } \mathbf{b} \in \mathcal{A}^\star, \tag{10.6}$$

where $\#0(\mathbf{b})$ and $\#1(\mathbf{b})$ denote the number of zeros and ones in $\mathbf{b}$. In order to simplify the notation, we define

$$P_t(\mathbf{b}) = (F^t\nu_\lambda)([\mathbf{b}]), \tag{10.7}$$

which will be informally referred to as "probability of occurrence of block $\mathbf{b}$ after t iterations of PCA rule F". With this notation, Eq. (10.4) can be written as

$$P_{t+1}(\mathbf{a}) = \sum_{\mathbf{b}\in\mathcal{A}^{|\mathbf{a}|+2r}} w(\mathbf{a}|\mathbf{b})P_t(\mathbf{b}), \tag{10.8}$$

for any $\mathbf{a} \in \mathcal{A}^\star$ and $t \in \mathbb{N}$. We will furthermore define

$$P_0(\mathbf{a}) = \nu_\lambda([\mathbf{a}]) = \lambda^{\#1(\mathbf{a})}(1-\lambda)^{\#0(\mathbf{a})} \tag{10.9}$$

for any $\mathbf{a} \in \mathcal{A}^\star$.

Elements of $\mathcal{A}^\star$ can be enumerated in lexicographical order, and corresponding probabilities arranged in an infinite column vector

$$\mathbf{P}_t = (P_t(0), P_t(1), P_t(00), P_t(01), P_t(10), P_t(11), P_t(000)\dots)^T. \tag{10.10}$$

Before we continue, note that not all these probabilities are independent. Due to additivity of measure, the following relationships, known as consistency conditions, are valid for any $\mathbf{a} \in \mathcal{A}^\star$,

$$P_t(\mathbf{a}) = P_t(\mathbf{a}0) + P_t(\mathbf{a}1) = P_t(0\mathbf{a}) + P_t(1\mathbf{a}). \tag{10.11}$$

These conditions will be frequently used in our subsequent considerations.

Since each $P_{t+1}(\mathbf{a})$, by the virtue of Eq. (10.8), is a linear combination of a finite number of $P_t(\mathbf{b})$ values, we can write

$$\mathbf{P}_{t+1} = \mathbf{M}\mathbf{P}_t, \tag{10.12}$$

where the infinite matrix $\mathbf{M}$ is defined by Eq. (10.8). This yields the following expression for probabilities of all finite words,

$$\mathbf{P}_t = \mathbf{M}^t\mathbf{P}_0, \tag{10.13}$$

where components of $\mathbf{P}_0$ are defined in Eq. (10.9). In theory, the above equation gives us a complete solution of the problem of determining the orbit of Bernoulli measure under iterations of a PCA rule. In practice, however, computing powers of an infinite matrix is a daunting, if not impossible, task.

In practical applications, however, we rarely need *all* probabilities $P_t(\mathbf{a})$, that is, all components of the vector $\mathbf{P}_t$. Sometimes we are interested only in one specific probability, for example, $P_t(1)$. For a binary PCA, the expected value of a given lattice site after t iterations of the rule is equal to $1 \cdot P_t(1) + 0 \cdot P_t(0) = P_t(1)$, and for that reason, $P_t(1)$ is sometimes referred to as an expected *density of ones*, to be denoted by ρ_t,

$$\rho_t = P_t(1). \tag{10.14}$$

Note that for Bernoulli measure ν_λ, we have $\rho_0 = \lambda$. Given ρ_0, could one find an explicit expression for ρ_t for a given PCA using Eq. (10.12)? This problem will be called a *density response problem*. Although it cannot be solved in a general case, we will demonstrate that for a sufficiently simple PCA, it is a doable task.

The idea is to set up a recursion similar to (10.12), but using a "smaller" set of block probabilities, for which the matrix $\mathbf{M}$ has somewhat simpler structure, lending itself to direct computation of $\mathbf{M}^t$. If we could then express ρ_t in terms of block probabilities from this "smaller" set, we would solve the density response problem.

Let us define the concept of the "smaller" set first. A set of words $\mathcal{A}^\star \supset C = \{\mathbf{a}_1, \mathbf{a}_2, \mathbf{a}_3, \ldots\}$ will be called *complete* with respect to a PCA rule F if for every $\mathbf{a} \in C$ and $t \in \mathbb{N}$, $P_{t+1}(\mathbf{a})$ can be expressed as a linear combination of $P_t(\mathbf{a}_1), P_t(\mathbf{a}_2), P_t(\mathbf{a}_3), \ldots$. We will show a concrete example of a complete set in the next section.

10.3 Example PCA Rule

As an example, we will consider a PCA rule which generalizes some of the CA rules investigated in [1]. This PCA can be viewed as a simple model for diffusion of innovations, spread of rumours, or a similar process involving transport of information between neighbours. We consider an infinite one-dimensional lattice where each site is occupied by an individual who has already adopted the innovation (1) or who has not adopted it yet (0). Once the individual adopts the innovation, he remains in state 1 forever. Individuals in state 0 can change their states to 1 (adopt the innovation) with probabilities depending on the state of nearest neighbours. All changes of states take place simultaneously. This process can be formally described as a radius 1 binary PCA with the following transition probabilities,

$$\begin{aligned} &w(1|000) = 0,\ w(1|001) = p,\ w(1|010) = 1,\ w(1|011) = 1, \\ &w(1|100) = q,\ w(1|101) = r,\ w(1|110) = 1,\ w(1|111) = 1, \end{aligned} \tag{10.15}$$

where p, q, r are fixed parameters of the model, $p, q, r \in [0, 1]$. In order to illustrate the difficulty of computing block probabilities for this rule, let us write Eq. (10.8) for blocks **a** of length 1 and 2,

$$\begin{aligned} P_{t+1}(0) &= P_t(000) + (1-p)P_t(001) + (1-q)P_t(100) + (1-r)P_t(101), \\ P_{t+1}(1) &= pP_t(001) + P_t(010) + P_t(011) + qP_t(100) + rP_t(101) + P_t(110) + P_t(111), \end{aligned}$$

$$\begin{aligned} P_{t+1}(00) &= P_t(0000) + (1-p)P_t(0001) + (1-q)P_t(1000) + (1-p)(1-q)P_t(1001), \\ P_{t+1}(01) &= pP_t(0001) + (1-p)P_t(0010) + (1-p)P_t(0011) + p(1-q)P_t(1001) \\ &\quad + (1-r)P_t(1010) + (1-r)P_t(1011), \\ P_{t+1}(10) &= (1-q)P_t(0100) + (1-r)P_t(0101) + qP_t(1000) + (1-p)qP_t(1001) \\ &\quad + (1-q)P_t(1100) + (1-r)P_t(1101), \\ P_{t+1}(11) &= pP_t(0010) + pP_t(0011) + qP_t(0100) + rP_t(0101) + P_t(0110) + P_t(0111) \\ &\quad + pqP_t(1001) + rP_t(1010) + rP_t(1011) + qP_t(1100) \\ &\quad + rP_t(1101) + P_t(1110) + P_t(1111). \end{aligned}$$

As we can see, in order to know $P_{t+1}(1)$, we need to know probabilities of blocks of length 3 at time step t, and in order to compute these, we would need probabilities of blocks of length 5 at time step $t-1$, etc.

We will now show, however, that for the PCA rule defined in Eq. (10.15), a complete subset of $\mathcal{A}^\star$ can be constructed. This subset consists of clusters of zeros bounded by 1 on each side, that is, of blocks of the type 10^n1, where $n \in \mathbb{N}$ and 0^n denotes n consecutive zeros.

Proposition 1 *The set* $C = \{101, 1001, 100001, \ldots\}$ *is complete with respect to the PCA rule defined in Eq. (10.15).*

In order to prove this, we need to show that every $P_{t+1}(10^n1)$ can be expressed as a linear combination of probabilities of the type $P_t(10^k1)$. Let us write Eq. (10.8) for

$\mathbf{a} = 10^n 1$. Two cases must be distinguished, $n = 1$ and $n > 1$. For $n = 1$, we have

$$\begin{aligned} P_{t+1}(101) = p(1-q)P_t(01001) + (1-r)P_t(01010) + (1-r)P_t(01011) \\ + pqP_t(10001) + (1-p)qP_t(10010) + (1-p)qP_t(10011) \\ + p(1-q)P_t(11001) + (1-r)P_t(11010) + (1-r)P_t(11011). \end{aligned}$$

By consistency conditions, $P_t(10010) + P_t(10011) = P_t(1001)$ and $P_t(11010) + P_t(11011) = P_t(1101)$, as well as $P_t(01001) + P_t(11001) = P_t(1001)$. This yields

$$\begin{aligned} P_{t+1}(101) =& (1-r)P_t(01010) + (1-r)P_t(01011) + pqP_t(10001) \\ &+ (1-p)qP_t(1001) + p(1-q)P_t(1001) + (1-r)P_t(1101), \end{aligned}$$

and further reduction is possible using $P_t(01010) + P_t(01011) = P_t(0101)$ and $P_t(0101) + P_t(1101) = P_t(101)$. The final result is

$$P_{t+1}(101) = (1-r)P_t(101) + (p - 2pq + q)P_t(1001) + pqP_t(10001). \tag{10.16}$$

For $n > 1$, using a similar procedure (omitted here), we obtain

$$P_{t+1}(10^n 1) = (1-p)(1-q)P_t(10^n 1) + (p - 2pq + q)P_t(10^{n+1} 1) + pqP_t(10^{n+2} 1). \tag{10.17}$$

Equations (10.16) and (10.17) clearly show that the set C is complete. □

10.4 Calculations of $P_t(10^n 1)$

Having a complete set of block probabilities, we can now write Eqs. (10.16) and (10.17) in matrix form,

$$\begin{bmatrix} P_{t+1}(101) \\ P_{t+1}(1001) \\ \vdots \\ P_{t+1}(10^n 1) \\ \vdots \end{bmatrix} = \begin{bmatrix} \tilde{a} & b & c & 0 & 0 & 0 & \cdots \\ 0 & a & b & c & 0 & 0 & \\ 0 & 0 & a & b & c & 0 & \\ 0 & 0 & 0 & a & b & c & \\ \vdots & & & & & & \ddots \end{bmatrix} \begin{bmatrix} P_t(101) \\ P_t(1001) \\ \vdots \\ P_t(10^n 1) \\ \vdots \end{bmatrix}, \tag{10.18}$$

where $a = (1-p)(1-q)$, $\tilde{a} = 1 - r$, $b = p - 2pq + q$, and $c = pq$.

Let us define

$$\mathbf{M} = \begin{bmatrix} \tilde{a} \; b \; c \; 0 \; 0 \; 0 \cdots \\ 0 \; a \; b \; c \; 0 \; 0 \\ 0 \; 0 \; a \; b \; c \; 0 \\ 0 \; 0 \; 0 \; a \; b \; c \\ \vdots \qquad\qquad \ddots \end{bmatrix}, \quad \mathbf{P}_t = \begin{bmatrix} P_t(101) \\ P_t(1001) \\ \vdots \\ P_t(10^n 1) \\ \vdots \end{bmatrix}. \tag{10.19}$$

We will use $\mathrm{diag}(x_1, x_2, x_3, \ldots)$ to denote an infinite matrix with $x_1, x_2, x_3, \ldots$ on the diagonal and zeros elsewhere. Similarly, $\mathrm{sdiag}(x_1, x_2, x_3, \ldots)$ will denote shifted diagonal matrix having $x_1, x_2, x_3, \ldots$ on the line above the diagonal and zeros elsewhere, and ${}^2\mathrm{sdiag}(x_1, x_2, x_3, \ldots)$ will denote doubly shifted diagonal matrix, with $x_1, x_2, x_3, \ldots$ on the second line above the diagonal and zeros elsewhere. With this notation, we have

$$\mathbf{M} = \mathbf{A} + \mathbf{B} + \mathbf{C}, \tag{10.20}$$

where

$$\begin{aligned} \mathbf{A} &= \mathrm{diag}(\tilde{a}, a, a, \ldots), \\ \mathbf{B} &= \mathrm{sdiag}(b, b, b, \ldots), \\ \mathbf{C} &= {}^2\mathrm{sdiag}(c, c, c, \ldots). \end{aligned}$$

Now,

$$\mathbf{P}_t = \mathbf{M}^t \mathbf{P}_0, \tag{10.21}$$

and we need to compute $\mathbf{M}^t$. We will do it by considering a special case first.

10.4.1 Special Case: $\tilde{a} = a$

When $\tilde{a} = a$, matrices **A**, **B**, and **C** pairwise commute, thus we can use the trinomial expansion formula,

$$\mathbf{M}^t = (\mathbf{A} + \mathbf{B} + \mathbf{C})^t = \sum_{i+j+k=t} \binom{t}{i, j, k} \mathbf{A}^i \mathbf{B}^j \mathbf{C}^k, \tag{10.22}$$

where

$$\binom{t}{i, j, k} = \frac{t!}{i! j! k!}. \tag{10.23}$$

Generalizing the previously introduced notation, let

$${}^n\mathrm{sdiag}(x_1, x_2, x_3, \ldots) \tag{10.24}$$

denote n-times shifted diagonal matrix, which has $x_1, x_2, x_3, \ldots$ entries on the nth line above the diagonal and zeros elsewhere. It is straightforward to prove that

$$\begin{aligned}\mathbf{A}^i &= \text{diag}(a^i, a^i, a^i, \ldots),\\ \mathbf{B}^j &= {}^j\text{sdiag}(b^j, b^j, b^j, \ldots),\\ \mathbf{C}^k &= {}^{2k}\text{sdiag}(c^k, c^k, c^k, \ldots),\end{aligned}$$

and, consequently,

$$\mathbf{A}^i\mathbf{B}^j\mathbf{C}^k = {}^{j+2k}\text{sdiag}(a^i b^j c^k, a^i b^j c^k, a^i b^j c^k, \ldots). \tag{10.25}$$

In the first row of the above matrix, the only nonzero element $(a^i b^j c^k)$ is in the column $1 + j + 2k$. In the second row, the only nonzero element $(a^i b^j c^k)$ is in the column $2 + j + 2k$, and so on. This means that

$$\mathbf{A}^i\mathbf{B}^j\mathbf{C}^k\mathbf{P}_0 = \begin{bmatrix} a^i b^j c^k P_0(10^{1+j+2k}1) \\ a^i b^j c^k P_0(10^{2+j+2k}1) \\ a^i b^j c^k P_0(10^{3+j+2k}1) \\ \vdots \end{bmatrix}. \tag{10.26}$$

Using the above and the fact that $P_0(10^n 1) = \rho_0^2(1 - \rho_0)^n$, we can now write

$$\mathbf{P}_t = \mathbf{M}^t\mathbf{P}_0 = \sum_{i+j+k=t} \binom{t}{i, j, k} \begin{bmatrix} \tilde{a}^i b^j c^k \rho_0^2(1 - \rho_0)^{1+j+2k} \\ a^i b^j c^k \rho_0^2(1 - \rho_0)^{2+j+2k} \\ a^i b^j c^k \rho_0^2(1 - \rho_0)^{3+j+2k} \\ \vdots \end{bmatrix}. \tag{10.27}$$

We finally obtain

$$\begin{aligned}P_t(10^l 1) &= \sum_{i+j+k=t} \binom{t}{i, j, k} a^i b^j c^k \rho_0^2(1 - \rho_0)^{l+j+2k}\\ &= \rho_0^2(1 - \rho_0)^l \sum_{i+j+k=t} \binom{t}{i, j, k} a^i [b(1 - \rho_0)]^j [c(1 - \rho_0)^2]^k\\ &= \rho_0^2(1 - \rho_0)^l \left(a + b(1 - \rho_0) + c(1 - \rho_0)^2\right)^t.\end{aligned} \tag{10.28}$$

10.4.2 General Case

We are now ready to handle the general case, without the $\tilde{a} = a$ assumption. Let us first note that tth powers of matrices

$$\begin{bmatrix} \tilde{a} & b & c & 0 & 0 & 0 & \cdots \\ 0 & a & b & c & 0 & 0 & \\ 0 & 0 & a & b & c & 0 & \\ 0 & 0 & 0 & a & b & c & \\ \vdots & & & & & & \ddots \end{bmatrix}^t, \quad \begin{bmatrix} a & b & c & 0 & 0 & 0 & \cdots \\ 0 & a & b & c & 0 & 0 & \\ 0 & 0 & a & b & c & 0 & \\ 0 & 0 & 0 & a & b & c & \\ \vdots & & & & & & \ddots \end{bmatrix}^t \tag{10.29}$$

differ only in their first row. This implies that the expression for $P_t(10^l1)$ given in Eq. (10.28) remains valid for $l > 1$ even if $\tilde{a} \neq a$. We only need to consider $l = 1$ case, that is, to compute $P_t(101)$. This can be done by writing Eq. (10.16) and replacing $P_t(1001)$ and $P_t(10001)$ by appropriate expressions obtained from Eq. (10.28),

$$P_{t+1}(101) = \tilde{a} P_t(101) + b\rho_0^2(1-\rho_0)^2 \left(a + b(1-\rho_0) + c(1-\rho_0)^2\right)^t + c\rho_0^2(1-\rho_0)^3 \left(a + b(1-\rho_0) + c(1-\rho_0)^2\right)^t. \tag{10.30}$$

This can be written as

$$P_{t+1}(101) = \tilde{a} P_t(101) + K\theta^t, \tag{10.31}$$

where

$$K = b\rho_0^2(1-\rho_0)^2 + c\rho_0^2(1-\rho_0)^3, \tag{10.32}$$

$$\theta = a + b(1-\rho_0) + c(1-\rho_0)^2. \tag{10.33}$$

Equation (10.31) is a first-order nonhomogeneous difference equation for $P_t(101)$, and, as such, it can be easily solved by standard methods [2]. The solution is

$$P_t(101) = P_0(101)\tilde{a}^t + K \sum_{i=1}^{t} \tilde{a}^{t-i}\theta^{i-1}. \tag{10.34}$$

The sum on the right-hand side is a partial sum of geometric series if $\tilde{a} \neq \theta$, or of an arithmetic series when $\tilde{a} = \theta$. Using appropriate formulae for partial sums of geometric and arithmetic series, one obtains

$$P_t(101) = \begin{cases} P_0(101)\tilde{a}^t + K(\tilde{a}^t - \theta^t)/(\tilde{a} - \theta) & \text{if } \tilde{a} \neq \theta, \\ P_0(101)\tilde{a}^t + K\tilde{a}^{t-1}t & \text{if } \tilde{a} = \theta. \end{cases} \tag{10.35}$$

Taking $P_0(101) = \rho^2(1-\rho)$ and replacing K and θ by their definitions we obtain, for $a - \tilde{a} + b(1-\rho_0) + c(1-\rho_0)^2 \neq 0$ (which is equivalent to $\tilde{a} \neq \theta$),

$$P_t(101) = \frac{{\rho_0}^2\,(1-\rho_0)^2\,(b+c-c\rho_0)}{a-\tilde{a}+b(1-\rho_0)+c(1-\rho_0)^2}\left(a+b(1-\rho_0)+c(1-\rho_0)^2\right)^t + \frac{{\rho_0}^2\,(1-\rho_0)\,(a-\tilde{a})}{a-\tilde{a}+b(1-\rho_0)+c(1-\rho_0)^2}\,\tilde{a}^t. \tag{10.36}$$

For $a-\tilde{a}+b(1-\rho_0)+c(1-\rho_0)^2=0$ (equivalent to $\tilde{a}=\theta$), the solution is slightly simpler,

$$P_t(101) = \rho_0^2\,(1-\rho_0)\left(\left(c{\rho_0}^2-(b+2\,c)\,\rho_0+b+c\right)t+\tilde{a}\right)\tilde{a}^{t-1}. \tag{10.37}$$

We now have expressions for $P_t(10^l1)$ for $l=1$ (Eqs. 10.36 and 10.37) and for $l>1$ (Eq. 10.28).

10.5 Cluster Expansion

We are finally ready to compute ρ_t. To do this, we will use the formula

$$P_t(0) = \sum_{k=1}^{\infty} k P_t(10^k1), \tag{10.38}$$

which we will refer to as "cluster expansion". Various proofs of this formula can be given (see, e.g., [12]), but we will show here that it is a direct consequence of additivity of measure.

Consider a cylinder set of a single zero anchored at i, $[0]_i$. A single zero must belong to a cluster of zeros of size k with possible values of k varying from 1 to infinity. If it belongs to a cluster of k zeros, then it must be the jth zero of the cluster, with possible values of j varying from 1 to k. Therefore,

$$[0]_i = \bigcup_{k=1}^{\infty}\bigcup_{j=1}^{k}[10^k1]_{i-j}. \tag{10.39}$$

Since all the cylinder sets on the right-hand side are mutually disjoint, their measures add up, thus

$$(F^t\nu_\lambda)([0]_i) = \sum_{k=1}^{\infty}\sum_{j=1}^{k}(F^t\nu_\lambda)([10^k1]_{i-j}). \tag{10.40}$$

The measure is shift-invariant, thus $(F^t\nu_\lambda)([10^k1]_{i-j}) = P_t(10^k1)$, and we obtain

$$P_t(0) = \sum_{k=1}^{\infty} \sum_{j=1}^{k} P_t(10^k 1), \tag{10.41}$$

which yields Eq. (10.38), as desired.

We can now compute $P_t(0)$ using the cluster expansion formula and Eqs. (10.36)–(10.28). We will first consider the case of $a - \tilde{a} + b(1-\rho_0) + c(1-\rho_0)^2 \neq 0$, that is, using Eq. (10.36) for $P_t(101)$.

$$P_t(0) = \sum_{l=1}^{\infty} l P_t(10^l 1) = \frac{{\rho_0}^2 (1-\rho_0)^2 (b+c-c\rho_0)}{a-\tilde{a}+b(1-\rho_0)+c(1-\rho_0)^2} \left(a+b(1-\rho_0)+c(1-\rho_0)^2\right)^t$$
$$+ \frac{{\rho_0}^2 (1-\rho_0)(a-\tilde{a})}{a-\tilde{a}+b(1-\rho_0)+c(1-\rho_0)^2} \tilde{a}^t + \sum_{l=2}^{\infty} l\rho_0^2 (1-\rho_0)^l \left(a+b(1-\rho_0)+c(1-\rho_0)^2\right)^t. \tag{10.42}$$

Since

$$\sum_{l=2}^{\infty} l(1-\rho_0)^l = \frac{(1+\rho_0)(1-\rho_0)^2}{{\rho_0}^2}, \tag{10.43}$$

we obtain

$$P_t(0) = \frac{{\rho_0}^2 (1-\rho_0)^2 (b+c-c\rho_0)}{a-\tilde{a}+b(1-\rho_0)+c(1-\rho_0)^2} \left(a+b(1-\rho_0)+c(1-\rho_0)^2\right)^t$$
$$+ \frac{{\rho_0}^2 (1-\rho_0)(a-\tilde{a})}{a-\tilde{a}+b(1-\rho_0)+c(1-\rho_0)^2} \tilde{a}^t + (1+\rho_0)(1-\rho_0)^2 \left(a+b(1-\rho_0)+c(1-\rho_0)^2\right)^t. \tag{10.44}$$

After substitution of $\tilde{a}, a, b, c$ and simplification, as well as taking $\rho_t = 1 - P_t(0)$, the following expression for ρ_t is obtained,

$$\rho_t = 1 - \frac{(1-\rho_0)^2 (r-(p-r+q)\rho_0)}{pq{\rho_0}^2 - (p+q)\rho_0 + r} ((\rho_0 q - 1)(\rho_0 p - 1))^t$$
$$- \frac{{\rho_0}^2 (1-\rho_0)((q-1)p - q + r)}{pq{\rho_0}^2 - (p+q)\rho_0 + r} (1-r)^t. \tag{10.45}$$

When $a - \tilde{a} + b(1-\rho_0) + c(1-\rho_0)^2 = 0$, similar calculations can be performed, but this time using Eq. (10.37) for $P_t(101)$. After simplification, this yields

$$\rho_t = 1 - (1-\rho_0)\left(\rho_0^2 (\rho_0 - 1)(pq\rho_0 - p + pq - q)t + 1 - r\right)(1-r)^{t-1}. \tag{10.46}$$

Let us summarize this in a more readable form, noticing that after substitution of $\tilde{a}, a, c, b$ by their definitions the condition $a - \tilde{a} + b(1-\rho_0) + c(1-\rho_0)^2 = 0$ becomes $pq\rho_0^2 - (p+q)\rho_0 + r = 0$. Our final expression for the density of ones can be written as

$$\rho_t = \begin{cases} 1 - E\left((\rho_0 q - 1)(\rho_0 p - 1)\right)^t - F(1-r)^t & \text{if } pq\rho_0^2 - (p+q)\rho_0 + r \neq 0, \\ 1 - (G + Ht)(1-r)^{t-1} & \text{if } pq\rho_0^2 - (p+q)\rho_0 + r = 0, \end{cases} \tag{10.47}$$

where definitions of E, F, G, H can be figured out by comparing the above to Eqs. (10.44) and (10.45).

We can see that in the nondegenerate case (when $pq\rho_0^2 - (p+q)\rho_0 + r \neq 0$), the limit $\rho_\infty = \lim_{t\to\infty} \rho_t$ always exists and that ρ_t approaches ρ_∞ exponentially fast, excluding special cases when $\rho_t = \text{const}$ (such as $\rho_0 = 0$ or $p = q = 0, r = 1$). In the degenerate case, ρ_∞ always exists as well, but the approach of ρ_t to ρ_∞ is linearly exponential.

It is worth noting that the existence of the degenerate case is a fairly subtle phenomenon and that it would be very difficult to discover the linearly exponential convergence by computer simulations alone. This illustrates the point that having a formula for ρ_t brings some advantages and that the search for such formulae is worthwhile.

As a separate remark, let us note that deterministic CA are nothing else but special cases of PCA, thus we can choose integer values of p, q, r and obtain relevant expression for ρ_t for a number of elementary CA rules (ECA), as follows.

ECA rule 206 ($p = 1, q = 0, r = 0$) or rule 220 ($p = 0, q = 1, r = 0$)

$$\rho_t = 1 - \rho_0(1-\rho_0) - (1-\rho_0)^{t+2}, \tag{10.48}$$

ECA rule 222 ($p = q = 1, r = 0$)

$$\rho_t = 1 + \frac{2(1-\rho_0)^{2t+1} + \rho_0(1-\rho_0)}{\rho_0 - 2}, \tag{10.49}$$

ECA rule 236 ($p = 0, q = 0, r = 1$)

$$\rho_t = 1 - (\rho_0 + 1)(1-\rho_0)^2, \tag{10.50}$$

ECA rule 238 ($p = 1, q = 0, r = 1$) or rule 252 ($p = 0, q = 1, r = 1$),

$$\rho_t = 1 - (1-\rho_0)^{t+1}, \tag{10.51}$$

ECA rule 254 ($p = q = r = 1$)

$$\rho_t = 1 - (1-\rho_0)^{2t+1}. \tag{10.52}$$

The above formulae agree with those derived informally in [1].

10.6 Conclusions

We presented a method for computing the density of ones in the orbit of the Bernoulli measure under the action of a probabilistic cellular automaton, using a simple PCA rule as an example. For this rule, we were able to construct a complete set of block probabilities, and then solve the resulting recurrence relations. By using the cluster expansion, we then obtained the required density of ones. Although this method is obviously applicable only to PCA rules with rather simple dynamics, it may be possible to find other PCA rules with complete sets, thus making the method useful for them. Generalization of the rule used in this paper to larger neighbourhood sizes comes to mind as a first possibility, and sufficiently simple deterministic rules, such as asymptotic emulators of identity investigated in [6], are also possible candidates.

Acknowledgements The author acknowledges partial financial support from the Natural Sciences and Engineering Research Council of Canada (NSERC) in the form of Discovery Grant. Some calculations on which this work is based were made possible by the facilities of the Shared Hierarchical Academic Research Computing Network (SHARCNET:www.sharcnet.ca) and Compute/Calcul Canada. The author thanks anonymous referees for suggestions leading to improvement of the article, including a simpler derivation of the cluster expansion formula.

References

1. Boccara, N., Fukś, H.: Modeling diffusion of innovations with probabilistic cellular automata. In: Delorme, M., Mazoyer, J. (eds.) Cellular Automata: A Parallel Model. Kluwer Academic Publishers, Dordrecht (1998)
2. Cull, P., Flahive, M., Robson, R.: Difference Equations. Springer, Berlin (2004)
3. Fukś, H.: Dynamics of the cellular automaton rule 142. Complex Syst. **16**, 123–138 (2006)
4. Fukś, H.: Probabilistic initial value problem for cellular automaton rule 172. DMTCS Proc. **AL**, 31–44 (2010)
5. Fukś, H.: Construction of local structure maps for cellular automata. J. Cell. Autom. **7**, 455–488 (2013)
6. Fukś, H., Gómez Soto, J.-M.: Exponential convergence to equilibrium in cellular automata asymptotically emulating identity. Complex Syst. **23**, 1–26 (2014)
7. Fukś, H., Skelton, A.: Response curves for cellular automata in one and two dimensions - an example of rigorous calculations. Int. J. Nat. Comput. Res. **1**, 85–99 (2010)
8. Fukś, H., Skelton, A.: Orbits of Bernoulli measure in asynchronous cellular automata. Discret. Math. Theor. Comput. Sci. **AP**, 95–112 (2011)
9. Kůrka, P.: On the measure attractor of a cellular automaton. Discret. Contin. Dyn. Syst. 524–535 (2005)
10. Kůrka, P., Maass, A.: Limit sets of cellular automata associated to probability measures. J. Stat. Phys. **100**, 1031–1047 (2000)
11. Pivato, M.: Conservation laws in cellular automata. Nonlinearity **15**(6), 1781 (2002)
12. Stauffer, D., Aharony, A.: Introduction to Percolation Theory. Taylor and Francis, London (1994)

Chapter 11
Statistical Equilibrium in Deterministic Cellular Automata

Siamak Taati

Abstract Some deterministic cellular automata have been observed to follow the pattern of the second law of thermodynamics: starting from a partially disordered state, the system evolves towards a state of equilibrium characterized by maximal disorder. This chapter is an exposition of this phenomenon and of a statistical scheme for its explanation. The formulation is in the same vein as Boltzmann's ideas, but the simple combinatorial set-up offers clarification and hope for generic mathematically rigorous results. Probabilities represent frequencies and subjective interpretations are avoided.

11.1 Introduction

The aim of statistical mechanics is to bridge between microscopic and macroscopic behaviour of systems consisting of a large number of interacting components. The prime example is a gas of particles moving and interacting according to the laws of mechanics, giving rise to macroscopic behaviour described in thermodynamics. The kinetic theory of gases, initiated by Clausius, Maxwell and Boltzmann, takes on the task of explaining the macroscopic behaviour of a gas on the basis of its microscopic description.

The main problem in kinetic theory is the derivation of the second law of thermodynamics (i.e. the tendency of an isolated thermodynamic system to evolve towards more disordered states). Starting from a collection of particles pictured as hard balls interacting through elastic collisions and using a simplifying (though erroneous) statistical assumption about the number of collisions of each type occurring in a small time interval (the *Stosszahlansatz*), Boltzmann was able to derive a version of the second law by showing that a certain quantity measuring disorder (Boltzmann's entropy) increases monotonically with time and is maximized precisely when the system is in equilibrium.

S. Taati (✉)
University of British Columbia, Vancouver, Canada
e-mail: siamak.taati@gmail.com

P.-Y. Louis and F.R. Nardi (eds.), *Probabilistic Cellular Automata*, Emergence, Complexity and Computation 27, https://doi.org/10.1007/978-3-319-65558-1_11

Although the second law of thermodynamics was originally formulated for thermodynamic systems, its applicability goes beyond a system of particles following the particular laws of (classical or quantum) mechanics. A mathematical understanding of the precise circumstances leading to the applicability of the more general law of tendency towards disorder is desirable but missing.

The purpose of this chapter is to demonstrate examples of results and experimental observations regarding the so-called randomization behaviour in cellular automata (going back to Miyamoto, Wolfram and Lind) that could be thought of as instances of this generalized version of the second law of thermodynamics. Notably, neither probabilistic hypotheses (i.e. incorporating intrinsic randomness in the model) nor subjective interpretations (see [15]) are needed — probabilities enter the picture only as intuitive means of representing statistical data. The combinatorial setting of cellular automata is simple enough that one could attempt to find generic mathematical conditions that guarantee the applicability of the second law. At the same time, the rich range of behaviour among cellular automata makes the challenge interesting and non-trivial.

The scenario is briefly as follows. Consider a configuration that is atypical of the maximally disordered state (so that there is a bias in the frequency of the patterns) but is not too rigidly regular either (e.g. it is not periodic). Over the time, a sufficiently chaotic cellular automaton shuffles such a configuration (albeit deterministically) in such a way that the bias gradually becomes undetectable. More specifically, the configuration of the system becomes more and more typical of the maximally disordered state, up to wider and wider ranges of observation.

11.2 Randomization Phenomenon: Examples

11.2.1 XOR Cellular Automaton

On the space of all bi-infinite sequences of symbols 0 and 1, consider a transformation T that maps a sequence x into another sequence Tx defined by $(Tx)_i \triangleq x_i + x_{i+1} \pmod 2$. In other words, T replaces each symbol with the sum (modulo 2) of that symbol and its right neighbour. The iteration of T defines a dynamical system on $\{0, 1\}^{\mathbb{Z}}$, which we refer to as the *XOR cellular automaton*.[1] Each sequence in $\{0, 1\}^{\mathbb{Z}}$ will be called a *configuration* of the system. A sample trajectory of this system is depicted in Fig. 11.1.

The map T is continuous with respect to the product topology. The *product topology* is the topology in which two configurations are considered "close" if they agree on a large region around the origin. Convergence in the product topology corresponds to site-wise eventual agreement. Another basic property of T is its translation symmetries. Namely, if σ^k denotes the *translation* by k (i.e. $(\sigma^k x)_i \triangleq x_{k+i}$), then

[1] XOR stands for "eXclusive OR".

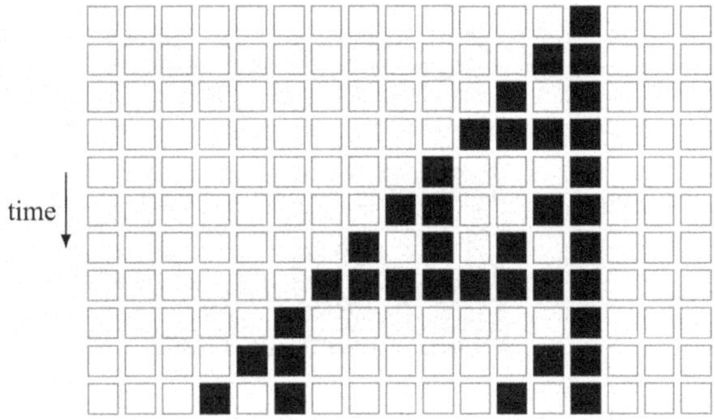

Fig. 11.1 A sample trajectory of the XOR cellular automaton starting with a configuration consisting of a single 1 at the origin and 0 everywhere else

$T\sigma^k = \sigma^k T$ for every $k \in \mathbb{Z}$. The map T is also *additive*, meaning $T(x + y) = Tx + Ty$, where the addition is performed site-wise and modulo 2. Although T is not invertible, it is onto and "almost one-to-one" in that every configuration $y \in \{0, 1\}^{\mathbb{Z}}$ has precisely 2 pre-images. Namely, choosing a symbol x_0 arbitrarily, we can find, recursively, unique values for the symbols x_i, for $i > 0$ and $i < 0$, such that $Tx = y$.

Slightly less obvious is the following *balance* property of T: if we choose the symbols in x by independent unbiased coin flips, the symbols in Tx will also be indistinguishable from independent unbiased coin flips. In other words, the *uniform Bernoulli measure* is invariant under T. To see this, take any block of n consecutive symbols $b_1 b_2 \ldots b_n$ and consider the probability that Tx takes values $b_1 b_2 \ldots b_n$ at positions k to $k + n - 1$. There are precisely two choices of values $a_1 a_2 \ldots a_{n+1}$ and $a'_1 a'_2 \ldots a'_{n+1}$ that, if on x at positions k to $k + n$, lead to the desired values $b_1 b_2 \ldots b_n$ on Tx at sites k to $k + n - 1$. Each of these two choices has probability $2^{-(n+1)}$ of appearing in independent flips of an unbiased coin. Therefore, $b_1 b_2 \ldots b_n$ appears at positions k to $k + n - 1$ of Tx with probability 2^{-n}.

Besides the balance property, the XOR cellular automaton has a wealth of other statistical regularities. For instance, if x is chosen according to a uniform Bernoulli measure (i.e. with independent unbiased coin flips), then for any n, the sequence of blocks $(T^t x)_{[k,k+n)}$, $t = k, k + 1, \ldots$ is independent of the block $x_{[k,k+n)}$. It follows from the law of large numbers that, almost surely, every pattern $a_1 a_2 \cdots a_n$ appears with the same frequency 2^{-n} in the space-time column $(T^t x)_{[k,k+n)}$, $t = 0, 1, 2, \ldots$. The same sort of eventual independence holds along any other "space-time direction": for every $a \in \mathbb{Z}^+$ and $b \in \mathbb{Z}$ and a sufficiently large t_0, the tilted column of blocks $(T^{at} x)_{[bt+k, bt+k+n)}$ with $t \geq t_0$ is independent of $x_{[k,k+n)}$.

Yet, the most remarkable property of the XOR cellular automaton for us is its *randomization effect*: if x is chosen using independent flips of a *biased* coin (say, with probability $p \in (0, 1)$ of having 1 at each site), then T will gradually *randomize* x, meaning $T^t x$ will be asymptotically indistinguishable (in distribution) from a configuration chosen using independent flips of an *unbiased* coin as $t \to \infty$, provided we ignore a negligible set of time steps t (Fig. 11.2).

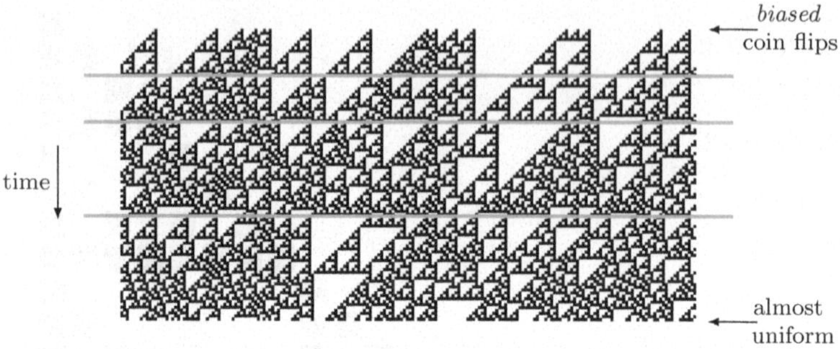

Fig. 11.2 Randomization effect of the XOR cellular automaton. The starting configuration is chosen by independent biased coin flips with probability $p = 0.1$ of having 1 at each site. Ignoring a negligible set of time steps (represented by *gray lines*), the configuration quickly becomes almost uniform

To state this more precisely, we need some notation and terminology. A (Borel) probability measure on $\{0, 1\}^{\mathbb{Z}}$ is uniquely identified by the probabilities it associates to the *cylinder* sets

$$[a_k a_{k+1} \dots a_{k+n}] \triangleq \{x \in \{0, 1\}^{\mathbb{Z}} : x_k x_{k+1} \dots x_{k+n} = a_k a_{k+1} \dots a_{k+n}\} .$$

For instance, for the *Bernoulli measure* with parameter p (the distribution of independent flips of a biased coin with probability p of having 1), which we will denote by μ_p, we have

$$\mu_p([a_k a_{k+1} \dots a_{k+n}]) = p^{\#_1(a)}(1-p)^{\#_0(a)}$$

for any block $a = a_k a_{k+1} \dots a_{k+n}$, where $\#_1(a)$ and $\#_0(a)$ denote, respectively, the number of 1s and 0s appearing in a. The image of a probability measure π under T is another probability measure $T\pi$ with $(T\pi)(E) = \pi(T^{-1}E)$ for any measurable set E. This is the distribution of Tx if x is chosen at random according to π. A sequence of probability measures $\nu_1, \nu_2, \dots$ is said to *converge weakly* to a measure π if the probabilities that ν_t associate to each fixed cylinder converge to the probability of that cylinder according to π.

By the above-mentioned balance property, $T\mu_{1/2} = \mu_{1/2}$. Miyamoto [26] and Lind [24] (following experimental observations made by Wolfram [38]) proved that

Theorem 1 *There is a set $J \subseteq \mathbb{N}$ of density 1 such that for every $p \in (0, 1)$, $T^t \mu_p \to \mu_{1/2}$ as $t \to \infty$ within J.*

Here, the *density* of a set of non-negative integers J is defined as

$$d(J) \triangleq \lim_{n\to\infty} \frac{|J \cap [0, n)|}{n}$$

when the limit exists. The theorem in particular implies that the Cesàro averages $\frac{1}{n}\sum_{t=0}^{n-1} T^t\mu_p$ converge to $\mu_{1/2}$ as $n \to \infty$.

The randomization behaviour of the XOR cellular automaton can be seen as an analog (or an instance) of the second law of thermodynamics: *the system evolves towards an equilibrium in a macroscopic state with highest degree of disorder*. Here, the term *macroscopic* is understood as synonymous with *statistical*: the *macroscopic state* of a configuration x consists of the frequency of occurrence of every finite word $a \in \{0, 1\}^*$ in x. This information is encapsulated conveniently in a translation-invariant probability measure π_x that is defined by those frequencies and which has x as a "typical element". The *equilibrium state* (the uniform Bernoulli measure) is the least presumptive (most random) state: every word of length n has the same frequency 2^{-n}. In Sects. 11.3 and 11.4, we shall make this interpretation more precise.

The starting configuration does not need to be Bernoulli for the XOR cellular automaton to randomize it. A random configuration which is a realization of a (bi-infinite) k-step Markov chain with positive transition probabilities is also randomized by the XOR cellular automaton. In other words, the conclusion of Theorem 1 remains true if μ_p is replaced with a full-support Markov measure [6]. More generally, randomization is known to occur as long as the starting measure is *harmonically mixing* [29, 30].[2] A complete characterization of the measures randomized by the XOR map is nevertheless missing.

11.2.2 A Reversible Cellular Automaton

The analogy with the second law of thermodynamics would have been stronger if the XOR cellular automaton was reversible[3] or symmetric under time reversal.[4] Consider now a different cellular automaton acting on the configurations of symbols from $S \triangleq \{0, 1\} \times \{0, 1\}$. Each site $i \in \mathbb{Z}$ of a configuration (x, y) carries two symbols x_i and y_i, and the cellular automaton map T is defined by $(T(x, y))_i \triangleq (y_i, x_i + y_{i+1})$, where the addition is again modulo 2.[5] Let us call this the *XOR-transpose* cellular

[2]For the definition of harmonic mixing and basic properties of the class of harmonically mixing measures, see [29, 30]. The result of [6] covers also the measures with *complete connections* and *summable decay*. These, however, turn out to be included in the class of harmonically mixing measures [13].

[3]A cellular automaton is said to be *reversible* if it is bijective and has another cellular automaton as inverse. This is equivalent to bijectivity, because the configuration space is compact and metric.

[4]A reasonable definition of *time-reversal symmetry* for cellular automata is that T is reversible and there is another reversible cellular automaton R such that $T^{-1} = R^{-1}TR$; see [8].

[5]A reader familiar with Kac's ring model (see [17], Sect. III.14) might recognize some similarity. The infinite version of Kac's model can be defined with $(T(x, y))_i \triangleq (x_i, x_{i+1} + y_{i+1})$. The first component represents the presence or absence of marks on each site and the second the colour of the balls.

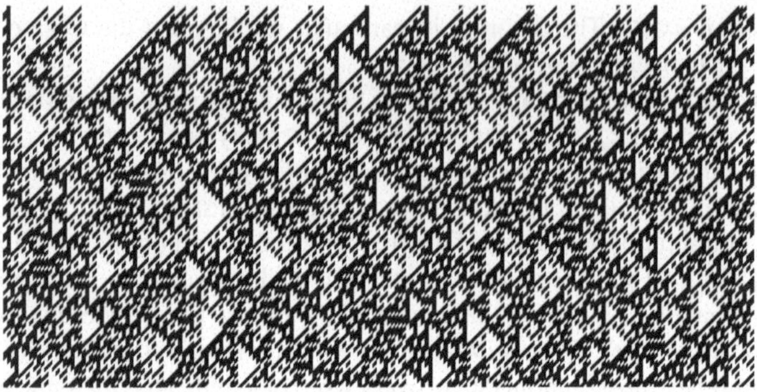

Fig. 11.3 Randomization effect of the reversible cellular automaton of Maass and Martínez. In the initial configuration, each component of the symbol at each site is chosen with an independent coin flip with probability 0.1 of having a 1

automaton.[6] As in the XOR example, the map T is continuous and translation-invariant.[7] It is also additive and has the balance property: the uniform Bernoulli measure μ on $S^{\mathbb{Z}}$ is invariant under T.[8] Unlike the XOR cellular automaton, the XOR-transpose is reversible and time-reversal symmetric: one can traverse backwards in time by switching the two symbols at each site.[9] Maass and Martínez [25] proved that the XOR-transpose cellular automaton has a similar randomization property as the XOR cellular automaton (Fig. 11.3):

Theorem 2 *Let π be the distribution of a single-step Markov chain on S with positive transition probabilities. Then, $\frac{1}{n}\sum_{t=0}^{n-1} T^t\pi$ converges to the uniform Bernoulli measure μ as $n \to \infty$.*

The convergence of the Cesàro averages implies the existence of a set $J \subseteq \mathbb{N}$ of density 1 of time steps t along which $T^t\pi$ converges to μ (see [16], Corollary 1.4), but the set J might, in principle, depend on the measure π.

---

[6]The name is suggested by the fact that the space-time diagrams of this cellular automaton are obtained from the space-time diagrams of a variant of the XOR cellular automaton with $(Tx)_i \triangleq x_{i-1} + x_{i+i} \pmod 2$ by exchanging the role of time and space.

[7]A *cellular automaton* may in fact be defined as a map on a symbolic configuration space $S^{\mathbb{Z}^d}$ that is continuous and invariant under translations. These are precisely the maps defined homogeneously using local update rules [11].

[8]The balance property is shared among all cellular automaton maps that are onto [11].

[9]More specifically, T has an inverse given by $\left(T^{-1}(x, y)\right)_i = (y_i + x_{i+1}, x_i)$. Setting $R(x, y) \triangleq (y, x)$, we can write the inverse map T^{-1} as $R^{-1}TR$.

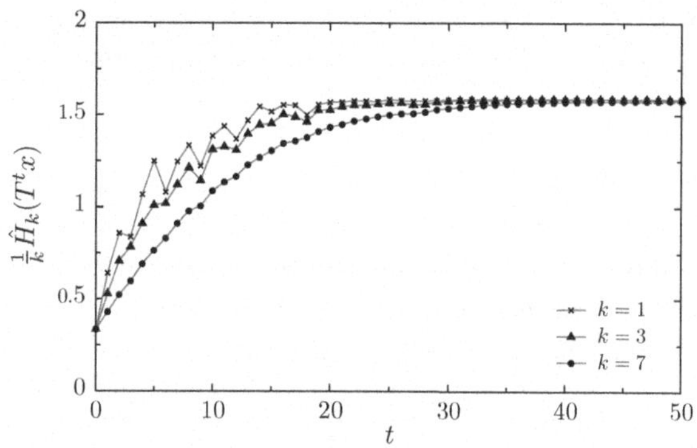

Fig. 11.4 Evidence of randomization in the bi-permutative cellular automaton defined in Eq. (11.1). The starting configuration x (on a ring of length 50,000) is chosen with independent flips of a biased 3-sided coin with distribution $(0 \mapsto 0.95, 1 \mapsto 0.025, 2 \mapsto 0.025)$

11.2.3 A Bi-permutative Cellular Automaton

As a third example, let us look at the cellular automaton with symbol set $S \triangleq \{0, 1, 2\}$, defined by

$$(Tx)_i \triangleq \varphi(x_{i-1}, x_i, x_{i+1}) \triangleq \begin{cases} x_{i-1} + x_{i+1} + 1 \pmod 3 & \text{if } x_i = 2, \\ x_{i-1} + x_{i+1} \pmod 3 & \text{otherwise.} \end{cases} \tag{11.1}$$

Unlike the last two examples, the map T is not additive. Nevertheless, the local rule φ is *bi-permutative*, which is to say both $a \mapsto \varphi(a, b, c)$ and $c \mapsto \varphi(a, b, c)$ are permutations.[10] It follows, similarly as in the case of the XOR cellular automaton, that the map T is 9-to-1. Like the last two examples, the uniform Bernoulli measure (i.e. the distribution of a configuration chosen at random by flipping an unbiased "3-sided coin"[11] independently for each site) is invariant under T. The bi-permutativity also ensures other statistical regularities for T, similar to those enjoyed by the XOR cellular automaton [32].

It is not known whether the above cellular automaton has a randomizing behaviour in the sense of Theorems 1 or 2. Nevertheless, there is experimental evidence suggesting that T indeed randomizes biased Bernoulli configurations (Fig. 11.4). The graphs in Fig. 11.4 depict the change in time of the empirical entropies of words of length 1, 3 and 7 in consecutive configurations of this cellular automaton, starting from a biased Bernoulli configuration. More specifically, a single pseudo-random configuration x is picked by simulating independent biased (3-sided) coin flips, and iterations of T on x are obtained for up to 50 time steps.[12] At each time step, the empirical entropy of words of length k (for $k = 1, 3, 7$) appearing in the current configuration is calculated as follows. For each word w of length k with symbols from

---

[10]Notice that the XOR cellular automaton is also bi-permutative.

[11]Or rolling a 3-sided die, if you wish.

[12]Instead of infinite configurations, configurations of symbols on a large ring (indexed by $\mathbb{Z}_N$ for N large) are used.

S, let $\overline{\zeta_w}(y)$ denote the frequency of appearance of the word w in configuration y. The *empirical entropy* of words of length k appearing in y is defined as

$$\hat{H}_k(y) \triangleq -\sum_{w \in S^k} \overline{\zeta_w}(y) \log \overline{\zeta_w}(y) .$$

Figure 11.4 shows that the empirical entropy $\hat{H}_k(T^t x)$ rapidly increases to reach its maximum at around $k \log 3$, where it stays.

The empirical entropy $\hat{H}_k(y)$ is a measure of disorder in y. It is maximized when all words of length k appear in y with approximately the same frequency. For instance, a configuration chosen using independent unbiased coin flips (which is considered to be maximally disordered) has, with probability 1, the maximum empirical entropy $\hat{H}_k$ for each k. The empirical entropy $\hat{H}_k(y)$ should be compared with Boltzmann's entropy (see below).[13] Although not exhaustive, the simulation in Fig. 11.4 suggests a gradual approach towards a maximally disordered state.

11.2.4 Rule 30

Yet another example where randomization seems to be present is the so-called *Rule 30* cellular automaton. The Rule 30 cellular automaton has the binary alphabet $\{0, 1\}$ as the symbol set and may be defined by the logical expression

$$(Tx)_i \triangleq \varphi(x_{i-1}, x_i, x_{i+1}) \triangleq x_{i-1} \text{ XOR } (x_i \text{ OR } x_{i+1}) ,$$

where XOR denotes "exclusive or". The local rule φ is not bi-permutative, but it is *left-permutative* (i.e. $a \leftrightarrow \varphi(a, b, c)$ is a bijection for each b and c). This still implies that the map T is onto, and that each configuration has at most 4 pre-images under T. It follows that the Rule 30 cellular automaton again has the balance property. Starting from an unbiased Bernoulli configuration, the Rule 30 cellular automaton enjoys similar statistical regularities as in the previous examples, along almost all space-time directions [33].[14]

This cellular automaton was first studied by Wolfram [39]. He noticed that even with a simple starting configuration, the iterations of the Rule 30 cellular automaton produce complex seemingly unpredictable patterns. He proposed a method for generation of pseudo-random sequences by initializing the Rule 30 cellular automaton with a "seed" and picking the symbols appearing on a particular site every few time steps, which he tested against standard statistical randomness tests.[15]

---

[13] For the interpretations of entropy, see e.g. [3, 10].

[14] More specifically, $T\sigma^k$ is an exact endomorphism unless $k = -1$.

[15] Rule 30 is in fact used in Mathematica as one of the methods for pseudo-random number generation.

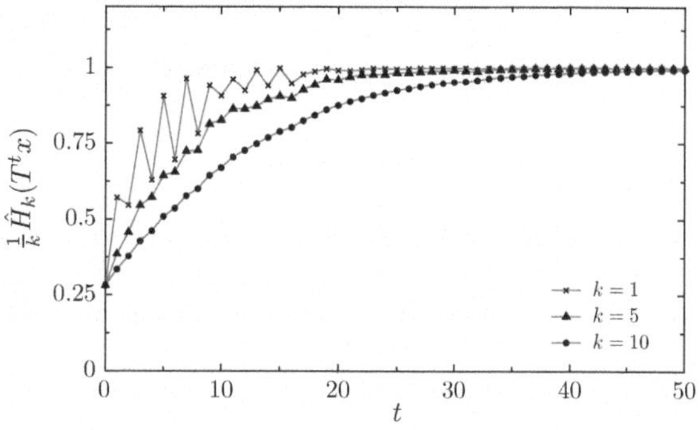

Fig. 11.5 Evidence of randomization in the Rule 30 cellular automaton. The starting configuration x (on a ring of length 50,000) is chosen with independent flips of a biased coin with probability $p = 0.05$ of having a 1

Figure 11.5 shows evidence for randomization in the Rule 30 cellular automaton starting from biased Bernoulli configurations. The empirical entropies are calculated as in the previous example.

11.2.5 Q2R Spin Dynamics

One feature that is common among the first three examples (and is suspected for the Rule 30 cellular automaton) is the absence of conserved energy-like quantities [7]. A non-trivial conserved quantity would partition the macroscopic states into unescapable fibers, hence preventing complete randomization. Nevertheless, one might still expect randomization within each fiber.

The next example is based on the configurations of the Ising model and was introduced by Vichniac [36]. The *Ising model* is a model of ferromagnetism: each site of an infinite two-dimensional square lattice (indexed by $\mathbb{Z}^2$) carries a symbol $\uparrow$ or $\downarrow$, representing two possible directions of a *magnetic spin*. The interaction between spins is modelled by assigning energy -1 or $+1$ to any pair of neighbouring sites whose symbols are, respectively, aligned or anti-aligned spins. The energy content of a region is the sum of the interaction energy of neighbouring sites in that region. Hence, lower energy in a region corresponds to an average tendency of neighbouring spins to be aligned.

The dynamics is through alternate applications of two maps T_0 and T_1: the first map updates the *even* sites (i.e. the sites (i, j) with $i + j$ even) and the second updates the *odd* sites. The updating is done in such a way that the energy is preserved: a spin is flipped if and only if the flipping does not change the total energy of the site and its four immediate neighbours. More specifically, let us say that a site (i, j) is *balanced* on a configuration x if half of the neighbouring spins $x_{i+1,j}$, $x_{i,j+1}$, $x_{i-1,j}$ and $x_{i,j-1}$ are upwards and the other half are downwards. For a spin $a \in \{\uparrow, \downarrow\}$, let $\overline{a}$ denote the spin with the opposite direction as a. Then,

$$(T_0x)_{i,j} \triangleq \begin{cases} \overline{x_{i,j}} & \text{if } i+j \text{ even and } (i,j) \text{ balanced on } x, \\ x_{i,j} & \text{otherwise,} \end{cases}$$

and similarly for T_1. The dynamical system defined by the composition $T \triangleq T_1 T_0$ is called the *Q2R* cellular automaton.[16]

The Q2R system is reversible and symmetric under time reversal in the sense that $T^{-1} = T_0 T_1$. By construction, it also conserves the energy. The *conservation of energy* can be formulated in various equivalent ways. For us, it suffices to say that T (indeed, each of T_0 and T_1) keeps the *average energy per site* invariant. Note that the average energy per site of a configuration x is a function of its macroscopic state π_x. The set of translation-invariant probability measures with a given average energy per site is convex and closed under the topology of weak convergence. Therefore, any limit or Cesáro limit of the measures $T^t \pi_x$ will have the same average energy per site as π_x.

As before, we consider the uniform Bernoulli measure on the configuration space $\{\uparrow, \downarrow\}^{\mathbb{Z}^2}$ to be a representation of a "maximally disordered" state, because it assigns the same probability $2^{-|A|}$ to all cylinder sets

$$[q_A] \triangleq \{x \in \{\uparrow, \downarrow\}^{\mathbb{Z}^2} : x_i = q_i \text{ for } i \in A\} .$$

Put another way, in a typical spin configuration obtained by independent unbiased coin flips, (translations of) each finite pattern $q_A : A \to \{\uparrow, \downarrow\}$ appears with the same frequency $2^{-|A|}$. Another way to express this is to say that the *entropy*

$$H_A(\mu) \triangleq - \sum_{q_A : A \to \{\uparrow, \downarrow\}} \mu([q_A]) \log \mu([q_A])$$

of any *finite window* $A \subseteq \mathbb{Z}^2$ has its maximum value $|A| \log 2$ if μ is the uniform Bernoulli measure.

The description of a "maximally disordered" state with a given average energy per site is more subtle. Indeed, since the constraint is not local, it might not be possible to maximize the entropy $H_A(\mu)$ simultaneously for all finite windows A. However, if $B \supseteq A$, a larger value for $H_B(\mu)$ is a better indication of disorder than a larger value for $H_A(\mu)$. Therefore, one may measure the disorder by the limit *entropy per site*

$$h(\mu) \triangleq \lim_{n \to \infty} \frac{H_{I_n}(\mu)}{|I_n|} ,$$

[16]Strictly speaking, this is not a cellular automaton with the common definition of the term, because the even and odd sites are treated differently. It can however be recoded into a standard cellular automaton.

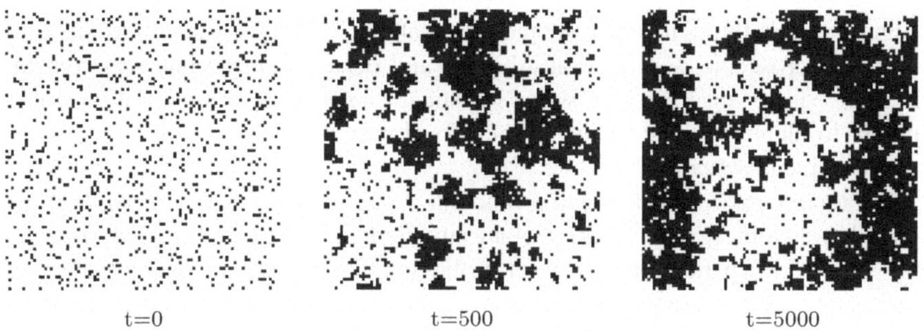

Fig. 11.6 Simulation of the Q2R cellular automaton starting from a coin-flip configuration with probability $p = 0.1$ of having $\uparrow$ (represented by *black*). After a relatively short while, the macroscopic look of the configuration seems to reach an equilibrium with upward and downward spins clustered together

where $I_n = [-n, n] \times [-n, n]$.[17] A maximally disordered state with a given average energy per site may therefore be identified with an ergodic translation-invariant measure that has the prescribed average energy per site and maximal entropy per site subject to the energy constraint. These are the ergodic *Gibbs measures* for the Ising model (see e.g. [14, 31]).[18]

Figure 11.6 shows few snapshots from a simulation of the Q2R cellular automaton starting from a biased Bernoulli configuration. At the beginning, the spins gradually cluster, even though the total length of the boundaries between upward and downward clusters remains constant. After a while, the macroscopic picture of the configuration appears to reach an equilibrium, which resembles a typical configuration chosen according to a Gibbs measure of the Ising model.[19] More elaborate simulations have shown numerical agreement with the Ising model (see e.g. [12, 35]), hence supporting the conjecture that Q2R indeed randomizes a coin-flip configurations within the corresponding average energy per site level.

In the next two sections, we attempt to make the concepts of macroscopic state and maximally disordered state more precise.

11.3 Macroscopic States

Let us fix a symbol set S and denote by $\mathcal{X} = \{x : \mathbb{Z}^d \to S\}$ the set of all d-dimensional configurations of symbols from S. A configuration $x \in \mathcal{X}$ is considered as a *microscopic state* of a system. The *macroscopic state* of x consists of all information in

---

[17]For translation-invariant measures, the limit exists and is equal to $\inf_n \frac{1}{|I_n|} H_{I_n}(\mu)$.

[18]In the standard Ising model, the ergodic Gibbs measures are considered to be suitable descriptions of the macroscopic states of equilibrium for the ferromagnetic material (see e.g. [9]).

[19]Again, the simulation is made on a torus $\mathbb{Z}_N^2$ rather than the infinite lattice $\mathbb{Z}^2$. The "equilibrium" configurations could be compared with a random configuration generated by a Gibbs sampler for the Ising model.

x that is visible through "macroscopic observations". Which observations are considered macroscopic is somewhat arbitrary and depends on the physical context. Here, we equate "macroscopic" with "statistical": a macroscopic observation would amount to identifying the frequency of a fixed pattern, or the spatial average of a "microscopic observation".

To be more specific, let us call a function $f : \mathcal{X} \to \mathbb{R}$ a *local observable* if the symbols x_i at finitely many sites $i \in A$ are sufficient to determine $f(x)$. For instance, if $q : A \to S$ is a pattern on a finite set A, the function $x \mapsto \zeta_q(x)$ that has value 1 if x agrees with q on A and 0 otherwise is a local observable. Furthermore, any local observable is a linear combination of observables of this type.

If f is a local observable, the spatial average

$$\overline{f}(x) \triangleq \lim_{n\to\infty} \frac{1}{|I_n|} \sum_{i\in I_n} f(\sigma^i x) \tag{11.2}$$

will be considered as a *macroscopic observable*. As before, $I_n \triangleq [-n, n]^d$, and σ^i denotes the translation by i. For instance, $\overline{\zeta_q}(x)$ is simply the frequency of the occurrence of (translations of) the pattern q in x. The limit in (11.2) may or may not exist. If well defined, the collection $\big(\overline{f}(x) : f \text{ local}\big)$ defines a unique translation-invariant probability measure π_x with

$$\pi_x(f) = \int f \mathrm{d}\pi_x = \overline{f}(x) \,,$$

describing the statistics of x. In particular, $\pi_x([q]) = \overline{\zeta_q}(x)$ for any finite pattern q.

Every translation-invariant probability measure on $\mathcal{X}$ arises from a configuration in the above fashion [28]. Nevertheless, not every translation-invariant probability measure should be considered as an unambiguous macroscopic state. To illustrate this, consider a one-dimensional configuration z with $z_i = 0$ for $i < 0$ and $z_i = 1$ for $i > 0$. Then $\pi_z = \frac{1}{2}(\delta_{\underline{0}} + \delta_{\underline{1}})$, where $\delta_{\underline{0}}$ and $\delta_{\underline{1}}$ are the point-mass measures at the uniform configurations $\underline{0}$ and $\underline{1}$. The measure $\frac{1}{2}(\delta_{\underline{0}} + \delta_{\underline{1}})$ however suggests an ambiguous situation in which we are uncertain about which of $\underline{0}$ and $\underline{1}$ is the real configuration of the system.[20] The ambiguity comes from the fact that the configuration z lacks homogeneity: its left and right tails have different macroscopic looks.[21]

Here, we focus on configurations that are homogeneous. We call a configuration x *homogeneous*[22] if

[20] See [9], Paragraph (7.8), for a similar reasoning.

[21] As an example in which inhomogeneity does not arise from left-right asymmetry, let $m_0 = 0$ and $m_k \triangleq 2(1 + 2^2 + \cdots + k^2)$, and construct a one-dimensional configuration $z : \mathbb{Z} \to \{0, 1\}$ with $z_i = 0$ if $m_k \leq |i| < m_k + (k+1)^2$ and $z_i = 1$ if $m_k + (k+1)^2 \leq |i| < m_{k+1}$. Then, again $\pi_z = \frac{1}{2}(\delta_{\underline{0}} + \delta_{\underline{1}})$.

[22] Such points are called *regular* in [27].

(i) π_x is well defined (i.e. the spatial average $\overline{f}(x)$ of every local observable f exists on x),
(ii) π_x cannot be written as a non-trivial convex combination of other translation-invariant measures (i.e. π_x is ergodic for the group of translations),[23] and
(iii) x is a *point of density* for π_x, which is to say, every finite pattern occurring in x occurs with positive frequency.

The measure π_x describing the statistical averages of a homogeneous configuration x will be called the *macroscopic state* of x. The set of homogeneous configurations sharing the same ergodic measure π as macroscopic state is called the *ergodic set* of π. The countability of the set of finite patterns together with the ergodic theorem implies that the ergodic set of any ergodic measure π has measure 1 with respect to π (see [27]).[24] Hence, one may think of a configuration in the ergodic set of π as a *typical* configuration with macroscopic state π.[25]

11.4 Maximally Disordered States

From the definition, it follows that, for any finite window $A \subseteq \mathbb{Z}^d$, the entropy $H_A(\pi)$ of a macroscopic state (i.e. a translation-ergodic measure) π agrees with the empirical entropy $\hat{H}_A(x)$ of any configuration x that is typical for π. The entropy $H_A(\pi)$ is a convex continuous function of π, taking its maximum value $|A| \log |S|$ only when π assigns equal probabilities to every cylinder with base A.

The limit entropy per site $h(\pi)$ is affine (hence convex) and takes its maximum value $\log |S|$ precisely when π is the uniform Bernoulli measure, that is the state with "maximum disorder". The map $\pi \mapsto h(\pi)$ is however not continuous. For example, for each $m > 0$, let $x(m)$ be a periodic configuration in $\{0, 1\}^{\mathbb{Z}}$ that has each word of length m exactly once in its period.[26] Then, the macroscopic states $\pi_{x(m)}$ have 0 entropy per site yet converge weakly to the uniform Bernoulli measure as $m \to \infty$.

---

[23]It might not be intuitively clear why π_x should be required to be ergodic in order for x to be called homogeneous. A perhaps more plausible condition equivalent to the ergodicity of π_x is that for every local observable f and each $\varepsilon > 0$, the upper density of the set

$$\left\{ a \in \mathbb{Z}^d : \left| \frac{1}{|I_m|} \sum_{i \in a + I_m} f(\sigma^i x) - \overline{f}(x) \right| > \varepsilon \right\}$$

in $\mathbb{Z}^d$ goes to 0 as $m \to \infty$ [27]. Note that for both examples of inhomogeneous configurations mentioned above, this condition fails for the function $f = \zeta_1$ with $f(x) = 1$ if $x_0 = 1$ and $f(x) = 0$ otherwise.

[24]In particular, the set of homogeneous configurations has measure 1 with respect to any translation-invariant probability measure.

[25]If need be, stronger notions of homogeneity and typicalness can be obtained by intersecting the ergodic set of π with other sets of measure 1.

[26]Such a configuration corresponds to an Eulerian circuit on the de Bruijn graph of order m.

In fact, every macroscopic state is a weak limit of macroscopic states of periodic configurations (which all have entropy 0). Nevertheless, entropy per site is upper semi-continuous: $\pi_n \to \pi$ implies $\limsup_{n\to\infty} h(\pi_n) \leq h(\pi)$ (see e.g. [37]).

Let us now consider a concept of energy as in the Ising model. Namely, let $f : \mathcal{X} \to \mathbb{R}$ be a local observable, representing the energy contribution of the symbol at the origin when interacting with the nearby symbols. For instance, for the Ising model, we can set

$$f(x) = \begin{cases} \frac{1}{2}\big(n_\downarrow(\partial_0 x) - n_\uparrow(\partial_0 x)\big) & \text{if } x_0 = \uparrow, \\ \frac{1}{2}\big(n_\uparrow(\partial_0 x) - n_\downarrow(\partial_0 x)\big) & \text{if } x_0 = \downarrow, \end{cases}$$

where $n_\uparrow(\partial_0 x)$ and $n_\downarrow(\partial_0 x)$ are the numbers of upward and downward spins among the four neighbours $x_{1,0}, x_{0,1}, x_{-1,0}$ and $x_{0,-1}$ of site 0. Then, $\overline{f}(x)$ represents the average energy per site of a configuration x, which is well defined if x is homogeneous, and agrees with $\pi_x(f)$.

Suppose e is a real number within the range of $\overline{f}$. Among all the macroscopic states π satisfying $\pi(f) = e$, those with maximum entropy per site $h(\pi)$ could be considered as the *most disordered*. These are the presumed *equilibrium states* of a system in which the energy f is conserved.[27] Applying the Lagrange multipliers method (Legendre transform), the optimization problem

$$\begin{aligned} &\text{maximize} && h(\pi) \\ &\text{subject to} && \pi(f) = e \end{aligned}$$

(for e in the range of $\overline{f}$) translates into the unconstrained problem

$$\text{maximize} \quad h(\pi) - \beta\pi(f) \tag{11.3}$$

(for $\beta \in \mathbb{R}$). The compactness of the space, the continuity of $\pi \mapsto \pi(f)$ and the upper semi-continuity of $\pi \mapsto h(\pi)$ imply that, in both problems, the maximums are achieved by some translation-invariant probability measures.[28] Dobrushin, Lanford and Ruelle proved that the macroscopic states solving the variational problem (11.3) are precisely the ergodic *Gibbs measures* at *inverse temperature* β.[29] See [14, 20, 31] for more information.

[27] A similar discussion applies if rather than a single notion of energy, we have a finite number of conserved quantities $f_1, f_2, \ldots, f_n$.

[28] However, the maximum in the first problem is not necessarily achieved by ergodic measures (i.e. macroscopic states). Such a situation corresponds to a first-order phase transition (see [14, 31]).

[29] In the current setting, Gibbs measures coincide with full-support Markov measures.

11.5 Boltzmann's Theory and Cellular Automata

Let us take a moment to draw parallels between the concepts in Boltzmann's gas theory and in cellular automata. We refer to the survey article of the Ehrenfests [4] and the book by Kac [17], which contain excellent accounts of Boltzmann's theory and related issues. See also [23] for a general discussion.

Boltzmann considered an isolated collection of n particles (identical hard spheres) interacting via elastic collisions.[30] The particles are assumed to be homogeneously distributed in (a bounded but large region of) the space. To be concrete, we may consider a cubic region with periodic boundary conditions. The focus is thus only on the velocity of the particles. Assuming that the number of particles is very large, we take $\rho(v, t)\mathrm{d}v$ to be the fraction of particles that, at time t, have velocities within an infinitesimal approximation $\mathrm{d}v$ of each value $v \in \mathbb{R}^3$. Using the assumption of spatial homogeneity, Boltzmann estimated the average number of collisions, in an infinitesimally small time interval $(t, t + \mathrm{d}t)$, among particles with velocities close to u and those with velocities close to v (the *Stosszahlansatz*).[31] The model of elastic collisions (the conservation of energy and momentum) could now be invoked to obtain the new distribution $\rho(v, t + \mathrm{d}t)\mathrm{d}v$ for the velocity of the particles at the end of the time interval $(t, t + \mathrm{d}t)$. This leads to an equation describing the time evolution of $\rho(v, t)$ known as the Boltzmann equation. Boltzmann used this equation to show that the quantity $-\int \rho(v, t) \log \rho(v, t)\mathrm{d}v$ is monotonically increasing in time, except at an equilibrium in which the velocities are distributed according to the Maxwell distribution $\rho(v) \sim \mathrm{e}^{-c|v|^2}$.

Boltzmann's derivation of the "law of increase in entropy" faced two major criticisms.[32] Loschmidt objected that a system governed by a reversible and time-reversal symmetric dynamics (like a system of particles interacting via elastic collisions) cannot possibly have an observable that is invariant under time reversal (like entropy) and is monotonically increasing in all situations. Zermelo's objection was based on Poincaré's recurrence theorem. According to Liouville's theorem, a Hamiltonian system (such as a system of particles) preserves the phase-space volume (i.e. the Lebesgue measure). Poincaré's theorem states that in a volume-preserving system whose phase space has finite volume (such as an isolated system of particles in a 3-dimensional torus), almost every trajectory eventually returns (infinitely many times) arbitrarily close to its starting point. This again implies that such a system cannot have a *continuous* observable that is monotonically increasing in time for *almost all* starting points.

In order to address these criticisms, Boltzmann later introduced another more refined framework.[33] In this new setting, each particle i is described by a state variable

[30] See Chap. I of [4] and Sects. III.1–2 of [17].

[31] More specifically, the *Stosszahlansatz* says that the frequencies of particles with different velocities that enter an infinitesimally small region at any given time are statistically independent.

[32] See Sect. 7 of [4] and Sect. III.7 of [17].

[33] See Chap. II of [4] and Sect. III.8 of [17].

x_i, which could, for instance, consist of the position as well as the velocity of the particle. The phase space of an individual particle (i.e. the range of values of x_i) is divided into small equally sized regions $A_1, A_2, \ldots$. Given a configuration of particles x, we can form the fraction $\rho_k = n_k/n$ of particles whose states are in region A_k. Conversely, given the macroscopic information $\rho = (\rho_1, \rho_2, \ldots)$, there corresponds a set $[\rho]$ consisting of all particle configurations that have fractions $\rho_1, \rho_2, \ldots$ of particles in regions $A_1, A_2, \ldots$. If the number of particles is very large, the volume of the set $[\rho]$ could be measured by the quantity $H(\rho) = -\sum_k \rho_k \log \rho_k$. Neglecting any interaction energy between particles, the energy of a configuration x could be written as $E(x) = n \sum_k \rho_k e_k$, where e_k is the (approximate) energy of a particle whose state is within A_k. Boltzmann now argued that a system with energy E at equilibrium is most likely to be found (at almost any moment of time) to have a state distribution ρ that maximizes the quantity $H(\rho)$ among all the state distributions with energy E, for this is the distribution for which $[\rho]$ takes the overwhelmingly largest portion of the set of all configurations with energy E. If n is large, this equilibrium distribution is (approximately) given by $\rho_k \sim \mathrm{e}^{-\beta e_k}$, where β is a Lagrange multiplier for tuning E.

The analogy with cellular automata should be clear. Rather than particles, the elementary pieces of information in cellular automata are carried by lattice sites representing discretized positions in the space. The symbol at site i should therefore be compared with the state of particle i. The model of elastic collisions governing the interaction between the particles is replaced with the local update rule describing the cellular automaton map $T : S^{\mathbb{Z}^d} \to S^{\mathbb{Z}^d}$. The fraction $\overline{\zeta_a}(x)$ of sites having symbol a is an elementary macroscopic observable analogous to the fraction n_k/n of particles whose states are in region A_k, but now it is clear that one must also take into account the macroscopic observables $\overline{\zeta_q}(x)$ (for larger patterns $q : A \to S$) which contain information about correlations between finite collections of sites. Boltzmann's entropy corresponds to the empirical entropy $\hat{H}_0(x)$ of symbols appearing in the configuration x, or more generally, the empirical entropy $\hat{H}_A(x)$, which measures lack of bias in the frequency of the patterns with support A occurring in x.

To understand Boltzmann's argument about the increase in entropy, consider the XOR cellular automaton (Sect. 11.2.1), and let x be a Bernoulli random configuration with parameter $p \in (0, 1)$ (i.e. a homogeneous configuration whose macroscopic state is described by the Bernoulli measure with parameter p). In particular, the words of length 2 have frequencies

$$\overline{\zeta_{00}}(x) = (1-p)^2\,, \quad \overline{\zeta_{01}}(x) = (1-p)p\,, \quad \overline{\zeta_{10}}(x) = p(1-p)\,, \quad \overline{\zeta_{11}}(x) = p^2$$

in x. It follows that the frequency $\overline{\zeta_1}(Tx)$ of occurrence of symbol 1 after one step is $\varphi(p) \triangleq 2p(1-p)$. If $\mathsf{H}(p) \triangleq -p \log p - (1-p)\log(1-p)$ denotes the binary entropy function, one can easily verify that $\mathsf{H}(\varphi(p)) \geq \mathsf{H}(p)$ with equality if and only if $p = \frac{1}{2}$. Therefore, it is indeed the case that the entropy $\hat{H}_0(Tx)$ is larger than $\hat{H}_0(x)$ unless $p = \frac{1}{2}$.[34]

Boltzmann's assumption about the number of collisions (the *Stosszahlansatz*) is analogous to the (invalid) assumption that the configuration Tx is also Bernoulli, so that the frequency of occurrence of a word $w = a_1 a_2 \ldots a_n$ in Tx is simply the product of the frequencies of $a_1, a_2, \ldots, a_n$. If true, this would lead to the conclusion that the entropy increases monotonically in the consecutive steps, that is, $H_0(x) \leq H_0(Tx) \leq H_0(T^2 x) \leq \cdots$ with the equalities only if $H_0(x) = \log 2$. The assumption is of course false.[35] Nevertheless, the randomization property of the XOR cellular automaton (Theorem 1) suggests a mathematically rigorous scenario that makes Boltzmann's conclusion essentially true. Indeed, the randomization implies that for any finite set $A \subseteq \mathbb{Z}$, the entropy $\hat{H}_A(T^t x)$ approaches (after ignoring a negligible set of time steps) to its maximum value $|A| \log 2$, even if this convergence might be non-monotonic.[36]

For cellular automata, the uniform Bernoulli measure plays the role of the Lebesgue measure on the phase space of a Hamiltonian system. The analog of Liouville's theorem is the balance property, which says that any surjective cellular automaton map $T : S^{\mathbb{Z}^d} \to S^{\mathbb{Z}^d}$ preserves the uniform Bernoulli measure. Hence, Poincaré's theorem applies to all surjective cellular automata. It says that starting from *almost every* configuration x (i.e. any configuration in a set of uniform Bernoulli measure 1), every finite pattern occurring on x (i.e. $q = x_A$ for some finite set $A \subseteq \mathbb{Z}^d$) reappears on the same position infinitely many times (i.e. $(T^t x)_A = q$ for infinitely many time steps t).[37] Note that this does not say anything about a starting configuration that is *not* typical for the uniform Bernoulli measure.

It is worth mentioning that surjective cellular automata preserve the limit entropy per site, that is, $h(T^t \pi) = h(\pi)$ for every macroscopic state π and $t = 1, 2, \ldots$ (see e.g. [19]). On the other hand, randomization implies a jump at the limit to $h(\mu) > \lim_{\substack{t \in J \\ t \to \infty}} h(T^t \pi) = h(\pi)$. This may be understood as follows.[38] A submaximum value for $H_A(\pi)$ expresses the presence of correlations among symbols with relative positions given by A. The convergence of the entropy $H_A(T^t \pi)$ to its maximal value $|A| \log |S|$ for larger and larger finite sets $A \subseteq \mathbb{Z}^d$ indicates that the correlations are gradually distributed over larger and larger regions, and are escaping to infinity as t grows to infinity.

---

[34]Similar (but more cumbersome) calculations lead to the same conclusion for the examples in Sects. 11.2.3 and 11.2.4. More generally, one can show that if $f : S \times S \to S$ is a function that is permutative on both its arguments, X and Y are independent S-valued random variables with distributions p and p', and $Z = f(X, Y)$, then $H(Z) \geq H(X)$ with equality if and only if p is uniform.

[35]For the XOR cellular automaton, Tx is Bernoulli only if $p = \frac{1}{2}$.

[36]Boltzmann's derivation for a system of particles was later made rigorous in a certain asymptotic regime (the Boltzmann–Grad limit) by the groundbreaking work of Lanford [22] and others [2].

[37]If the map T is ergodic (e.g. if T is the XOR map or XOR-transpose map), the average time between two consecutive reappearances is $2^{|A|}$ by Kac's recurrence theorem.

[38]For simplicity, we assume that T has no non-trivial conserved quantity.

11.6 How Far the Second Law Goes?

The phenomenon described by the second law of thermodynamics extends to all physical systems. What constitutes a physical system and what is the exact statement of the second law are much less clear. Let us discuss few prerequisites for the presence of the randomization effect. For simplicity, we focus on the case that the cellular automaton has no non-trivial conserved quantity.

To fix the terminology, let us say that a cellular automaton $T : S^{\mathbb{Z}^d} \to S^{\mathbb{Z}^d}$ *randomizes* a translation-invariant probability measure π if the Cesàro averages $\frac{1}{n}\sum_{t=0}^{n-1} T^t\pi$ converge weakly to the uniform Bernoulli measure μ as n goes to infinity. Equivalently, T randomizes π if $T^t\pi \to \mu$ along a subsequence $J \subseteq \mathbb{N}$ of density 1 of time steps (see [16], Corollary 1.4).

An obvious case in which randomization fails is when T is not surjective. Lack of surjectivity (or reversibility) has been suggested as a mechanism behind the contrasting phenomenon of self-organization (see e.g. [38]). The requirement for T to be surjective is a relaxation compared to reversibility (let alone time-reversal symmetry), which is common among most microscopic physical theories. Yet, surjectivity already guarantees the invariance of the uniform Bernoulli measure. Moreover, surjective cellular automata are in some way close to being injective: if two configurations x, y differ on at most finitely many positions, then Tx and Ty are distinct (see e.g. [18]).[39]

Besides surjectivity, the cellular automaton requires to have certain degree of chaos in order for randomization to occur. For instance, for a one-dimensional surjective cellular automata with equicontinuity points,[40] the Cesàro averages $\frac{1}{n}\sum_{t=0}^{n-1} T^t\pi$ with a Bernoulli starting measure π converge but not necessarily to the uniform Bernoulli measure [1]. Such a cellular automaton typically has too many distinct conserved quantities, resulting in failure of any thermodynamic behaviour (see [19, 34]).

Another obstacle for randomization is too much regularity in the starting configuration. The simplest type of regularity is periodicity. Note that a spatially periodic configuration is also temporally periodic.[41] Therefore, no cellular automaton can randomize a periodic configuration. A spatially periodic configuration has zero entropy per site. As a more sophisticated example of a regularity obstructing randomization, consider the XOR cellular automaton. The XOR cellular automaton has the following

[39] A non-surjective cellular automaton may still act surjectively on a natural subspace (e.g. a mixing subshift of finite type). Randomization within such a subspace may still occur.

[40] A cellular automaton T is *equicontinuous* (or *stable*) at a configuration x if for each finite set A, there is a finite set B such that for any configuration y that agrees with x on B, $T^t x$ and $T^t y$ agree on A for every $t \geq 0$. A cellular automaton with equicontinuity points is not sensitive, hence not chaotic (see [21]).

[41] A configuration x is said to be *spatially periodic* if its translation orbit is finite, or equivalently, if there are d linearly independent elements $q_1, q_2, \ldots, q_d \in \mathbb{Z}^d$ such that $x_{a+nq_i} = x_a$ for all $a \in \mathbb{Z}^d$ and $n \in \mathbb{Z}$. A configuration x is *temporally periodic* if $T^p x = x$ for some $p > 0$. Observe that every spatially periodic configuration is homogeneous.

self-similarity property, which can be verified by induction: $(T^{2^n}x)_i = x_i + x_{i+2^n}$ (mod 2) for every $n \geq 0$. Define the *duplicate* of a configuration x as the configuration Dx with $(Dx)_{2i} \triangleq (Dx)_{2i+1} \triangleq x_i$. It follows from self-similarity that $DTx = T^2Dx$, or more generally $D^nTx = T^{2^n}D^nx$. For the uniform Bernoulli measure μ, in particular, we find that $T^{2^n}D^n\mu = D^n\mu$. Note that if π is a translation-ergodic measure, the measure $\overline{D}\pi \triangleq \frac{1}{2}(D\pi + \sigma D\pi)$ is also translation-ergodic and has entropy per site $\frac{1}{2}h(\pi)$. Moreover, if $T^k\pi = \pi$, we get $T^{2k}\overline{D}\pi = \overline{D}\pi$. Therefore, we have an infinite sequence $\overline{D}\mu, \overline{D}^2\mu, \overline{D}^3\mu, \ldots$ of distinct translation-ergodic measures (i.e. macroscopic states) with positive entropy that are not randomized by the XOR cellular automaton.[42]

In summary, randomization is expected only if the cellular automaton is surjective and "sufficiently chaotic", and the starting configuration does not have "too much regularity".

Acknowledgements Research supported by ERC Advanced Grant 267356-VARIS of Frank den Hollander. I would like to thank Aernout van Enter, Nazim Fatès and Ville Salo for helpful comments and discussions. This article is dedicated with love and appreciation to my teacher Javaad Mesgari.

References

1. Blanchard, F., Tisseur, P.: Some properties of cellular automata with equicontinuity points. Annales de l'Institut Henri Poincaré, Probabilités et Statistiques **36**(5), 569–582 (2000)
2. Cercignani, C., Illner, R., Pulvirenti, M.: The Mathematical Theory of Dilute Gases. Springer, Berlin (1994)
3. Cover, T.M., Thomas, J.A.: Elements of Information Theory. Wiley, New York (1991)
4. Ehrenfest, P., Ehrenfest, T.: The conceptual foundations of the statistical approach in mechanics. Cornell University Press (1959). Originally appeared in 1912 as volume IV, part 32 of the Encyklopädie der mathematischen Wissenschaften
5. Einsiedler, M.: Invariant subsets and invariant measures for irreducible actions on zero-dimensional groups. Bull. Lond. Math. Soc. **36**(3), 321–331 (2004)
6. Ferrari, P.A., Maass, A., Martínez, S., Ney, P.: Cesàro mean distribution of group automata starting from measures with summable decay. Ergod. Theory Dyn. Syst. **20**(6), 1657–1670 (2000)
7. Formenti, E., Kari, J., Taati, S.: On the hierarchy of conservation laws in a cellular automaton. Nat. Comput. **10**(4), 1275–1294 (2011)
8. Gajardo, A., Kari, J., Moreira, A.: On time-symmetry in cellular automata. J. Comput. Syst. Sci. **78**(4), 1115–1126 (2012)
9. Georgii, H.O.: Gibbs Measures and Phase Transitions. Walter de Gruyter, Berlin (1988)
10. Georgii, H.O.: Probabilistic aspects of entropy. In: Greven, A., Keller, G., Warnecke, G. (eds.) Entropy, pp. 37–54. Princeton University Press, Princeton (2003)
11. Hedlund, G.A.: Endomorphisms and automorphisms of the shift dynamical system. Math. Syst. Theory **3**, 320–375 (1969)
12. Herrmann, H.J.: Fast algorithm for the simulation of Ising models. J. Stat. Phys. **45**(1/2), 145–151 (1986)

[42] In fact, the XOR cellular automaton has a continuum of distinct translation-ergodic measures with positive entropy per site that are not randomized [5].

13. Host, B., Maass, A., Martínez, S.: Uniform Bernoulli measure in dynamics of permutative cellular automata with algebraic local rules. Discret. Contin. Dyn. Syst. — Ser. A **9**(6), 1423–1446 (2003)
14. Israel, R.B.: Convexity in the Theory of Lattice Gases. Princeton University Press, Princeton (1979)
15. Jaynes, E.T.: Information theory and statistical mechanics. Phys. Rev. **106**(4), 620–630 (1957)
16. Johnson, A., Rudolph, D.J.: Convergence under $\times_q$ of $\times_p$ invariant measures on the circle. Adv. Math. **115**(1), 117–140 (1995)
17. Kac, M.: Probability and Related Topics in Physical Sciences. Interscience Publishers, New York (1959)
18. Kari, J.: Theory of cellular automata: a survey. Theor. Comput. Sci. **334**, 3–33 (2005)
19. Kari, J., Taati, S.: Statistical mechanics of surjective cellular automata. J. Stat. Phys. **160**(5), 1198–1243 (2015)
20. Keller, G.: Equilibrium States in Ergodic Theory. Cambridge University Press, Cambridge (1998)
21. Kůrka, P.: Topological and symbolic dynamics. Cours Spécialisés, vol. 11. Société Mathématique de France (2003)
22. Lanford III, O.E.: Time evolution of large classical systems. In: Moser, J. (ed.) Dynamical Systems, Theory and Applications. Lecture Notes in Physics, vol. 38, pp. 1–111 (1975)
23. Lebowitz, J.L.: Statistical mechanics: a selective review of two central issues. Rev. Mod. Phys. **71**(2) (1999)
24. Lind, D.A.: Applications of ergodic theory and sofic systems to cellular automata. Phys. D: Nonlinear Phenom. **10**(1–2) (1984)
25. Maass, A., Martínez, S.: Time averages for some classes of expansive one-dimensional cellular automata. Cellular Automata and Complex Systems, Nonlinear Phenomena and Complex Systems, vol. 3, pp. 37–54. Kluwer Academic Publishers, Dordrecht (1999)
26. Miyamoto, M.: An equilibrium state for a one-dimensional life game. J. Math. Kyoto Univ. **19**, 525–540 (1979)
27. Oxtoby, J.C.: Ergodic sets. Bull. Am. Math. Soc. **58**(2), 116–136 (1952)
28. Oxtoby, J.C.: On two theorems of Parthasarathy and Kakutani concerning the shift transformation. In: Wright, F.B. (ed.) Ergodic Theory, pp. 203–215. Academic Press, New York (1963)
29. Pivato, M., Yassawi, R.: Limit measures for affine cellular automata. Ergod. Theory Dyn. Syst. **22**, 1269–1287 (2002)
30. Pivato, M., Yassawi, R.: Limit measures for affine cellular automata II. Ergod. Theory Dyn. Syst. **24**, 1961–1980 (2004)
31. Ruelle, D.: Thermodynamic Formalism, 2nd edn. Cambridge University Press, Cambridge (2004)
32. Shereshevsky, M.A.: Ergodic properties of certain surjective cellular automata. Monatshefte für Mathematik **114**(3–4), 305–316 (1992)
33. Shereshevsky, M.A.: K-property of permutative cellular automata. Indag. Math. **8**(3), 411–416 (1997)
34. Takesue, S.: Reversible cellular automata and statistical mechanics. Phys. Rev. Lett. **59**(22), 2499–2502 (1987)
35. Toffoli, T., Margolus, N.: Cellular Automata Machines: A New Environment for Modeling. MIT Press, Cambridge (1987)
36. Vichniac, G.Y.: Simulating physics with cellular automata. Phys. D **10**, 96–116 (1984)
37. Walters, P.: An Introduction to Ergodic Theory. Springer, Berlin (1982)
38. Wolfram, S.: Statistical mechanics of cellular automata. Rev. Mod. Phys. **55**, 601–644 (1983)
39. Wolfram, S.: Random sequence generation by cellular automata. Adv. Appl. Math. **7**, 123–169 (1986)

Chapter 12
Epidemic Automaton and the Eden Model: Various Aspects of Robustness

Lucas Gerin

Abstract The two-dimensional probabilistic cellular automaton *Epidemic* models the spread of an epidemic without recovering on graph. We discuss some well-known and less well-known properties of Epidemic on a finite grid and its analogous on the infinite square lattice: the Eden model. This survey is intended for non-probabilists and gives a detailed study of the robustness of a cellular automaton with respect to several sources of randomness.

12.1 Introduction

We discuss here several random perturbations of a particular (yet very interesting) 2D probabilistic cellular automaton: *Epidemic*. This is a "toy model" for the propagation in a graph of an epidemic without recovering. The goal of this article is to analyse the behaviour of Epidemic with respect to several sources of randomness:

- randomness in the updating scheme,
- randomness in the initial configuration,
- randomness (or defaults) of the graph.

We will discuss theses aspects on two variants of Epidemic: on a finite square grid (in Sect. 12.2) and the analogous rule on the infinite square lattice (Sect. 12.3). The latter variant is often called *Eden model* by physicists and probabilists.

The main questions that we will discuss are

- On a finite grid: how long does it take for the spread to occupy the whole grid?
- On $\mathbb{Z}^2$: what is the *typical* shape of the spread after a large time?
- For both models: how do these behaviours depend on the different parameters of the models?

Beyond its own interest, we believe that Epidemic is a good candidate to study robustness of Cellular Automata with respect to randomness. Its behaviour is rich

L. Gerin (✉)
CMAP, École Polytechnique, Route de Saclay, 91128 Palaiseau Cedex, France
e-mail: gerin@cmap.polytechnique.fr

P.-Y. Louis and F.R. Nardi (eds.), *Probabilistic Cellular Automata*, Emergence, Complexity and Computation 27, https://doi.org/10.1007/978-3-319-65558-1_12

enough to reveal some interesting phenomena and simple enough to allow rigorous analysis. Some results stated here are all more or less *folklore*, but the statements are not so easy to find in literature, our goal was to present them in a self-contained way. The results on Epidemic are all rigorously proved; the discussion on the Eden model is more thought as a reading-guide in the probabilistic literature.

12.2 Epidemic on a Finite Grid

12.2.1 The Model

Let $L \geq 3$ be an integer, we denote by Λ the square grid $L \times L$, with torical boundary conditions (*i.e.* we identify Λ with $\mathbb{Z}/L\mathbb{Z} \times \mathbb{Z}/L\mathbb{Z}$). Let $n = L^2$ be the number of *cells* in Λ, n will be our scale for the asymptotics.

We endow Λ with the usual graph distance, the *ball* $B(c, r)$ with centre c and radius r being the set of cells c' such that there exists a *path* of $0 \leq \ell \leq r$ neighbouring cells

$$c = c_0 \to c_1 \to c_2 \cdots \to c_\ell = c'.$$

A *configuration* σ is one of the 2^n elements of $\mathbb{Q}^\Lambda$. For $c \in \Lambda$, $\sigma_c \in \mathbb{Q}$ is the *state* of cell c in configuration σ.

For c in Λ, $\mathcal{N}(c)$ is the so-called *Von Neumann* neighbourhood of c:

$$\mathcal{N}(c) = \{c, c + (1, 0), c + (-1, 0), c + (0, 1), c + (0, -1)\},$$

where $+$ stands for addition modulo L. In other words, $\mathcal{N}(c) = B(c, 1)$.

We now can define Epidemic as a stochastic dynamical system. Each cell in state 0 (healthy), when it is updated, turns into state 1 (infected) if one at least of its neighbours is infected. There is no recovering: a 1 remains 1 forever. Besides, updating is random and independent from each other.

More formally:

Definition 1 (*Epidemic in the α-synchronous dynamics*) Let $\alpha \in (0, 1)$, the α-synchronous Epidemic cellular automaton is the stochastic process $\left(\sigma^t\right)_{t\geq 0}$ with values in $\{0, 1\}^\Lambda$ such that $\sigma^0 \in \{0, 1\}^\Lambda$ and whose evolution is given in the following way.

For every $t \geq 0$, given σ^t at time t, the configuration σ^{t+1} is defined as follows: each cell in Λ is updated independently with probability α (independently from the past and from the $n - 1$ other cells) ;

- If c is updated, $\sigma_c^{t+1} = 1$ if and only if at least one cell in $\mathcal{N}(c)$ is in state 1 at time t;
- Otherwise, $\sigma_c^{t+1} = \sigma_c^t$.

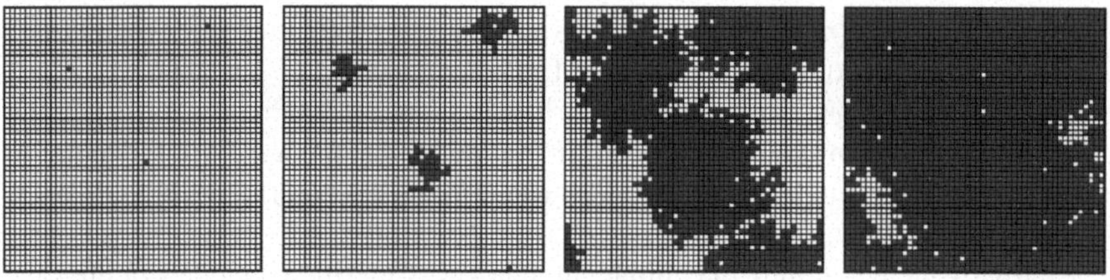

Fig. 12.1 A simulation of α-synchronous Epidemic with $\alpha = 0.1$ and $L = 50$, σ^0 has only 3 cells in state 1. Simulations were performed with FiatLux: http://fiatlux.loria.fr/

The sequence (σ^t) is then a discrete-time Markov chain with values in $\{0, 1\}^\Lambda$. Obviously, it eventually reaches one of the two fixed configurations 0^Λ our 1^Λ (Fig. 12.1). The configuration 0^Λ is *isolated*:

$$\sigma^t \text{ reaches } 0^\Lambda \Leftrightarrow \sigma^0 = 0^\Lambda.$$

From now on, we exclude the trivial case $\sigma^0 = 0^\Lambda$

Definition 2 (*Convergence time*) For an initial configuration $\sigma^0 \neq 0^\Lambda$, the *convergence time* $T_n(\sigma^0)$ is the first time at which all cells are infected:

$$T_n(\sigma^0) = \min\left\{t \geq 0 \text{ such that } \sigma^t = 1^\Lambda\right\}.$$

In this section, we will focus on the asymptotic behaviour of the expectation of this random variable: $\mathbb{E}\left[T_n(\sigma^0)\right]$, in *worst* and *typical* cases.

12.2.2 The Worst Expected Convergence Time

We first consider the *worst* expected convergence time (WECT) for Epidemic, i.e. the mean convergence time when $\sigma^0 = \sigma^{\mathrm{w}}$, where σ^{w} is such that

$$\mathbb{E}\left[T_n(\sigma^{\mathrm{w}})\right] = \max_{\substack{\sigma^0 \in \{0,1\}^\Lambda \\ \sigma^0 \neq 0^\Lambda}} \mathbb{E}\left[T_n(\sigma^0)\right].$$

Obviously such σ^{w}'s are exactly the n configurations with a single 1.

Before stating the result, let us motivate the analysis of the WECT:

1. In dimension one, Fatès et al. [9] have studied the WECT of the Elementary Cellular Automata with two quiescent states (see their article for the definitions). Their work revealed that the asymptotic behaviour of the WECT provides a relevant classification of 1D cellular automata. Precisely, they have shown that these rules may be classified into 5 families, according to whether the WECT is $\Theta(n \log n)$,

$\Theta(n^2), \Theta(n^3), \Theta(n2^n)$ or infinite.[1] This approach was extended in [8] for a family of 2D cellular automata (in particular, for Epidemic).
2. Another motivation comes from algorithmic complexity theory, since cellular automata are often thought as model in computability theory. With this point of view, it is natural to ask what happens when the system starts from the "worse" configuration.
3. Alternatively, if we think of cellular automata as (simplified!) models of physical or biological systems, studying the WECT provides us with an estimation of the maximum time needed to go back to equilibrium when a perturbation is applied.

Theorem 1 (Worst Expected Convergence Time) *For Epidemic on a finite grid with n cells, for all $\alpha \in (0, 1)$, if n is large enough,*

$$\frac{\sqrt{n}}{8\alpha} \leq \mathbb{E}\left[T_n(\sigma^{\mathrm{w}})\right] \leq 3\frac{\sqrt{n}}{\alpha}.$$

Remark 1 • In [8], a very similar result (but less precise, because of $\log(n)$ terms in both sides of the inequality) was proved in *asynchronous* dynamics.
• We believe that $\mathbb{E}\left[T_n(\sigma^{\mathrm{w}})\right]/\sqrt{n}$ converges to a constant, and more precisely that the sequence of random variables $T_n(\sigma^{\mathrm{w}})/\sqrt{n}$ converges in probability. We have not been able to prove so, and the usual tool (*subadditivity* theory, see [2] for instance) to deal with similar problems does not seem to work here.

Proof **Lower bound.** Let c be the only cell in state $\mathtt{1}$ in σ^{w}, and let c' be one of the cells of Λ which is at distance $\lfloor L/2 \rfloor$ from c, where $\lfloor x \rfloor$ is the integer part of x. It is enough to prove that $\mathbb{E}[\tau_{c'}] \geq \frac{L}{8\alpha}$, with

$$\tau_{c'} = \min\left\{t \geq 0,\ \sigma^t_{c'} = \mathtt{1}\right\},$$

since obviously $T_n(\sigma^{\mathrm{w}}) \geq \tau_{c'}$.

Set $k = \lfloor L/10\alpha \rfloor$, we will prove that if L is large enough

$$\mathbb{P}(\tau_{c'} < k) \leq 1/2. \tag{12.1}$$

This proves the upper bound since then

$$\mathbb{E}\left[T_n(\sigma^{\mathrm{w}})\right] \geq \mathbb{E}[\tau_{c'}] \geq \sum_{r=1}^{k} \mathbb{P}(\tau_{c'} \geq r) \geq 1 + \sum_{r=2}^{k} 1/2 = \left(\lfloor L/10\alpha \rfloor + 1\right)/2 \geq \frac{\sqrt{n}}{20\alpha}.$$

We focus on (12.1). The proof is not very difficult but quite technical, here is the general strategy (Fig. 12.2). As often in this kind of optimization problem, there is a trade-off between two phenomena:

---

[1] We write $f_n = \Theta(g_n)$ when there exist two positive numbers c_1, c_2 such that, for n large enough, $c_1 g_n \leq f_n \leq c_2 g_n$.

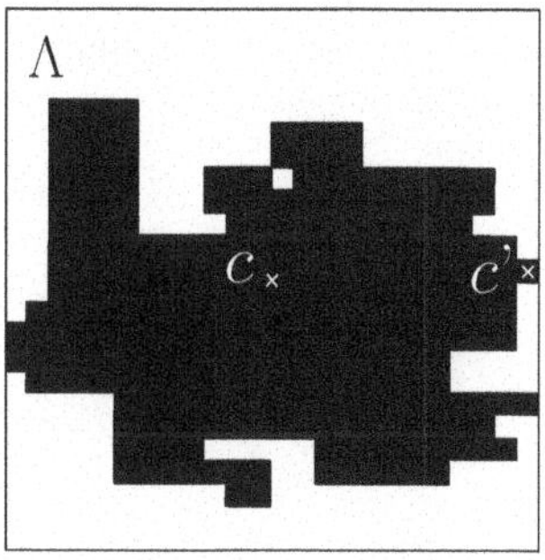

Fig. 12.2 The configuration at time $\tau_{c'}$

- There is a huge number of paths (we will bound this number by 3^j in the proof) going from c to c' along which successive updatings would infect cell c';
- On the other hand, if we fix such a path, it is very unlikely if k is well-chosen (this will be our Lemma 1 below) that its cells are updated in the proper order before k.

We now go into the details. Assume that $\tau_{c'} \leq k$, then there is a j with $\lfloor L/2 \rfloor \leq j \leq k$ and a path $\mathcal{P}$ made of j disjoint cells, going from c to c':

$$\mathcal{P} = \left\{c = c_0 \to c_1 \to c_2 \to \cdots \to c_j = c'\right\}$$

such that, during the k first time units, cells $c_1, \ldots, c_j$ are updated *in this order*. This gives

$$\mathbb{P}(\tau_{c'} \leq k) = \mathbb{P}\left(\bigcup_{j=\lfloor L/2 \rfloor}^{k} \bigcup_{\mathcal{P}, |\mathcal{P}|=j} \left\{c_1, c_2, \ldots, c_j \text{ are updated in this order and before } k\right\}\right)$$
$$\leq \sum_{j=\lfloor L/2 \rfloor}^{k} \sum_{\mathcal{P}, |\mathcal{P}|=j} \mathbb{P}\left(c_1, c_2, \ldots, c_j \text{ are updated in this order and before } k\right),$$

(here, we used $\mathbb{P}(\cup) \leq \sum \mathbb{P}$, often called *union bound*). The second sum runs over all paths of j cells going from c to c'. Fix such a $\mathcal{P}$ and bound the last probability. Among times $\{1, 2, \ldots, k\}$, there are times $t_1 < t_2 < \cdots < t_j$ such that, at time t_j, cell c_j is updated. Each updating being independent, we have

$$\mathbb{P}\left(c_1, c_2, \ldots, c_j \text{ are updated in this order and before } k\right) = \mathbb{P}\left(\text{Binom}(k, \alpha) \geq j\right),$$

where $\text{Binom}(k, \alpha)$ has the binomial distribution with parameters k, α and has expectation $k\alpha$. If $k\alpha \ll j$ then this last probability is small, the following lemma is useful, this is for instance (2.5) in [14].

Lemma 1 (Right-deviations for the binomial) *For all* $j \geq k\alpha$,

$$\mathbb{P}\left(\text{Binom}(k,\alpha)\geq j\right)\leq \exp\left(-3\frac{(j-k\alpha)^2}{2k\alpha+j}\right).$$

There are less than 3^j paths of length j going from c to c' (this is a rough bound but sufficient here), we have

$$\mathbb{P}(\tau_{c'}\leq k)=\sum_{j=\lfloor L/2\rfloor}^{k} 3^j \exp\left(-3\frac{(j-k\alpha)^2}{2k\alpha+j}\right).$$

One can check that $j\mapsto 3^j\exp\left(-3\frac{(j-k\alpha)^2}{2k\alpha+j}\right)$ is non-increasing for $L/2\leq j\leq k$ (recall $k\alpha\approx L/10$) and thus, skipping the integer parts in order to lighten notations,

$$\begin{aligned}\mathbb{P}(\tau_{c'}\leq k)&\leq k\times\max_j\left\{3^j\exp\left(-3\frac{(j-k\alpha)^2}{2k\alpha+j}\right)\right\}\\&\leq k\exp\left(\log(3)L/2-3\frac{(L/2-k\alpha)^2}{2k\alpha+L/2}\right)\\&\leq\frac{L}{10\alpha}\exp\left(\log(3)L/2-3L\frac{(1/2-1/10)^2}{2/10+1/2}\right)\leq\frac{L}{10\alpha}\exp(-0.2\times L),\end{aligned}$$

and therefore is less than $1/2$ if L is large (depending on α). We have proved (12.1).
Upper bound. We will prove that for L large enough and $k\geq 2L/\alpha$

$$\mathbb{P}(T_n(\sigma^{\mathrm{w}})>k)\leq L^2\exp(-k\alpha/32). \tag{12.2}$$

This yields the upper bound since

$$\begin{aligned}\mathbb{E}\left[T_n(\sigma^{\mathrm{w}})\right]&=\sum_{k\geq 1}\mathbb{P}(T_n(\sigma^{\mathrm{w}})\geq k)\\&\leq 2\sqrt{n}/\alpha+\sum_{k\geq 2\sqrt{n}/\alpha}\mathbb{P}(T_n(\sigma^{\mathrm{w}})\geq k)\\&\leq 2\sqrt{n}/\alpha+L^2\sum_{k\geq 2\sqrt{n}/\alpha}\exp(-k\alpha/4)\\&\leq 2\sqrt{n}/\alpha+n\frac{\exp(-\frac{\alpha}{4}\frac{2\sqrt{n}}{\alpha})}{1-e^{-\alpha/4}}\\&=2\sqrt{n}/\alpha+\mathrm{o}(\sqrt{n})\leq 3\sqrt{n}/\alpha\end{aligned}$$

Let us prove (12.2). First, for each $c'\neq c$, we choose (in a non-random way) a path $\mathcal{P}_{c'}$ among all shortest paths $c\to c'$: $\mathcal{P}_{c'}$ can be written

$$\mathcal{P}_{c'}=\left\{c=c_0\to c_1\to c_2\to\cdots\to c_j=c'\right\},$$

where $j \leq L/2$ is the distance between c and c'.

If $T_n(\sigma^{\mathrm{w}}) > k$, then there is c' which is still in state 0 at time k. In particular, there is $j \leq L/2$ and a cell c' at distance j such that cells $c_1, c_2, \ldots, c_j$ of its associated path $\mathcal{P}_{c'}$ have not been updated in this order before time k:

$$\mathbb{P}(T_n(\sigma^{\mathrm{w}}) > k) = \mathbb{P}\left(\bigcup_{c' \in \Lambda} \{c_1, c_2, \ldots, c_j \text{ are not updated in this order and before } k\}\right)$$
$$\leq \mathrm{card}(\Lambda) \max_{c' \in \Lambda} \mathbb{P}\left(c_1, c_2, \ldots, c_j \text{ are updated in this order and before } k\right).$$

and by the same argument as for the lower bound

$$\mathbb{P}\left(c_1, c_2, \ldots, c_j \text{ are updated in this order and before } k\right) = \mathbb{P}\left(\mathrm{Binom}(k, \alpha) < j\right),$$

now we need the following bound, this is (2.6) in [14].

Lemma 2 (Left-deviations for the binomial) *For all $j \leq k\alpha$,*

$$\mathbb{P}\left(\mathrm{Binom}(k, \alpha) < j\right) \leq \exp\left(-\frac{(k\alpha - j)^2}{2k\alpha}\right).$$

For all $j \leq L/2$, we have $j \leq k\alpha$ (recall $k \geq 2L/\alpha$) and we can apply the lemma:

$$\begin{aligned}
\mathbb{P}(T_n(\sigma^{\mathrm{w}}) > k) &\leq L^2 \max_{1 \leq j \leq L/2} \exp\left(-\frac{(k\alpha - j)^2}{2k\alpha}\right) \\
&\leq L^2 \exp\left(-\frac{(k\alpha - L/2)^2}{2k\alpha}\right) \\
&\leq L^2 \exp\left(-\frac{(k\alpha - 3k\alpha/4)^2}{2k\alpha}\right) \qquad \text{since} (k\alpha - L/2)^2 \geq k\alpha/2 \\
&\leq L^2 \exp\left(-k\alpha/32\right),
\end{aligned}$$

we have proved (12.2).

12.2.3 *Typical Convergence Time*

We now estimate the *typical* expected convergence time, when σ^0 is drawn uniformly at random in $\{0, 1\}^\Lambda$.

$$\mathrm{Typ}_n := \frac{1}{2^n} \sum_{\sigma^0 \in \{0,1\}^\Lambda} \mathbb{E}\left[T_n(\sigma^0)\right].$$

As expected, Typ_n is much smaller than in the worst case.

Theorem 2 (Typical expected convergence time) *For n large enough,*

$$\frac{1}{4\alpha}\log\, n \le \mathrm{Typ}_n \le \frac{6}{\alpha}(\log\, n)^{3/2}.$$

Proof We closely follow ([11], Chap. 2).
Lower bound. The number of cells in state 0 in σ^0, which is a Binomial $(n, 1/2)$, is larger than $n/2$ with more than 50% chance. For such σ^0, the convergence takes more time than the time needed to update all these cells at least once. Thus

$$\mathbb{E}\left[T_n(\sigma^0)\right] \ge \mathbb{E}\left[\max\left\{\mathcal{G}_1, \mathcal{G}_2, \dots, \mathcal{G}_{n/2}\right\}\right],$$

where $\mathcal{G}_i$ are i.i.d geometric random variables with mean $1/\alpha$. For large k, we have (see [18] for instance)

$$\frac{2\log(k)}{3\alpha} \le \mathbb{E}\left[\max\left\{\mathcal{G}_1, \mathcal{G}_2, \dots, \mathcal{G}_k\right\}\right] \le \frac{2\log(k)}{\alpha} \tag{12.3}$$

and then, for large enough n, $\mathbb{E}\left[T_n(\sigma^0)\right] \ge \frac{2}{3\alpha}\log(n/2) \ge \frac{1}{2\alpha}\log(n)$ when σ^0 has more than $n/2$ cells in state 0. Now,

$$\begin{aligned}\mathrm{Typ}_n = \frac{1}{2^n}\sum_{\sigma^0\in\{0,1\}^\Lambda}\mathbb{E}\left[T_n(\sigma^0)\right] &\ge \frac{1}{2^n}\sum_{\substack{\sigma^0 \text{ with}\\ \text{more than } n/2\, 0\text{'s}}}\mathbb{E}\left[T_n(\sigma^0)\right]\\ &\ge \frac{1}{2^n}\mathrm{card}\left\{\sigma^0 \text{ with more than } n/2\, 0\text{'s}\right\}\frac{1}{2\alpha}\log(n)\\ &\ge \frac{1}{2^n}\frac{2^n}{2}\frac{1}{2\alpha}\log(n),\end{aligned}$$

hence the lower bound.
Upper bound. The first step is to show that with high probability there is no ball of radius $3\sqrt{\log(L)}$ which is full of 0's in the initial configuration σ^0.
Precisely, set $a(L) = \lfloor 3\sqrt{\log L}\rfloor$, introduce the vent

$$\begin{aligned}A &= \bigcup_{c\in\Lambda}\{\text{ the ball of centre } c \text{ and radius } a(L) \text{ is full of } 0 \text{ 's at time } 0.\}\\ &= \bigcup_{c\in\Lambda}\left\{\forall c' \in B(c, a(L)), \sigma^0_{c'} = 0\right\}.\end{aligned}$$

By the union bound,

$$\mathbb{P}(A) \le \sum_{c\in\Lambda}\mathbb{P}\left(\forall c' \in B(c, a(L)), \sigma^0_{c'} = 0\right).$$

Each ball $B(c, a(L))$ contains more than $2a(L)^2$ cells, it is full of 0's with probability less than $(1/2)^{2a(L)^2}$. We get (we skip integer parts once more)

$$\mathbb{P}(A) \leq \text{card}(\Lambda)(1/2)^{2a(L)^2} \leq L^2(1/2)^{2a(L)^2} \leq \exp\left(2\log L - 2\times 3^2 \log(L)\log 2\right) \leq 1/L = 1/\sqrt{n}$$

(for large n). Let us write

$$\begin{aligned}
\frac{1}{2^n}\sum_{\sigma^0}\mathbb{E}[T(\sigma^0)] &= \frac{1}{2^n}\sum_{\substack{\sigma^0 \text{ such that}\\ A \text{ is true}}}\mathbb{E}[T(\sigma^0)] + \frac{1}{2^n}\sum_{\substack{\sigma^0 \text{ such that}\\ A \text{ is false}}}\mathbb{E}[T(\sigma^0)],\\
&\leq \mathbb{P}(A)\max_{\substack{\sigma^0 \text{ such that}\\ A \text{ is true}}}\mathbb{E}[T(\sigma^0)] + \mathbb{P}(\text{not } A)\max_{\substack{\sigma^0 \text{ such that}\\ A \text{ is false}}}\mathbb{E}[T(\sigma^0)],\\
&\leq 1/\sqrt{n}\max_{\substack{\sigma^0 \text{ such that}\\ A \text{ is true}}}\mathbb{E}[T(\sigma^0)] + 1\times\max_{\substack{\sigma^0 \text{ such that}\\ A \text{ is false}}}\mathbb{E}[T(\sigma^0)]. \qquad (12.4)
\end{aligned}$$

We bound both terms:

- $\max_{\substack{\sigma^0 \text{ such that}\\ A \text{ is true}}}\mathbb{E}[T(\sigma^0)]$ is a $\mathcal{O}(\sqrt{n})$ by the upper bound of the WECT;
- If A is false, then every 0 is less than $a(L)$ away from a 1. The configuration has thus converged before the time at which each cell has been updated $a(L)$ times. By (12.3), it takes less than $2\log(n)/\alpha$ in average to update the n cells at least once. Then

$$\max_{\sigma^0; A \text{ is false}}\mathbb{E}[T(\sigma^0)] \leq a(L)\frac{2}{\alpha}\log(n).$$

And (12.4) yields

$$\text{Typ}_n \leq c^{\text{st}}\sqrt{n}\frac{1}{\sqrt{n}} + 3\sqrt{\log(\sqrt{n})}\frac{2}{\alpha}\log(n) \leq \frac{6}{\alpha}\log(n)^{3/2},$$

for large enough n.

Remark 2 It is in fact possible (but tedious) using (12.2) to improve the upper bound from $\mathcal{O}(\log(n)^{3/2})$ to $\mathcal{O}(\log(n))$. The idea is that a ball of 0's of radius $\log(L)$ is filled with 1's in less than $\mathcal{O}(\log(L))$ time steps (with high probability).

Discussion

Our aim here was to present with self-contained proofs some quantitative results that did not seem to appear in literature. It is worth noting that many natural questions still remain open: in particular, the order of magnitude of the variance of Variance$(T_n(\sigma^{\text{w}})$ is still unknown.

Fig. 12.3 The Eden model with $\alpha = 0.02$, at different times up to 10^6

12.3 Epidemic in $\mathbb{Z}^2$: The Eden Model

We now consider the analogous of Epidemic on the infinite lattice $\mathbb{Z}^2$, it is usually referred to as the *Eden model* [7] or *Richardson's model* [17]. Let $\alpha > 0$, we consider the stochastic process $\left(\sigma^t\right)_{t\geq 0}$ with values in $\{\mathtt{0}, \mathtt{1}\}^{\mathbb{Z}^2}$ defined as follows:

- $\sigma^0_{\mathbf{0}} = \mathtt{1}$ and $\sigma^0_c = \mathtt{0}$ for $c \neq \mathbf{0}$, where $\mathbf{0}$ is the origin of $\mathbb{Z}^2$.
- At time $t + 1$, each $\mathtt{0}$ that has a neighbour in state $\mathtt{1}$ in σ^t turns into $\mathtt{1}$ with probability α, independently from the past and the other cells.

We are interested in the asymptotic behaviour $\left(\sigma^t\right)_{t\geq 0}$. We observe on simulations (Fig. 12.3) that the component of $\mathtt{1}$'s seems to grow like a particular shape. Richardson [17] has proved that it has indeed a *limiting shape*, in the following sense (Fig. 12.3).

Theorem 3 (Limiting shape theorem for the Eden model) *Let B_t be the set of cells in state $\mathtt{1}$ at time t. There is a non-random set $B_\star \subset \mathbb{R}^2$ which is compact, convex and non void such that for every $\varepsilon > 0$,*

$$\mathbb{P}\left(B_\star(1-\varepsilon) \subset \frac{B_t}{t} \subset B_\star(1+\varepsilon)\right) \overset{t\to+\infty}{\to} 1.$$

This result was further improved by [4] into an almost-sure convergence (Fig. 12.4).

12.3.1 The Link with First-Passage Percolation

The Eden model is a dynamical model of growth process but in fact it can be seen as a static model. To do so, set as before, for $c' \in \mathbb{Z}^2$,

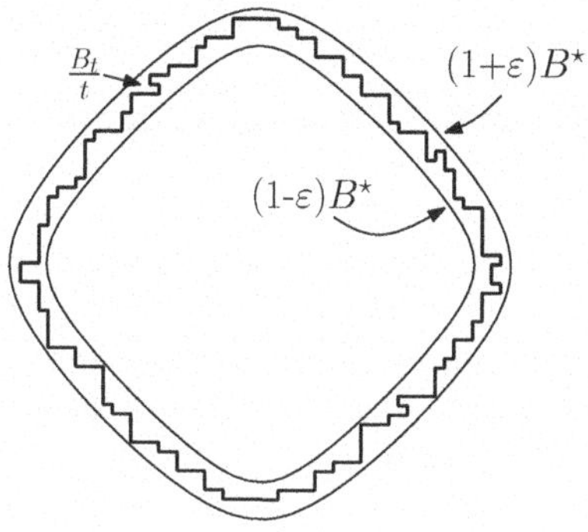

Fig. 12.4 What Theorem 3 says, for large t

$$\tau_{c'} = \min\left\{t \geq 0,\ \sigma^t_{c'} = 1\right\}.$$

As in the previous section, $\tau_{c'} \leq k$ if and only if there is a path $\mathcal{P}$ of $j \leq k$ neighbouring cells going from $\mathbf{0}$ to c'

$$\mathcal{P} = \left\{\mathbf{0} \to c_1 \to c_2 \to \cdots \to c_j = c'\right\}$$

such that successive updating along $\mathcal{P}$ turn c' into 1. Then one can show that, for each fixed c',

$$\tau_{c'} \overset{\text{(d)}}{=} \min_{\mathcal{P}:\mathbf{0}\to c'} \sum_{i=1}^{|\mathcal{P}|} g_{c_i}, \tag{12.5}$$

where $\left\{g_c,\ c \in \mathbb{Z}^2\right\}$ is a family of i.i.d. geometric random variables with mean $1/\alpha$ and $\overset{\text{(d)}}{=}$ means "are equal in distribution". Here, g_{c_i} is the time needed to update c_i, once one of its neighbours is 1.

Thus, a way to construct $\tau_{c'}$ is to draw for each cell in $\mathbb{Z}^2$ some independent random times g_{c_i}, and then $\tau_{c'}$ is the sum of these times over the path $\mathbf{0} \to c'$ such that the sum is minimal. This model is known as *First-passage percolation* (FPP) and has been studied for the first time by Hammersley and Welsh [13]. We refer to [2] for a modern introduction to FPP and its connections with growth models.

The full connection between $\tau_{c'}$'s in the Eden model and first-passage percolation can be written as follows:

Proposition 1 (Eden model is FPP) *Let $\left\{g_c,\ c \in \mathbb{Z}^2\right\}$ be a family of i.i.d. geometric random variables with mean $1/\alpha$. Then*

$$\left\{\tau_c\right\}_{c\in\mathbb{Z}^2} \overset{(d)}{=} \left\{\min_{\mathcal{P}:\mathbf{0}\to c} \sum_{i=1}^{|\mathcal{P}|} g_{c_i}\right\}_{c\in\mathbb{Z}^2}.$$

where the min *is taken over all paths going from* $\mathbf{0}$ *to* c*:*

$$\mathcal{P} = \left(\mathbf{0} = c_0 \to c_1 \to \dots c_{|\mathcal{P}|} = c\right)$$

and $|\mathcal{P}|$ *is the number of cells of* $\mathcal{P}$.

The connection between Eden model and FPP is usually attributed to Richardson, even if first-passage percolation is not clearly mentioned in [17]. Surprisingly enough, it seems that there is no rigorous proof of Proposition 1 available in literature. It is often considered as folklore, but it is not so easy to write down a complete proof (the main difficulty is to establish the equality for the whole family of τ_c's and not only for a fixed c).

12.3.2 Influence of the Lattice

The Eden model being a toy model for propagation, one might wonder if the properties proved in this particular model are *robust* under various perturbations of the lattice. This question is not clearly understood.

Since [17], the following conjecture is attributed to Eden:

Conjecture 1 (Eden conjecture) For the Eden model in continuous time, the set of cells in state 1 is asymptotically shaped as a disc: $B_\star$ is a euclidian ball.

The Eden model in continuous time is defined as before, except that the updating times are exponentially distributed. There are some simulations in [7], but the conclusions are not so clear:

As yet the samples of configurations computed in this way appear to be too few to justify anything more than a few qualitative statements. It is to be seen that the colony is essentially circular in outline.

In 1984, H.Kesten had the intuition that this conjecture should be false, at least in high dimensions, for geometrical reasons. He disproved the conjecture for $d > 600000$ (see [15]), since then [6] and [3] improved the result up to dimension $d > 35$.

Of course, we are far from a physical or biological model, yet this result says something interesting: the asymptotic properties of the Eden model strongly depend on the lattice on which it is constructed. We know that this is not the case for the position of a standard random walk on a regular lattice, whose asymptotic law does not depend on the lattice and is the normal distribution. It seems that the Eden model is sensible to the microscopic structure of the lattice.

12.3.3 Eden Model and Random Defaults

What happens in Theorem 3 if some proportion of cells is *immunised* against infection? How does it change the growth of B_t?

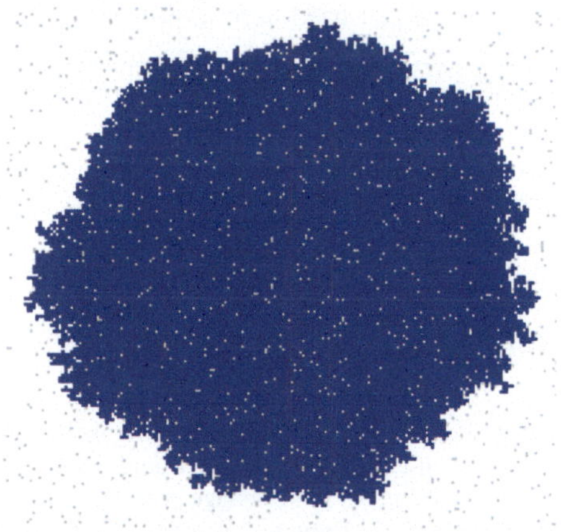

Fig. 12.5 The Eden model with immunised cells (*grey*)

Assume that each cell is originally immunised independently with probability p (Fig. 12.5). Obviously B_t cannot grow infinitely if the following event E occurs:

- either $\mathbf{0}$ is immunised,
- or there is a path of immunised cells surrounding $\mathbf{0}$.

Of course, $\mathbb{P}(E) \geq \mathbb{P}(\mathbf{0} \text{ is immunised}) \geq p > 0$ but we can prove (see [12] for instance) that if p is small enough then $\mathbb{P}(E) < 1$. In the case where E does not occur, it is possible than B_t grow infinitely and the growth is linear, as in the initial model. This has been proved rigorously by [10], we need a few notations to state the result.

Let $\mathbf{n}$ be the cell of coordinates $(n, 0)$. Let $\mathcal{A}$ be the (random) set of integers n such that there is a path of non-immunised cells going from the origin to $\mathbf{n}$. If E does not occur then there are infinitely many cells than can be infected and if $n \in \mathcal{A}$, the first time $\tau_{\mathbf{n}}$ at which $\mathbf{n}$ is in B_t is finite (almost surely).

Theorem 4 (Linear growth for the Eden model with defaults) *There exists* $\mu > 0$ *such that, if E does not occur,*

$$\lim_{\substack{n\to+\infty,\\ n\in\mathcal{A}}} \frac{\tau_{\mathbf{n}}}{n} = \mu \quad \textit{almost surely},$$

where the limit is taken along the random subsequence $\{n \in \mathcal{A}\}$.

Discussion

Much is known now about the quantitative properties of the Eden model. In particular, many efforts have been made in order to understand the dependence of the limiting shape with respect to the different parameters of the model: dependence with respect to α [5, 16] and to p [1].

We have just tried here to present a few results for non-probabilists, we refer to [2] for a nice and modern introduction to this topic. In particular, it is discussed of the variant in which there is a *competition* between two epidemics.

Acknowledgements I am grateful to two anonymous referees for their careful reading of the first version of this article.

References

1. Basdevant, A.L., Enriquez, N., Gerin, L., Gouéré, J.B.: The shape of large balls in highly supercritical percolation. Electron. J. Probab. **19**, 1–14 (2014)
2. Blair-Stahn, N.: First passage percolation and competition models (2010). arXiv:1005.0649
3. Couronné, O., Enriquez, N., Gerin, L.: Construction of a short path in high dimensional first-passage percolation. Electron. Commun. Probab. **16**, 22–28 (2011)
4. Cox, J., Durrett, R.: Some limit theorems for percolation processes with necessary and sufficient conditions. Ann. Probab. **9**(4), 583–603 (1981)
5. Cox, J., Kesten, H.: On the continuity of the time constant of first-passage parcolation. J. Appl. Probab. **18**, 809–819 (1981)
6. Dhar, D.: First passage percolation in many dimensions. Phys. Lett. A **130**, 308–310 (1988)
7. Eden, M.: A two-dimensional growth process. In: Proceedings of the Fourth Berkeley Symposium on Mathematical Statistics and Probability, vol. IV, pp. 223–239. University of California Press, Berkeley, Calif (1961)
8. Fatès, N., Gerin, L.: Examples of fast and slow convergence of 2d asynchronous cellular systems. J. Cell. Autom **4**, 323–337 (2009)
9. Fatès, N., Morvan, M., Schabanel, N., Thierry, E.: Fully asynchronous behavior of double-quiescent elementary cellular automata. Theoretical Comput. Sci. **362**, 1–16 (2006)
10. Garet, O., Marchand, R.: Asymptotic shape for the chemical distance and first-passage percolation on the infinite Bernoulli cluster. ESAIM Probab. Stat. **8**, 169–199 (2004) (electronic)
11. Gerin, L.: Aspects probabilistes des automates cellulaires, et d'autres problèmes en informatique théorique. (2008). Thèse de l'Université Nancy 1
12. Grimmett, G.: Percolation, vol. 321, 2nd edn. Springer, Berlin (1999)
13. Hammersley, J.M., Welsh, D.J.A.: First-passage percolation, subadditive processes, stochastic networks, and generalized renewal theory. In: Proceedings of the International Research Seminar Statistical Laboratory, University of California, Berkeley, Calif pp. 61–110. Springer, New York (1965)
14. Janson, S., Łuczak, T., Rucinski, A.: RanDom Graphs. In: Wiley-Interscience Series in Discrete Mathematics and Optimization. Wiley-Interscience, New York (2000)
15. Kesten, K.: Aspects of first passage percolation. In: École d'été de probabilités de Saint-Flour, XIV—1984, Lecture Notes in Mathematics, vol. 1180, pp. 125–264. Springer, Berlin (1986)
16. Marchand, R.: Strict inequalities for the time constant in first passage percolation. Ann. Appl. Probab. **12**, 1001–1038 (2002)
17. Richardson, D.: Random growth in a tessellation. In: Proceedings of the Cambridge Philosophical Society **74**, 515–528 (1973)
18. Szpankowski, W., Rego, V.: Yet another application of a binomial recurrence order statistics. Computing **43**, 401–410 (1990)

Chapter 13
Convergence Time of Probabilistic Cellular Automata on the Torus

Lorenzo Taggi

Abstract Many probabilistic cellular automata (PCA) exhibit a transition from an ergodic to a non-ergodic regime. Namely, if the free parameter is above a certain critical threshold, the process converges to a state that does not depend on the initial state (ergodicity), whereas if the free parameter is below the threshold, then the process converges to a state that depends on the initial state (non-ergodicity). If one considers the corresponding model on a finite space, such a transition is not observed (the process is always ergodic), nevertheless the convergence time is "small" when the corresponding process on infinite space is ergodic and "large" when the corresponding process on infinite space is non-ergodic. We analyse this correspondence for Percolation PCA, a class of probabilistic cellular automata which are closely related to oriented percolation.

13.1 Introduction and Overview

Probabilistic cellular automata (PCA) are discrete-time Markov processes modelling the evolution of a multicomponent system. Every component has a finite number of states, and it interacts with its neighbours according to a certain probabilistic rule. States of components are updated synchronously. This is the main difference from interacting particle systems, where different components are updated at different times.

There are several reasons to consider PCA. One of them is that they appear in many contexts of applied sciences, e.g. biology, economics, population dynamics, as models of systems, whose evolution is characterized by local interactions between their components (see e.g. [1, 10, 25, 30, 46]). Another reason is that, in the context of the classification of (deterministic) cellular automata (Wolfram's program), robustness to random errors can be used as a discriminating criterion [17].

L. Taggi (✉)
Max Planck Institute for Mathematics in the Sciences, Inselstrasse 22, Leipzig, Germany
e-mail: taggi@mis.mpg.de

L. Taggi
TU Darmstadt, Darmstadt, Deutschland

P.-Y. Louis and F.R. Nardi (eds.), *Probabilistic Cellular Automata*, Emergence, Complexity and Computation 27, https://doi.org/10.1007/978-3-319-65558-1_13

In this chapter, we consider PCA from the perspective of non-equilibrium statistical mechanics. Probabilistic cellular automata are relevant models in statistical physics as, together with interacting particle systems, they constitute favourable models to study non-equilibrium phenomena. Indeed, on the one hand their definition is simple, as space of configurations is discrete and interactions are local. On the other hand, they exhibit very complex phenomena. One example of such phenomena is the transition from an *ergodic* regime to a *non-ergodic* regime. Such transition is also often referred to as a *phase transition*, as it occurs *abruptly* as a free parameter is tuned above or below a certain critical value. More precisely, we say that a PCA is *ergodic* if it admits a unique invariant probability measure and if such probability measure is attractive. Namely, the process converges to a measure that does not depend on the initial state. Alternatively, if more than one invariant measure exist, the process is non-ergodic. This means that the long-time limit state of the dynamics preserves part of the information on the initial state.

Some of most challenging mathematical problems about PCA involve the proof of occurrence of a phase transition (see [29, 45] for extensive surveys and [18, 35] for some interesting examples), the analysis of the long-time limit probability measure of the process (see e.g. [9, 11, 12, 14, 21, 26, 34]), and the study of the convergence behaviour to the invariant measure (see e.g. [5, 11, 12, 28, 34, 39]). Typical questions addressed by mathematicians working in probability involve the estimation of the probability of certain events in the long-time limit [6, 8, 11, 21, 31, 34], the comparison between the asymptotic probability measure and the Gibbs measure [12, 21], and the estimation of the speed of convergence to the asymptotic state [5, 12, 39]. Another relevant problem is to understand how asynchrony in the updating scheme affects the dynamics of the process. A variation of the updating rule can affect in a non-trivial way the long-time behaviour of the system and induce a phase transition (see [20] for a survey on asynchronous cellular automata and [17] for extensive experimental studies). Despite PCA have been intensively studied in the last 50 years, many problems have remained open for long time. For example, a proof of the existence of a phase transition for the symmetric majority model in two dimensions, which is supported by extensive numerical investigations [24], is still missing (see e.g. [2, 3, 22] for some partial results). A list of open problems can be found in the surveys [42, 45]. Only few of them have been solved since the publication of these works (e.g. [7, 39]).

The problem we consider in this chapter involves the convergence time of probabilistic cellular automata on the finite lattice. While many probabilistic cellular automata on infinite lattice show a transition between ergodicity and non-ergodicity, the corresponding processes on finite lattice are always ergodic. Nevertheless, the process on finite space exhibits different convergence regimes providing information on the behaviour of the corresponding model on infinite lattice. Indeed, it is conjectured that the convergence time to the asymptotic state is *small* when the model on infinite space is ergodic and *large* otherwise, where "small" and "large" have a precise mathematical formulation. This means that the transition point between the two convergence regimes on finite space coincides precisely with the transition point between ergodic and non-ergodic regime on infinite space. This conjecture corresponds to *Unsolved Problem* 3.5.1 in [42] in the context of probabilistic cellular automata. This problem is important, as computer simulations refer directly to finite systems. Hence, whenever we interpret results of numerical simulations as telling us something about ergodicity and non-ergodicity of infinite systems, we need to know how the transition points are related for the model on finite and infinite lattice.

We address this problem from a mathematically rigorous point of view by considering a class of PCA that in [42] is referred to as *Percolation Systems*. In this chapter we call them *Percolation PCA*, the same as in [39]. For these systems, the above-mentioned conjecture has been proved in [38–40]. In Percolation PCA, the state of every component is synchronously updated according to the following rule: if the state of all sites in the neighbourhood of the component is one, then the state at the next time step is one with probability one; whereas, if the state of some neighbour is zero, then the state at the next time step is one with probability ϵ. It is well known [38, 40, 45] that for every finite neighbourhood there exists a unique, positive critical value ϵ_c such that the process is ergodic and converges to the state "all ones" for every $\epsilon > \epsilon_c$ and it admits an infinite number of invariant measures for every $\epsilon < \epsilon_c$. This class of processes undergo a so called *absorbing-state phase transition* (see [23] for a wide overview). This means that they undergo a transition from a regime of almost sure convergence to an *absorbing state* (a realization that cannot be left once it is reached) to a fluctuating regime. On a finite lattice, the process reaches the absorbing state with probability one for all parameter values.

The main problem discussed in this chapter involves the average time the process takes to reach the absorbing state. The average absorbing time grows logarithmically (resp. exponentially) with the system size when the process on infinite space is ergodic (resp. non-ergodic). This fact has been proved in [38–40]. In [38, 40], the existence of the two convergence regimes has been proved for ϵ small and large enough, respectively. In [39], the existence of such regimes has been proved for all ϵ. This chapter is an introduction to the mathematical methods employed for the proof of such convergence regimes in [38–40]. Some related open problems are discussed as well. One of the reasons to consider Percolation PCA is that, as the name suggests, they can be interpreted as a kind of site percolation, where the parameter $1 - \epsilon$ plays the same role of p in percolation. Hence, mathematical techniques developed for the study of percolation models can be employed for the analysis of the transition from ergodic to non-ergodic regime in Percolation PCA. The same methods do not naturally extend to the study of other classes of probabilistic cellular automata.

Percolation PCA belong to a class of models introduced by Toom [41]. These models are defined as a random perturbation of monotone (deterministic) cellular automata presenting an "erosion property". Namely, these cellular automata erase in a finite time any finite island of impurities in a predominantly homogeneous configuration. A recent extensive overview on these models can be found in the Ph.D. thesis of Ponselet [34]. Results proved for the class of models discussed in [34] hold also for Percolation PCA. The high-noise regime of these models is quite well understood. The noise weakens the interaction between neighbouring components, and this leads to an exponential convergence speed to the unique invariant measure and to an exponential decay of correlations [26]. The low-noise regime is still open to investigations. See [4, 13, 30, 48] for numerical results and [5, 11, 12, 21] for theoretical studies. One of the most studied Percolation PCA is the *Stavskaya's process*. This process has been introduced by Stavskaya and Piatetski-Shapiro in [38] more than half-century ago and since that time it has been intensively studied (see e.g. [11, 12, 32, 36, 37, 42, 43, 45]).

We end this introductory section presenting the structure of the chapter. In Sect. 13.2, we define probabilistic cellular automata and Percolation PCA. In Sect. 13.3, we describe the properties of Percolation PCA. In particular, in Sect. 13.3.1 we discuss monotonicity with respect to the noise parameter and in Sect. 13.3.2 we discuss percolation properties. In Sect. 13.4 we study the convergence time of Percolation PCA presenting the mathematical techniques employed in [38–40, 45].

13.2 Probabilistic Cellular Automata

A probabilistic cellular automaton (PCA) is a discrete-time Markov process on a discrete (finite or countable infinite) realization space. A realization of the process is a vector, whose components correspond to the states of cells located on a grid S, that we call *space*. In this chapter we consider $S \in \{\mathbb{Z}_n, \mathbb{Z}\}$, where $\mathbb{Z}_n$ is the one-dimensional torus of size n. We consider the case of binary PCA, namely the states of cells can be 0 or 1. Every realization of the process corresponds to an element of the *space of realizations* $\Sigma = \{0, 1\}^S$. We denote realizations by $\eta \in \Sigma$. For every site $x \in S$ and every subset $\Lambda \subset S$, we denote by η_x the state of the cell located at $x \in S$ and by η_Λ the set of states of the cells located on the sites of $\Lambda \subset S$.

The range of interaction between cells is finite and it is defined by a *neighbourhood* $\mathcal{U}$,

$$\mathcal{U} = \{s_1, s_2, \ldots, s_k\} , \tag{13.1}$$

where s_1, s_2, ..., s_k are distinct elements of S and k is a finite integer. For every $x \in S$, we define the neighbours of x as the set of sites,

$$\mathcal{U}(x) := \mathcal{U} + x , \tag{13.2}$$

that means assuming translation invariance of the neighbourhood. We also define the neighbourhood of a set $\Lambda \subset S$ as $\mathcal{U}(\Lambda) = \bigcup_{x \in \Lambda} \mathcal{U}(x)$.

The evolution of the process is defined by probabilistic interaction rules that involve neighbour cells. These are represented by the *transition probabilities*,

$$\{ \ p(a \,|\, \xi) \ \}_{a \in \{0,1\}, \xi \in \{0,1\}^{\mathcal{U}}} , \tag{13.3}$$

Given that $p(\,\cdot \,|\, \xi\,)$ is a probability, the normalization condition holds, i.e. $p(1 \,|\, \xi\,) + p(0 \,|\, \xi\,) = 1$. In the specific case of Percolation PCA, the transition probabilities are,

$$p(1 \mid \xi) = \begin{cases} 1 & \text{if } \xi = \mathbf{1}_{\mathcal{U}} \\ \epsilon & \text{otherwise,} \end{cases} \tag{13.4}$$

where $\mathbf{1}_{\mathcal{U}} \in \{0, 1\}^{\mathcal{U}}$ is the realization with all states equal to "one".

Hence, we introduced all the ingredients necessary to define a stochastic dynamics in the product space Σ. Such a dynamics can be represented by introducing a *transfer operator* $\mathcal{P} : \mathcal{M}(\Sigma) \to \mathcal{M}(\Sigma)$, which acts on the space of probability measures of Σ.

We denote by $\mu\mathcal{P}$ the probability measure obtained by applying $\mathcal{P}$ to $\mu \in \mathcal{M}(\Sigma)$. More formally, denoting by $C_{\eta'_K}$ the cylinder set $C_{\eta'_K} = \{\eta \in \Sigma : \eta_K = \eta'_K\}$, with $K \subset S$, the probability of the set $C_{\eta'_K}$ is defined as,

$$\mu\mathcal{P}(C_{\eta'_K}) = \sum_{\eta_{\mathcal{U}(K)} \in \{0,1\}^{\mathcal{U}(K)}} \mu(C_{\eta_{\mathcal{U}(K)}}) \prod_{x \in K} p(\eta'_x \mid \eta_{\mathcal{U}(x)}). \tag{13.5}$$

As on the right-hand side the product over all sites appear, conditioning on the realization at time $t-1$, realizations of cells at time t are independent one from the other. From a mathematically rigorous point of view, Eq. (13.5) does not define the probability measure $\mu\mathcal{P}$ on the whole σ-algebra generated by subsets of Σ, but only on cylinder sets. A well known theorem in measure theory, the Kolmogorov Extension Theorem (see e.g. [15, p. 410]), ensures that this is enough, as there is a unique and well defined extension of such a measure to the whole sigma-algebra.

Space-Time Evolution

In order to characterize the time evolution of PCA it is useful to introduce the set of *space-time realizations*, $\tilde{\Sigma} = \{0, 1\}^V$, where $V = S \times \mathbb{N}$ is the *space-time* set. The elements of $\tilde{\Sigma}$ are the realizations of the process at all times, $\tilde{\eta} = (\eta^t)_{t=0}^{\infty} \in \tilde{\Sigma}$. We shall then introduce a directed graph $\mathcal{G}_{\mathcal{U}} = (V, \mathbf{E}_{\mathcal{U}})$, whose edges connect any vertex $(x, y) \in V$ to the vertices $(z, y-1) \in V$, where $z \in \mathcal{U}(x)$, and they are oriented. See Fig. 13.1 for an example. The vertices reachable from $(x, y) \in V$ through a path on $\mathcal{G}_{\mathcal{U}}$ are those which can influence the state of the variable η^t_s: denoting by $\mathcal{U}^{\ell}(s)$, $s \in S, t \in \mathbb{N}$,

$$\mathcal{U}^{\ell}(s) = \overbrace{\mathcal{U} \circ \mathcal{U} \circ \cdots \circ \mathcal{U}}^{\ell}(s),$$

any η^t_s depends in a probabilistic way on $\eta^{t-1}_{\mathcal{U}(s)}$, which depends on $\eta^{t-2}_{\mathcal{U}^2(s)}$, and so on. We end this section providing some definitions.

Definition 1 (*Evolution Cone*) Consider a vertex $(y, t) \in V$. The set of vertices (x, z), such that $z \in \{0, 1, \ldots, t\}$ and $x \in \mathcal{U}^{t-z}(y)$ constitute the *evolution cone* of (y, t). See Fig. 13.1 for an example.

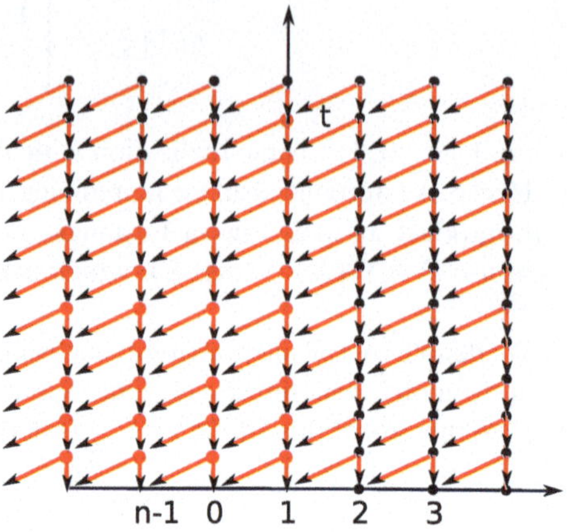

Fig. 13.1 Representation of (part of) the graph $\mathcal{G}_{\mathcal{U}}$ in the case of space $S = \mathbb{Z}_n$ and of neighbourhood $\mathcal{U} = \{-1, 0\}$. In the example, vertices coloured by *red* belong to the evolution cone of $(1, t)$

Definition 2 (*Space line*) We define for any $t \in \mathbb{N}$ the space line set $S^t = \{(s, t) \in V,\ s \in S\}$, which is a subset of V.

The next definition will be employed in Sect. 13.4.

Definition 3 (*Evolution measure*) Consider the space of probability measures on the evolution space $\mathcal{M}(\tilde{\Sigma})$, where $\tilde{\Sigma}$ is the set of space-time realizations previously defined. We define the evolution measure $\tilde{\mu} \in \mathcal{M}(\tilde{\Sigma})$ as the joint probability distribution of

$$\delta_{\mathbf{0}},\ \delta_{\mathbf{0}}\mathcal{P},\ \delta_{\mathbf{0}}\mathcal{P}^2,\ \delta_{\mathbf{0}}\mathcal{P}^3, \ldots$$

Namely, the evolution measure $\tilde{\mu}$ measures the probability of events in the space-time realization set in the case of Percolation PCA starting from the initial state "all zeros".

Definition 4 (*Expectation*) We define $\mathbb{E}(\,\cdot\,)$ as the expectation with respect to the evolution measure $\tilde{\mu}$.

Observe that all the previous definitions are consistent both in the case of infinite space and in the case of finite space with periodic boundary conditions. So far, we denoted the space by S and we did not specify if $S = \mathbb{Z}$ or $S = \mathbb{Z}_n$.

13.3 Phase Transition of Percolation PCA

In this section, we discuss two properties of Percolation PCA: monotonicity with respect to the noise and its interpretation as an oriented percolation model.

13.3.1 Monotonicity

The operator $\mathcal{P}$ of Percolation PCA is *monotone*. Monotonicity of $\mathcal{P}$ means that it preserves partial order among elements of $\mathcal{M}(\Sigma)$. We first introduce partial order "$\prec$" in Σ defining for any two configurations $\eta, \eta' \in \Sigma, \eta \prec \eta' \Leftrightarrow \forall s \in S\ \eta_s \leq \eta'_s$. Hence, we introduce the functions $\varphi : \Sigma \longmapsto \mathbb{R}$ dependent only on a finite number of sites. We call φ *monotone iff* for any $\eta, \eta' \in \Sigma,\ \eta \prec \eta' \Rightarrow \varphi(\eta) \leq \varphi(\eta')$. Then, we introduce partial order in $\mathcal{M}(\Sigma)$ defining $\mu \prec \mu' \Leftrightarrow$ for any *monotone* function φ, $\int \varphi\, d\mu \leq \int \varphi\, d\mu'$. Observe that any measure $\mu \in \mathcal{M}(\Sigma)$ satisfies,

$$\delta_{\mathbf{0}} \prec \mu \prec \delta_{\mathbf{1}}, \tag{13.6}$$

where $\delta_{\mathbf{0}}$ (resp. $\delta_{\mathbf{1}}$) is the probability measure that assigns probability one to the realization "all zeroes" (resp. "all ones").

Definition 5 (*Monotone Operator*) An operator $\mathcal{P} : \mathcal{M}(\Sigma) \longmapsto \mathcal{M}(\Sigma)$ is called monotone if for any pair of measures $\mu, \mu' \in \mathcal{M}(\Sigma)$, $\mu \prec \mu' \Rightarrow \mu\mathcal{P} \prec \mu'\mathcal{P}$.

For the transition operator of the Percolation PCA, this property is a consequence of the fact that the transition probability (13.4) preserves order locally, i.e. for any $\xi_1, \xi_2 \in \{0, 1\}^{\mathcal{U}}$,

$$\xi_1 \prec \xi_2 \ \Rightarrow\ p(1 \mid \xi_1) \leq p(1 \mid \xi_2), \tag{13.7}$$

(see for example [45, p. 28] for a proof). Monotonicity of $\mathcal{P}$ implies that the probability measure,

$$\nu_\epsilon := \lim_{t \to \infty} \delta_{\mathbf{0}}\, \mathcal{P}^t, \tag{13.8}$$

is well defined.

Definition 6 (*Critical Noise*) We define the critical noise for Percolation PCA as,

$$\epsilon_c(\mathcal{U}) = \inf_{\epsilon \in [0,1]} \{\nu_\epsilon = \delta_{\mathbf{1}}\}. \tag{13.9}$$

It is immediate to see that the Dirac measure $\delta_{\mathbf{1}}$, where $\mathbf{1} = (1, 1, 1, \ldots)$, is stationary for all $\epsilon \in [0, 1]$, i.e. $\delta_{\mathbf{1}} = \delta_{\mathbf{1}}\mathcal{P}$. By definition, for any $\epsilon > \epsilon_c$ the process is ergodic, i.e. there exists a unique stationary measure, which is attractive. In fact, if $\nu_\epsilon = \delta_{\mathbf{1}}$, then by monotonicity and by (13.6) it follows that for every $\mu \in \mathcal{M}(\Sigma)$, $\lim_{t \to \infty} \mu\, \mathcal{P}^t = \delta_{\mathbf{1}}$. Alternatively, if $\epsilon < \epsilon_c$, then $\nu_\epsilon \neq \delta_{\mathbf{1}}$. This means that the process admits at least two invariant probability measures, $\delta_{\mathbf{1}}$ and ν_ϵ. As any convex combination of two invariant probability measures is still an invariant probability measure (see e.g. [34, p. 22]), it follows that if $\epsilon < \epsilon_c$, the process admits an infinite number of invariant probability measures i.e. for all $\alpha_1, \alpha_2 \in [0, 1]$ such that $\alpha_1 + \alpha_2 = 1$, the probability measure $\alpha_1\nu_\epsilon + \alpha_2\delta_{\mathbf{1}}$ is invariant. Further results by Vaserstein and Leontovitch [47] state that, for all $\epsilon < \epsilon_c$, all the homogeneous invariant measures correspond to any of such convex combinations.

Fig. 13.2 Critical noise values ϵ_c for Percolation PCA with neighbourhood $\mathcal{U} = \{0, 1, 2, \ldots, k\}$

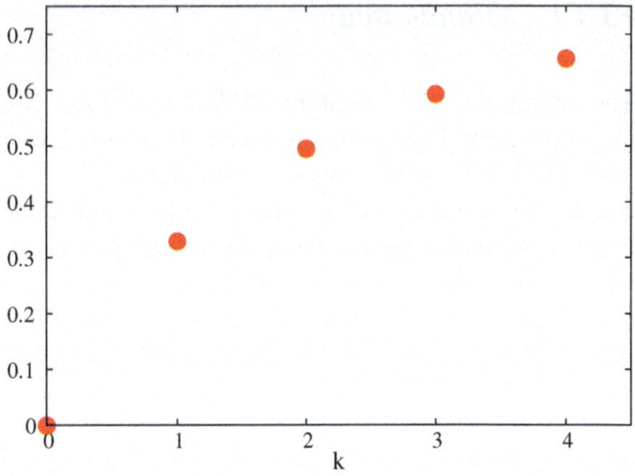

The original proof of the fact that $\epsilon_c(\mathcal{U}) \in (0, 1)$ for Percolation PCA on the infinite one-dimensional lattice with finite neighbourhood and such that $|\mathcal{U}| \geq 2$ is due to [36]. A different proof, based on the counting path method and on the Peierls argument, can be found in [38, 40]. The value of ϵ_c is not known exactly, and it depends on the neighbourhood. In the case of one dimension and of neighbourhood $\mathcal{U} = \{-1, 0\}$, the analytical estimation $0.09 < \epsilon_c < 0.323$ is due to [40]. A sharp numerical estimation based on a Monte Carlo simulation is provided by [32], $\epsilon_c = 0.29450(5)$. In the case of neighbourhood $\{-1, 0, 1\}$, the numerical estimation $\epsilon_c = 0.462$ comes from [33, 39]. In Fig. 13.2, numerical estimations of critical noise values for several neighbourhoods are considered. These are provided by [39], together with analytical upper bounds for ϵ_c.

Remark 1 Consider Percolation PCA with neighbourhood

$$\mathcal{U} = \{-1 + k, k, 1 + k\}.$$

It is possible to prove (or to verify by means of numerical simulations) that the critical noise value for the corresponding Percolation PCA is the same for all $k \in \mathbb{N}$. Namely, the critical noise value is invariant under translation. Consider Percolation PCA with neighbourhood

$$\mathcal{U} = \{-a, 0, a\},$$

where a is a positive integer. Again, it is possible to prove that the critical noise value for the corresponding Percolation PCA is the same for all $a \in \mathbb{N}$. Namely, a change of scale does not affect the critical noise. On the contrary, the critical noise value of Percolation PCA with neighbourhood $\mathcal{U} = \{-1, 0, 2\}$ is different from the one with neighbourhood $\mathcal{U} = \{-1, 0, 1\}$. Hence, critical thresholds do not simply depend on the cardinality of the neighbourhood, but they also depend on its structure and they are invariant under certain transformations. Indeed, the probability law of the state of a cell x at time t depends on the states of cells in the evolution cone of (x, t) and the structure of such an evolution cone depends on the definition of neighbourhood.

- **Open Problem 1: equivalence classes for neighbourhoods**. Identify all classes of neighbourhoods whose corresponding Percolation PCA have the same critical noise value. Are translation and change of scale the only transformations that preserve the critical noise value?
- **Open Problem 2: order relation between neighbourhoods**. Identify order relations between neighbourhoods. We say that $\mathcal{U} \prec \mathcal{U}'$ if $\epsilon_c(\mathcal{U}) < \epsilon_c(\mathcal{U}')$ and $\mathcal{U} \equiv \mathcal{U}'$ if $\epsilon_c(\mathcal{U}) = \epsilon_c(\mathcal{U}')$. If $\mathcal{U}$ and $\mathcal{U}'$ have the same cardinality, when is $\mathcal{U} \prec \mathcal{U}'$?

Remark 2 As the number n of cells is finite, the model does not undergo any phase transition, i.e. $\epsilon_c = 0$ for all neighbourhoods. Indeed, for every $t \in \mathbb{N}$, the probability of reaching an absorbing state given the previous state is uniformly bounded from below by ϵ^n. This implies that, independently on the initial state of the dynamics, with probability one there exists a finite time τ_n such that at that time the process "falls" in the absorbing state and that the average of τ_n is bounded from above by $\frac{1}{\epsilon}^n$.

13.3.2 Relation with Oriented Percolation

Definition 7 We introduce the probability

$$\Theta(\epsilon) = \nu_\epsilon(\eta_0 = 0),$$

omitting the dependence on the neighbourhood, where ν_ϵ is defined in (13.8) and corresponds to the probability measure reached by the process at infinite time starting from the state "all zeros".

By definition of ϵ_c, on the infinite lattice $\mathbb{Z}$, $\Theta(\epsilon)$ is positive for every $\epsilon < \epsilon_c$ and it is zero for every $\epsilon > \epsilon_c$. On the contrary, on $\mathbb{Z}_n$ $\Theta(\epsilon) = 0$ for all $\epsilon \in [0, 1]$ (recall Remark 2).

The function $\Theta(\epsilon)$ can be interpreted as a percolation probability for the infinite graph $\mathcal{G}_\mathcal{U} = (S \times \mathbb{Z}^+, \mathbf{E}_\mathcal{U})$. Indeed, consider a new model where every vertex of $\mathcal{G}_\mathcal{U}$ is *open* with probability $1 - \epsilon$ and *closed* with probability ϵ independently. Then, $\Theta(\epsilon)$ *equals the limit* $t \longrightarrow \infty$ *of the probability that the vertex* $(0, t)$ *belongs to a path of open vertices connecting* $(0, t)$ *to* S^0 (recall Definition 2). Hence, $1 - \epsilon$ plays the same role played in percolation by the parameter p. If $\epsilon < \epsilon_c$ ($p > p_c$ in percolation) we say that the process is *supercritical*; if $\epsilon > \epsilon_c$ ($p < p_c$ in percolation) we say that the process is *subcritical*. An overview on a similar percolation model can be found in [16].

In this section, we describe the percolation properties of Percolation PCA from a mathematically rigorous point of view, following [39, 40, 44, 45]. We introduce an auxiliary space $\Omega = \{0, 1\}^{S \times \mathbb{N}}$, we denote its elements by ω and we introduce a Bernoulli product measure $\mathbb{P}_\epsilon$. Namely, the state of every component $\omega_{x,y}$, where $(x, y) \in S \times \mathbb{N}$, is 1 with probability ϵ and 0 with probability $1 - \epsilon$ independently.

We represent the Percolation PCA starting from an initial realization $\eta^i \in \Sigma$ by introducing a deterministic mapping

$$D : \Omega \times \Sigma \longrightarrow \tilde{\Sigma}.$$

For every $(x, t) \in V$, the component $D^t_x : \Omega \times \Sigma \to \{0, 1\}$ of D is defined as

$$D^t_x := \begin{cases} \min\{\omega_{x,t-1}, \max_{k \in \mathcal{U}(x)} \{D^{t-1}_k\}\}, & \text{if } t \in \mathbb{Z}_+ \\ \eta^i_x, & \text{if } t = 0, \end{cases} \tag{13.10}$$

where $(\omega_{x,t})_{x \in S, y \in \mathbb{N}}$ are elements of Ω. This mapping defines any $D^T_z, z \in S, T \in \mathbb{Z}_+$, as a function of the variables $\omega_{x,y}$ associated to vertices belonging to the evolution cone of $(z, T) \in V$, and of the initial realization η^i.

Definition 8 (*Open Path*) Consider a path in $\mathcal{G}_\mathcal{U}$, i.e. a sequence of sites (x_0, t), $(x_1, t-1)$, $(x_2, t-2)$, ..., $(x_t, 0)$, such that $x_i \in \mathcal{U}(x_{i-1})$ for all $i \in \{1, 2, \ldots, t\}$. The path is *open* if all its vertices are open.

Definition 9 We denote the event "(x, t) is connected by an open path to some site $(y, 0)$ s.t. $y \in S^0$", by $\{(x, t) \longrightarrow S^0\}$.

The next proposition relates the probability of certain "percolation" events in the auxiliary space with the probability of certain events in the original space. Its proof can be found in [39, Sect. 3.1].

Proposition 1 *Consider Percolation PCA with space $S \in \{\mathbb{Z}_n, \mathbb{Z}\}$, represented by the operator $\mathcal{P} : \mathcal{M}(\Sigma) \to \mathcal{M}(\Sigma)$. Then,*

$$\delta_{\mathbf{0}}\mathcal{P}^t(\eta_x = 0) = \mathbb{P}_\epsilon(\omega \in \Omega \text{ s.t. } D^t_x(\omega, \mathbf{0}) = 0) = \mathbb{P}_\epsilon((x, t) \longrightarrow S^0) \tag{13.11}$$

Remark 3 Definitions 8, 9 and Proposition 1 hold both for finite space with periodic boundaries $\mathbb{Z}_n$ and for infinite space $\mathbb{Z}$. In the case of periodic boundaries, the event

$$\{(x, t) \longrightarrow S^0\}$$

occurs also if the open path connecting the site (x, t) to the line $y = 0$ leaves one of the vertical boundaries ($x = 0$ or $x = n - 1$) from one side and it re-appears at the same height on the other side (see path $a \cup c$ in Fig. 13.3).

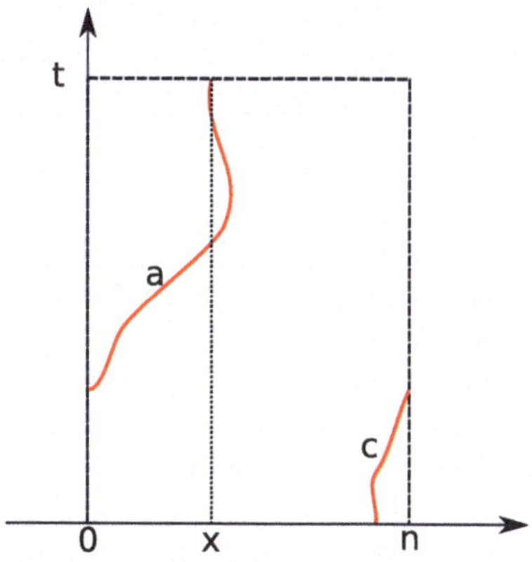

Fig. 13.3 Representation of an open path

13.4 Absorption Time of Percolation PCA on Finite Space

In this section, we consider Percolation PCA on a finite space with periodic boundaries, $S = \mathbb{Z}_n$, as defined in Sect. 13.2. Theorem 1 provides an estimation for the convergence time to the invariant measure as a function of n. We present some of the techniques employed for the proof of this theorem following [40, 45]. We consider the case of $\mathcal{U} = \{-1, 0\}$ (Stavskaya's process), as represented in Fig. 13.1. We refer to [39] for the complete proof in the case of general neighbourhood.

We introduce the average *absorption time* τ_n.

Definition 10 (*Absorption time*) We define the absorption time τ_n of Percolation PCA on space $S = \mathbb{Z}_n$ as the first time the process reaches the configuration *all ones*. Namely, $\tau_n : \tilde{\Sigma} \to \mathbb{N}$ defined as,

$$\tau_n := \min_{t \in \mathbb{N}} \{\forall x \in [0, n-1], \quad \eta_x^t = 1\}. \tag{13.12}$$

The following theorem has been proved in [38–40]. Recall from Definitions 3 and 4 that the average absorption time of Percolation PCA starting from initial state "all zeros" is denoted by $\mathbb{E}[\tau_n]$.

Theorem 1 *Consider Percolation PCA on $\mathbb{Z}_n$ with finite neighbourhood $\mathcal{U}$ and periodic boundary conditions. For every ϵ, there exist some positive constants A, B, C, D, a, b, c, d dependent on ϵ, and an integer $n_0 \in \mathbb{N}$ large enough such that the two following propositions hold for every $n > n_0$.*

(a) *if* $\epsilon > \epsilon_c, \quad A \log(a\,n) < \mathbb{E}[\tau_n] < B \log(b\,n)$,
(b) *if* $\epsilon < \epsilon_c, \quad C \exp(c\,n) < \mathbb{E}[\tau_n] < D \exp(d\,n)$.

The proof of the right side of (b) is trivial (see Remark 2 in Sect. 13.3). The proof of the left side of (a) is simple and does not require techniques from percolation. For an arbitrary integer d and for any n multiple of d, the interval $[0, n]$ is divided into $\frac{n}{d}$ smaller intervals of length d. The probability that all components of a smaller interval have state 1 at time t is bounded from above by $1-(1-\epsilon)^{d\cdot t}$. Hence, one can show that the probability that the process is in the absorbing state at time t is bounded from above by $[1-(1-\epsilon)^{d\cdot t}]^{\frac{n}{d}}$, i.e. the product over the n/d intervals of the probability that the interval is in the absorbing state. From this estimation and from the definition of expectation, one concludes after some computations that $\mathbb{E}[\tau_n]$ is at least $O(log(n))$ for every ϵ. The complete proof can be found in [38, Sect. 2].

In the next pages, we consider the Stavskaya's process (Percolation PCA with neighbourhood $\mathcal{U}=\{0,1\}$). First, we present the proof of the right side of (a) for ϵ close enough to 1. The original proof can be found in [40]. Then, we present the proof of the left side of (b) for ϵ close enough to 0. The original proof can be found in [38]. The complete proof of the right side of (a) and of the left side of (b) can be found in [39]. We denote by $\tilde{\mu}$ the evolution measure of the process starting from initial state "all zeros" (recall Definition 3). By definition of τ_n,

$$\tilde{\mu}(\tau_n > t) = \tilde{\mu}(\exists x \in [0, n-1] \text{ s.t. } \eta_x^t = 0).$$

Hence, using Proposition 1, $\tau_n > t$ if and only if there exists an open path that connects one of the vertices of the set $S^t = \{(x,t) \text{ s.t. } x \in \mathbb{Z}_n\}$ to one of the vertices of $S^0 = \{(y,0) \text{ s.t. } y \in \mathbb{Z}_n\}$. The next expression follows from the definition of expectation and from Proposition 1.

$$\mathbb{E}(\tau_n) = \sum_{t=0}^{\infty} \tilde{\mu}(\tau > t) = \sum_{t=0}^{\infty} \mathbb{P}_\epsilon(\exists y \in \{0,1,\ldots,n-1\} \text{ s.t. } (y,t) \longrightarrow S^0)\,. \tag{13.13}$$

High-Noise Regime

We provide a lower bound for $\mathbb{P}_\epsilon(\exists x \in \{0,1,\ldots,n-1\} \text{ s.t. } (x,t) \rightarrow S^0)$ in the case of ϵ close to 1. First, observe that from the union bound and from translation invariance the following equation holds. Namely,

$$\begin{aligned}\mathbb{P}_\epsilon(\exists x \in \{0,1,\ldots,n-1\} \text{ s.t. } (x,t) \longrightarrow S^0) &\le \sum_{x=0}^{n-1} \mathbb{P}_\epsilon((x,t) \rightarrow S^0) \\ &= n \cdot \mathbb{P}_\epsilon((0,t) \longrightarrow S^0)\,.\end{aligned} \tag{13.14}$$

Observe that with different boundary conditions translation invariance wouldn't hold. We use then a technique widely used in percolation theory called *counting path method*. Call then $C_{n,t}$ the set of possible paths connecting the vertex $(0,t)$ to one of the vertices in S^0 and call $N_{n,t}$ the total number of such paths. Observe that,

$$\begin{aligned}&\mathbb{P}_\epsilon((0,t) \rightarrow S^0) = \\ &\mathbb{P}_\epsilon\left(\bigcup_{c \in C_{n,t}} \{\, c \text{ is open }\}\right) \le \sum_{c \in C_{n,t}} \mathbb{P}_\epsilon(\, c \text{ is open }),\end{aligned} \tag{13.15}$$

where in the second inequality we used the union bound. Observe that the probability that a path is open equals the probability that all its vertices are open. As a path that connects the point $(0, t)$ to S^0 has t vertices, this probability is $(1-\epsilon)^t$. As every step has two possible states, down-right, down-left, the total number of possible paths connecting $(0, t)$ to S^0 is $N_{n,t} \leq 2^t$. Recalling (13.13) and (13.14), we conclude that,

$$\begin{aligned} \mathbb{E}[\,\tau_n\,] &\leq \sum_{t=1}^{\infty} \min\{1, n \cdot \mathbb{P}((0,t)\longrightarrow S^0)\} \\ &\leq \sum_{t=1}^{\infty} \min\{1, n \cdot e^{\log[2(1-\epsilon)]t}\} \qquad (13.16) \\ &\leq \frac{\log[n]}{\log[2(1-\epsilon)]} + K\,, \end{aligned}$$

where K is a positive constant. The last inequality holds if $\epsilon > \frac{1}{2}$. This proves the statement (a) of Theorem 1 for ϵ large enough. It is not difficult to adapt this estimation to the case of different neighbourhoods.

Low-Noise Regime

We provide an upper bound for $\tilde{\mu}(\tau > t)$ in the case of ϵ small enough. The technique used for this purpose is called *Peierls argument*, and it is often employed in statistical physics and in the analysis of random spatial processes. The Peierls argument is a sort of counting path method on the dual graph of $\mathcal{G}_{\mathcal{U}}$. We denote the dual graph of $\mathcal{G}_{\mathcal{U}}$ by $\mathcal{G}_{\mathcal{U}}^D$. In the case of neighbourhood $\mathcal{U} = \{-1, 0\}$, the construction of $\mathcal{G}_{\mathcal{U}}^D$ is due to Toom [40]. The dual graph is represented in Fig. 13.4 (right). Its edges have three

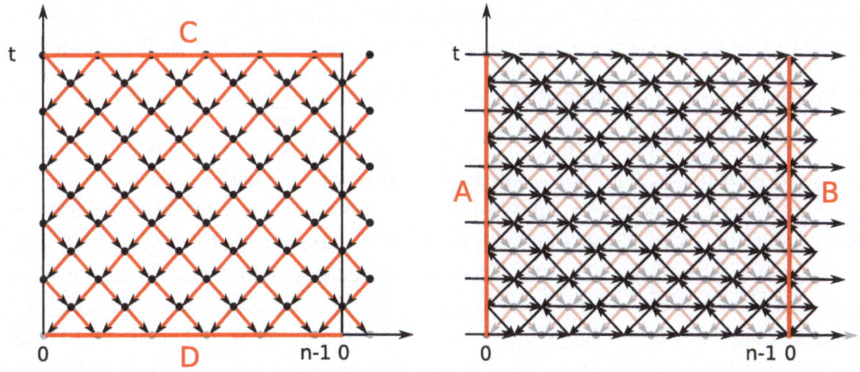

Fig. 13.4 *Left* representation of the graph $\mathcal{G}_{\mathcal{U}}$ with $\mathcal{U} = \{-1, 0\}$, similar to Fig. 13.1. In this representation sites at odd times have been translated to the *right* of $\frac{1}{2}$ and the horizontal axis has been rescaled by $\frac{1}{2}$. *Right* graph $\mathcal{G}_{\mathcal{U}}^D$, as defined in the text, in case $\mathcal{U} = \{-1, 0\}$. The dual graph is superimposed to the graph represented on the *left*. Edges of the dual graph are *arrows* pointing to the *right*, *up-left* and *down-left*

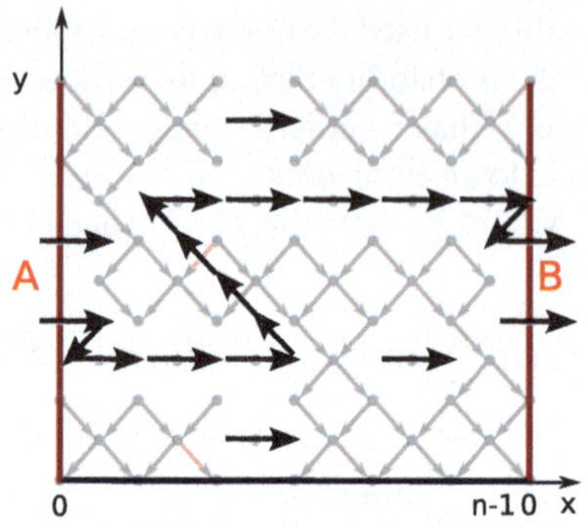

Fig. 13.5 Representation of a realization. Open vertices in the dual lattice (i.e. closed vertices in $\mathcal{G}_{\mathcal{U}}$) are marked by an *arrow* pointing to the *right*. *Arrows* "down-right" and "down-left" connect open vertices in $\mathcal{G}_{\mathcal{U}}$ (i.e. closed vertices in $\mathcal{G}_{\mathcal{U}}^D$). In the figure, an open path in the dual lattice ("right", "down-left" and "up-left" edges) connects the segment A to the segment B. Indeed, as stated in Proposition 2, there are no open paths in $\mathcal{G}_{\mathcal{U}}$ connecting the line $y = t$ to the line $y = 0$

possible directions: *right*, *down-left* and *up-left*. Every edge "right" is located over a site of $\mathcal{G}_{\mathcal{U}}$. We declare edges "right" *open* if and only if the corresponding site of $\mathcal{G}_{\mathcal{U}}$ is closed. Down-left and up-left edges are always open. The following proposition connects percolation on $\mathcal{G}_{\mathcal{U}}$ with percolation on its dual.

Proposition 2 *Call A the segment connecting the point* $(0, 0)$ *to the point* $(0, t)$ *and B the segment connecting the point* $(n, 0)$ *to the point* (n, t). *An open path in* $\mathcal{G}_{\mathcal{U}}$ *connects some vertex* (x, t), $x \in \mathbb{Z}_n$, *to some vertex* $(y, 0)$, $y \in \mathbb{Z}_n$, *if and only if there is no open path in* $\mathcal{G}_{\mathcal{U}}^D$ *connecting A to B.*

See Fig. 13.5. Then, providing a lower bound for the probability of the event $\{\exists x \in \{0, 1, \ldots, n-1\}$ s.t. $(x, t) \longrightarrow S^0\}$ is equivalent to providing an upper bound for the probability of the existence of an open path in $\mathcal{G}_{\mathcal{U}}^D$ connecting A to B. We denote this event by $\{A \longrightarrow B\}$. Observe that $\mathbb{P}_\epsilon(A \longrightarrow B) = \mathbb{P}_\epsilon(\bigcup_{y=0}^{t} \{(y, 0) \longrightarrow B\}) \le \sum_{y=0}^{t} \mathbb{P}_\epsilon((y, 0) \longrightarrow B)$. Call $G_{n,t,y,r}$ the set of paths having r horizontal edges and connecting $(y, 0)$ to B in $\mathcal{G}_{\mathcal{U}}^D$. Call $T_{n,t,y,r}$ the total number of such paths. Call $d\ell$ and $u\ell$ the number of steps, respectively, "down-left" and "up-left" of a path. Observe that any path connecting A to B must be such that $r \ge n$ and such that $2r - d\ell - u\ell = n$. Hence, the total number of steps of a path connecting A to B and having r steps "right" is $r + d\ell + u\ell = 3r - n$, with $r \ge n$. As every edge can be only in one of 3 possible states, it follows that $T_{n,t,y,r} \le 3^{3r-n}$. By the union bound as in (13.14) and the fact that for every $g \in G_{n,t,y,r}$, $\mathbb{P}_\epsilon(\text{the path } g \text{ is open}) = \epsilon^r$, we derive the next expression,

$$\mathbb{P}_\epsilon(A \longrightarrow B) = \mathbb{P}_\epsilon\left(\bigcup_{y=0}^{t}\{(y,0)\longrightarrow B\}\right) =$$

$$\mathbb{P}_\epsilon\left(\bigcup_{y=0}^{t}\bigcup_{g\in G_{n,t,y,r}}\{\text{the path } g \text{ is open}\}\right) \le \sum_{y=0}^{t}\sum_{g\in G_{n,t,y,r}} \mathbb{P}_\epsilon(\text{the path } g \text{ is open})$$

$$< t\sum_{r=n}^{\infty} T_{n,t,y,r}\epsilon^r < t\sum_{r=n}^{\infty} 3^{3r-n}\epsilon^r\ , \tag{13.17}$$

which converges to $t\frac{3^n\epsilon^{2n}}{1-3\epsilon^3}$ if $\epsilon < \frac{1}{27}$. Recalling (13.13) and using the previous expression, for all $\epsilon < \frac{1}{27}$ we derive the following lower bound for the expected absorption time,

$$\mathbb{E}[\tau_n] \ge \sum_{t=1}^{\infty} t\left(1 - t\frac{3^n\epsilon^{2n}}{1-3\epsilon^3}\right) \ge j\left(1 - j\frac{3^n\epsilon^{2n}}{1-3\epsilon^3}\right), \tag{13.18}$$

for any positive integer j. By choosing $j = [\frac{1}{2}\frac{1-3\epsilon^3}{3^n\epsilon^{2n}}]$, the statement **(a)** of Theorem 1 follows for every $\epsilon < \frac{1}{27}$.

13.5 Discussion

The transition from "fast" to "slow" convergence regime on finite space and its relation with the transition from ergodic to non-ergodic behaviour on infinite space is very far from being understood from a mathematically rigorous point of view. In particular, such correspondence does not involve only Percolation PCA, but it involves a wide class of processes exhibiting a non-equilibrium phase transition. In this section, we report some relevant questions.

- **Open problem 3: effect of the boundaries**. Theorem 1 describes the average absorption time of Percolation PCA on finite space with periodic boundaries. What is the behaviour of the model in the case of different boundaries? For example, in the case of non-symmetric neighbourhoods "typical" open paths from the line $y = t$ have a drift (dependent on ϵ), and they intersect the boundary before reaching S^0, as t is large. Hence, if one considers 1 boundary (i.e. cells at the boundary sites are one at any time independently on the past), the absorption time will be substantially smaller. This could substantially affect the convergence regimes of the process and perhaps introduce a new regime, as the slope of open paths depends on ϵ.
- **Open Problem 4: critical behaviour**. What is the convergence behaviour when $\epsilon = \epsilon_c$? The critical behaviour is usually the most difficult to understand. For a similar interacting particle system, the *contact-process* (see for example [27]), the average absorption time has been proved in [19] to grow polynomially with n at

criticality. It is reasonable to expect the same for Percolation PCA and techniques employed in [19] could be adapted to analyse Percolation PCA.

- **Open Problem 5: necessary conditions for Theorem** 1. Are there examples of processes undergoing an absorbing-state phase transition for which the transition point between ergodicity and non-ergodicity on infinite space does not coincide with the transition point between "fast" and "slow" convergence regime on finite space?

References

1. Bagnoli, F., Rechtman, R.: Topological bifurcations in a model society of reasonable contrarians. Phys. Rev. E **88**, 062914 (2013)
2. Balister, P., Bollobás, B., Kozma, R.: Large deviations for mean field models of probabilistic cellular automata. Random Struct. Algorithms **29**(3), 399–415 (2006)
3. Balister, P., Bollobás, B., Johnson, J., Walters, M.: Random majority percolation. Random Struct. Algorithms **36**(3), 315–340 (2010)
4. Bennet, C., Grinstein, G.: Role of irreversibility in stabilizing complex and nonergodic behavior in locally interacting discrete systems. Phys. Rev. Lett. **55**(7), 657–666 (1985)
5. Berezner, S., Krutina, M., Malyshev, V.: Exponential convergence of Toom's probabilistic cellular automata. J. Stat. Phys. **73**(5–6), 927–944 (1993)
6. Bigelis, S., Cirillo, E.N.M., Lebowitz, J.L., Speer, E.R.: Critical droplets in metastable states of probabilistic cellular automata. Phys. Rev E. **59**, 3935–3941 (1999)
7. Chassaing, P., Mairesse, J.: A non ergodic probabilistic cellular automaton with a unique invariant measure. Stoch. Process. Appl. **121**(11), 2474–2487 (2011)
8. Cirillo, E.N.M., Nardi, F.R., Spitoni, C.: Metastability for reversible probabilistic cellular automata with self-interaction. J. Stat. Phys. **132**, 431–447 (2008)
9. Dai Pra, P., Louis, P.-Y., Roelly, S.: Stationary measures and phase transition for a class of probabilistic cellular automata. ESAIM: Probab. Stat. **6**, 89–104 (2002)
10. Dai Pra, P., Sartori, E., Tolotti, M.: Strategic interaction in trend-driven dynamics. J. Stat. Phys. **152**(4), 724–741 (2013)
11. de Maere, A., Ponselet, L.: Exponential decay of correlations for strongly coupled Toom probabilistic cellular automata. J. Stat. Phys. **147**(3), 634–652 (2012)
12. Depoorter, J., Maes, C.: Stavskaya's measure is weakly Gibbsian. Markov Process. Relat. Fields **12**(4), 791–804 (2006)
13. Diakonova, M., MacKay, R.: Mathematical examples of space-time phases. Int. J. Bifurc. Chaos **21**(8), 791–804 (2006)
14. Dobrushin, R.: Markov processes with a large number of locally interacting components: existence of a limit process and its ergodicity. Probl. Inf. Transm. **7**(2), 1490164 (1071)
15. Durrett, R.: Oriented percolation in two dimensions. Ann. Probab. **12**(4), 929–1227 (1984)
16. Durrett, R.: Probability: Theory and Examples, 4th edn. Cambridge University Press, Cambridge (2010)
17. Durrett, R., Schonmann, R.H., Tanaka, N.I.: The contact process on a finite set. III: the critical case. Ann. Probab. **17**(4), 1303–1321 (1989)
18. Fatès, N.: Asynchronism induces second-order phase transitions in elementary cellular automata. J. Cell. Autom. **4**(1), 21–38 (2009)
19. Fatès, N.: A guided tour of asynchronous cellular automata. J. Cell. Autom. **9**, 387–416 (2014)
20. Fatès, N., Morvan, M., Schabanel, N., Thierry, É.: Fully asynchronous behavior of double-quiescent elementary cellular automata. Theor. Comput. Sci. **362**(1–3), 1–16 (2006)
21. Fernández, R., Toom, A.: Non-Gibbsianness of the invariant measure of non-reversible cellular automata with totally asymmetric noise. Asthérisque **287**, 71–87 (2003)

22. Gray, L.: The critical behaviour of a class of simple interacting systems - a few answers and a lot of questions. In: Durret, R. (ed.) Particle Systems, Random Media and Large Deviations. Contemporary Mathematics, vol. 41, pp. 149–160. AMS, Providence (1985). Asthérisque **287**, 71–87 (2003)
23. Hinrichsen, H.: Nonequilibrium Critical Phenomena and Phase Transitions into Absorbing States. Lectures Held at the International Summer School on Problems in Statistical Physics XI. Leuven, Belgium (2005)
24. Kozma, R., Puljic, M., Balister, P., Bollobas, B., Freeman, W.: Phase transitions in the neuropercolation model for neural population with mixed local and non-local interactions. Biol. Cybern. **92**, 367–379 (2005)
25. Landman, K.A., Binder, B.J., Newgreen, D.F.: Modeling development and disease in our "second" brain. Cell. Autom. Lect. Notes Comput. Sci. **7495**, 405–414 (2012)
26. Lebowitz, J., Maes, C., Speer, E.: Statistical mechanics of probabillistic cellular automata. J. Stat. Phys. **59**, 117–170 (1990)
27. Liggett, T.M.: Interacting Particle Systems, 2nd edn. Springer, Berlin (2005)
28. Louis, P.Y.: Ergodicity of PCA: equivalence between spatial and temporal mixing conditions. Electron. Commun. Probab. **9**, 119–131 (2004)
29. Mairesse, J., Marcovici, I.: Around probabilistic cellular automata. J. Theor. Comput. Sci. **559**, 42–72 (2014)
30. Makowiec, D.: Modeling heart pacemaker tissue by a network of stochastic oscillatory cellular automata. In: Mauri, G., et al. (eds.) UCNC 2013. LNCS, vol. 7956, pp. 138–149 (2013)
31. Manzo, F., Nardi, F.R., Olivieri, E., Scoppola, E.: On the essential features of metastability: tunnelling time and critical configurations. J. Stat. Phys. **115**(1–2), 591–642 (2004)
32. Mendoça, J.: Monte Carlo investigation of the critical behavior of Stavskaya's probabilistic cellular automaton. Phys. Rev. E **83**(1), 012102 (2011)
33. Pearce, C.E.M., Fletcher, F.K.: Oriented site percolation phase transitions and probability bounds. J. Inequal. Pure Appl. Math. **6**(5), 135 (2005)
34. Ponselet, L.: Phase transitions in probabilistic cellular automata. Ph.D. thesis (2013). arXiv:1312.3612
35. Regnault, D.: Proof of a phase transition in probabilistic cellular automata. Developments in Language Theory, pp. 433–444 (2013)
36. Shnirman, M.: On the problem of ergodicity of a Markov chain with infinite sets of states. Probl. Kibern. **20**, 115–124 (1968)
37. Stavskaja, O.N.: Gibbs invariant measures for Markov chains on finite lattices with local interaction. Mat. Sbornik **21**, 395 (1976)
38. Stavskaya, O., Piatetski-Shapiro, I.: On homogeneous nets of spontaneously active elements. Syst. Theory Res. **20**, 75–88 (1971)
39. Taggi, L.: Critical probabilities and convergence time of percolation probabilistic cellular automata. J. Stat. Phys. **159**(4), 853–892 (2015)
40. Toom, A.: A family of uniform nets of formal neurons. Sov. Math. Dokl. **9**, 1338–1341 (1968)
41. Toom, A.: Stable and attractive trajectories in multicomponent systems. In: Dobrushin, R., Sinai, Y. (eds.) Multicomponent Random Systems. Advanced Probability Related Topics, vol. 6, pp. 549–575. Dekker, New York (1980)
42. Toom, A.: Cellular automata with errors: problems for students of probability. In: Snell, L. (ed.) Topics in Contemporary Probability and Its Applications. Probability and Stochastics Series. CRC Press, Boca Raton (1995)
43. Toom, A.: Contours, convex sets, and cellular automata. Notes for a Course Delivered at the 23th Colloquium of Brazilian Mathematics, Rio de Janeiro (2004)
44. Toom, A.: Ergodicity of cellular automata. Notes for a Course Delivered at Tartu University, Estonia (2013)
45. Toom, A., Vasilyev, N.B., Stavskaya, O.N., Mityushin, L.G., Kurdyumov, G.L., Pirogov, S.A.: Discrete local Markov systems. Stochastic Cellular Systems: Ergodicity, Memory, Morphogenesis. Manchester University Press, Manchester (1990)

46. Tomé, T., de Carvalho, K.C.: Stable oscillations of a predator-prey probabilistic cellular automaton: a mean-field approach. J. Phys. A.: Math. Theor. **40** (2007)
47. Varerstein, L., Leontovitch, A.: Invariant measures of certain Markov operators describing a homogeneous random medium. Probl. Inf. Transm. **6**(1), 61–69 (1970)
48. Vasilyev, N., Petrovskaya, M., Piatetski-Shapiro, I.: Modelling of voting with random errors. Autom. Remote Control **10**, 1632–1642 (Translated from Russian) (1970)

Chapter 14
Percolation Operators and Related Models

Piotr Słowiński

Abstract Here we review the theory and methods of the two basic stochastic models with absorbing states: percolation operators and contact processes. We explore connections between them by studying discrete-time approximations of a continuous-time contact processes. In particular, we look at the approximations based on both synchronous and asynchronous updating schemes. Additionally, we go on to discuss several individual-based models, which are commonly used to model different biological phenomena. Specifically, we focused on models with absorbing states that, have spatially non-homogenous stationary states, or that have shown to be bi-stable. More generally, we aim to demonstrate the challenges associated with reconciling different mathematical descriptions of natural phenomena.

14.1 Introduction

Percolation operators (PO) are spatially extended, discrete-time Markov processes that are synchronously updated (that is, all sites are updated simultaneously in each computation step) [38, 39]. Contact processes (CP) are continuous-time analogs of the POs [22, 27]. Both are well understood and used as a starting point for the development of more complicated models in multiple fields of science, e.g. biology, physics and social sciences [13, 16].

POs and CPs are stochastic models consisting of units placed on a graph Γ. Γ is often an integer lattice $\mathbb{Z}^d$. Each unit can be in one of two individual states, say 0 and 1. The interpretation of the two states depends on the model application, for example they could refer to: "empty" and "occupied"; "infected" and "healthy"; "grass" and "tree", and so forth. The state of the unit can change in one of two random transitions. These are commonly called "death" ($1 \to 0$) and "birth" ($0 \to 1$). Both

P. Słowiński (✉)
College of Engineering, Mathematics and Physical Sciences, Harrison Building, Streatham Campus, University of Exeter, Exeter EX4 4QF, UK
e-mail: p.m.slowinski@exeter.ac.uk

P.-Y. Louis and F.R. Nardi (eds.), *Probabilistic Cellular Automata*, Emergence, Complexity and Computation 27, https://doi.org/10.1007/978-3-319-65558-1_14

transitions, $1 \to 0$ and $0 \to 1$, are conditionally independent given the previous steps. Additionally, births ($0 \to 1$) are conditional on the presence of a neighbour in the state 1. In other words, the POs and CPs are the simplest spatially extended models of local facilitation (activation).

The feature that makes the POs and CPs particularly interesting is their proven non-ergodicity [27, 39]. This means that for some parameter values there are two different stationary states that can be admitted by the system. Namely, the 'all 0s' state and the 'mixture of 0s and 1s' state. The 'all 0s' state is absorbing, i.e. once it is reached it is impossible to leave it. That is, if the initial state of the lattice is 'all 0s' it will stay 'all 0s' forever. Since the final state of the system depends on both the initial state and parameters of the system, POs and CPs correspond to real systems that partially remember their initial state.

Proofs of non-ergodicity of POs and CPs work only for infinite systems; on a graph Γ of finite size the POs and CPs are always ergodic [9, 38]. For a system of finite size, the absorption in the 'all 0s' state is inevitable. This is due to a fact that series of events in which all 1s die has a small, but strictly positive, probability. Nevertheless, non-ergodicity of the infinite system has consequences that can be observed in a system of finite size. Specifically, the convergence time to absorption for POs and CPs grows as an exponent of the system size in the non-ergodic regime. Whereas, in the ergodic regime the convergence time to absorption grows as a logarithm of the system size; details for CPs can be found in Sect. 2 p. 334 of [13] and references therein, and for POs see Theorem 2.2 in [36], discussion in [37] or Chap. 6 in [39].

The inverse of this statement might, however, not be true. That is to say, in some systems the time to absorption can be very long even when the system is proved to be ergodic; such behaviour is called metastability. In such cases the convergence time to a stationary state is often related to the probability of nucleation of a droplet with critical size [33]. A thorough review of results for stochastic systems with absorbing states and individual state space consisting of two states can be found in [23].

Although many studies of POs and CPs were motivated by biological processes observed in ecology, epidemiology and neurosciences [13, 16, 28, 38], it was quickly realised that the basic POs and CPs are too simple to reproduce the more complicated phenomena observed in nature, e.g. complex spatial patterns or multi-stability. A simple method to overcome the limitations of POs and CPs is to extend the individual state space of the units from two to many states. Models with multiple individual states are able to exhibit a much wider range of behaviours, however they are more difficult to study [16]. In fact, it is only recently that an introduction to the theory of non-ergodicity of a big class of stochastic systems with three individual states has been published; in [8].

In Sect. 14.2 of this chapter we discuss the properties of the POs; we compare them to discrete-time contact processes and show that they are "eroders". In Sect. 14.3 we review continuous-time contact processes and their discrete-time approximations. Finally, in Sect. 14.4 we present a short survey of models based on POs and CPs, that are being used in biology and ecology. A summary of this chapter is presented in Sect. 14.5.

14.2 Percolation Operators

Percolation operators, as defined in [38], consist of synchronously updated units placed on a graph Γ. Their dynamics can be described by a discrete-time Markov process, with independent update probabilities for the state x_r on site r, given the state on a neighbourhood x_η of site r. Each unit can be in one of two states, $x_r = \{0, 1\}$. The neighbourhood η of site r is a finite subset of the graph Γ, and it includes site r.

The transition probabilities of the POs are defined by the composition of a deterministic operator $D_{\max}$ and a random operator R_p. The deterministic operator $D_{\max}$ is a maximum function $f : \{0, 1\}^n \rightarrow \{0, 1\}$ defined for the neighbourhood η of site r. The random operator R_p turns each 1 into 0 with probability p, and does not affect sites in the state 0. Operators for different sites are independent, conditionally of the past. In other words, at each time step, each site observes the maximum of its neighbourhood and either, updates its state according to the outcome of the deterministic operator $D_{\max}$, or makes an error with probability p.

Figure 14.1 shows a schematic action of a 1D $PO_{n=3}$, with neighbourhood $\eta = \{r - 1, r, r + 1\}$. The arrow labeled D indicates an action of the deterministic operator and the arrow labeled R_p indicates one of the possible outcomes of a random part of the $PO_{n=3}$. The blue and red triangles illustrate the fact that description of the deterministic part of the percolation operator, $D_{\max}$, can be expressed either, as an action of the neighbourhood on the site (this is illustrated by the red triangle), or as an action of the site on its neighbourhood (this is illustrated by the blue triangle). In other words, descriptions "site r assumes the state of $D_{\max}(x_\eta)$" and "each site s in the neighbourhood of r assumes state given by $\max(x_r, x_s)$" are equivalent.

A representation of a stochastic models as the composition of deterministic and stochastic operators, as presented in [39], is very helpful in understanding their evolution. For example, to see how a PO acts on a state of the whole lattice, we can start by considering separately the actions of the operators $D_{\max}$ and R_p. On one hand, under an action of the $D_{\max}$ alone any site in state 1 will "reproduce", and as time tends to infinity, it will fill the whole lattice (a part of the space if the neighbourhood is asymmetric). That also means, that the $D_{\max}$ is an "eroder", i.e. in finite time it "erodes" any finite distortion (set of 0s) of the stationary state 'all 1s'.

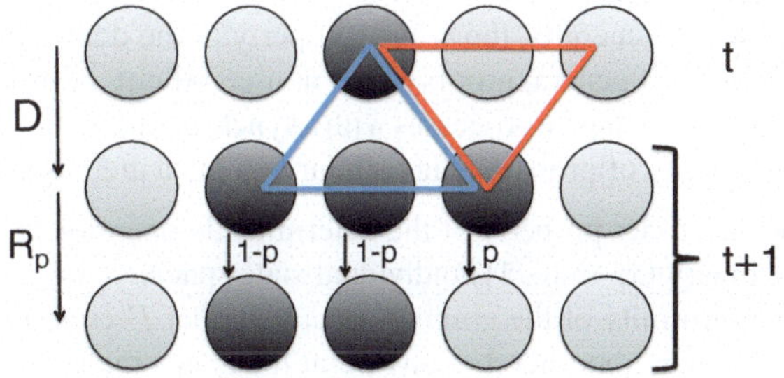

Fig. 14.1 Discrete time synchronous 1D PO with n = 3. *Black circles* indicate sites in state 1

On the other hand, if the operator R_p acts alone, then as time tends to infinity, any initial state will turn into the 'all 0s' state. If we compose actions of both R_p and $D_{\max}$ then the evolution of the process depends on the noise intensity p as well as on the initial state.

Since the POs are eroders, their non-ergodicity follows from the Theorem 3.1 in [38]. In practice, the non-ergodicity of the POs means that on an infinite lattice all POs with neighbourhood of size $n \geq 2$ have a critical value $p^* > 0$ for which if $p > p^*$ they are ergodic and for $p < p^*$ they are non-ergodic. That is, if $p > p^*$ the process initialised with any initial state gets absorbed in the 'all 0s' state, whereas for $p < p^*$ probability of absorption is less than 1. We note that, for any value of p, if a PO is initialised in the 'all 0s' state it will stay in that state forever.

The lower and upper bounds for p^* are given as:

$$0.09 < p^* \leq 1 - \frac{1}{n}. \tag{14.1}$$

The lower bound of Eq. (14.1) is based on a comparison of the probability of absorption with the probability of percolation in the oriented site percolation process. The upper bound of Eq. (14.1) is derived from considerations of the POs on tree-like graphs. Due to the monotonicity of the POs, the bounds depend only on the size of the neighbourhood and not on the dimension nor on the exact structure of the graph Γ. More precisely, the bounds are proved for: graphs in which site r belongs to its neighbourhood, uniform graphs and tree-like graphs; see Chap. 8 of [39] for details. For proof of bounds in Eq. (14.1) and of the non-ergodicity of the POs, we refer the reader to [38, 39]. For a demonstration that the analogue result is true for the convergence time of finite systems we refer the reader to [37, 39].

For the simplest 1D $PO_{n=2}$ with neighbourhood, $\eta = \{r, r+1\}$, the bounds are $0.09 < p^*_{PO,n=2} \leq 0.323$, [38, 39] and the numerically estimated critical value is equal to $p^*_{PO,n=2} \approx 0.29450(5)$ [31]. Interestingly, the estimate of the critical value in [31] is only a small amount greater than $1 - 0.7058 = 0.2943 < p^*_{PO,n=2}$; that is the estimate of $p^*_{PO,n=2}$ based on the critical value of the oriented site percolation equal to 0.7058 ± 1, which was obtained by means of the renormalisation group approach in [26]. For $PO_{n=3}$ the bounds are $0.09 < p^*_{PO,n=3} \leq 0.495$, and the numerically estimated critical value is equal to $p^*_{PO,n=3} \approx 0.462$ [36]. Note that, $PO_{n=2}$ corresponds to the Stavskaya model introduced in [34], but with inverted roles of 0s and 1s.

More generally, the eroder property of the deterministic operator is an important element of general proofs of the non-ergodicity of probabilistic cellular automata (a class of stochastic processes with a synchronous updating rule). The proofs of the non-ergodicity of probabilistic cellular automata are based on the following properties:

- the eroder property of the deterministic operator D,
- binary ($x_r = \{0, 1\}$) individual state space,
- uniformity of the graph Γ (meaning that Γ commutates with shifts of space, i.e. the neighbourhood of any point looks just the same as that of any other point),
- uniformity and monotonicity of operator D (as defined in [39]),

- uniformity and independence of the random operator R.

The theorems, their proofs and counterexamples (operators and processes that do not have one or more of the above properties) can be found in [38, 39].

14.2.1 Discrete-Time Contact Process

In this Section we present the discrete-time contact process (DTCP) defined in [13]. This is as follows:

- Particle (a site in state 1) dies with probability γ and survives with probability $1-\gamma$.
- If particle survives, with probability λ it gives birth to a new particle at each of its neighbouring sites.
- If at the immediately preceding time-step a new particle appeared at site r due to birth, or if a particle at site r survived, then site r is occupied at the next time step; otherwise the site r is empty.
- All sites are updated synchronously.
- The neighbourhood η of site r consists of all its nearest neighbours (the set of points at a Manhattan distance of 1) and the site r itself. E.g. in 1D $\eta = \{r-1, r, r+1\}$ and in 2D η is the von Neumann neighbourhood.

Similarly to the POs, DTCPs can be defined as a composition of the operator $D_{\max}$ and the random noise operator $\widetilde{R}_p$. However, in case of the DTCPs, the $\widetilde{R}_p$ depends on the state of the site x_r^t and on the outcome of the operator $D_{\max}(x_\eta^t)$. Note that, the random operator $\widetilde{R}_p$ is uniform and independent for each site, conditionally of the past. For $\lambda = 1$ the DTCPs simplify to the POs. Transition probabilities of the DTCPs (see Table 14.1) show that the operator with $\lambda < 1$ is bounded from above by the operator with $\lambda = 1$; in the sense that $P_{\lambda<1}(1|x_\eta) \leq P_{\lambda=1}(1|x_\eta)$ for any state of the neighbourhood x_η [39]. This means that critical value $\gamma^*(\lambda)$ of DTCP is less than or equal to $\gamma^*(\lambda = 1)$. Intuitively, for $\lambda < 1$ there are less births which counteract the absorption in the state 'all 0s'. The bounds for the critical death probability $\gamma^*(\lambda, d)$

Table 14.1 Transition probabilities of the discrete-time synchronous contact process. For $\lambda < 1$ the probability of error depends on the initial state of the site x_r^t and the outcome of the operator $D_{\max}(x_\eta^t)$

x_r^t	$D_{\max}(x_\eta^t)$	$\widetilde{R}_p(x_r^t)$ for $\lambda < 1$	$\widetilde{R}_p(x_r^t)$ for $\lambda = 1$	x_r^{t+1}
1	1	$1-\gamma$	$1-\gamma$	1
		γ	γ	0
0	1	$\lambda(1-\gamma)$	$1-\gamma$	1
		$\lambda\gamma+1-\lambda$	γ	0
	0	1	1	0

of the DTCPs depend on the lattice dimension d and on the probability of births λ [13] and are given as:

$$\min_{k\geq 1}(1-(0.819+2(1-\lambda)^{dk})^{(1/(2k-1))}) \leq \gamma^*(\lambda,d) \leq \frac{2d\lambda}{2d\lambda+1}. \tag{14.2}$$

The lower bound in Eq. (14.2) uses the estimate of the upper bound of a critical value for the oriented site percolation equal to 0.819 [11], and the upper bound in Eq. (14.2) is a function of a number of nearest neighbours of site r, which for a DTCP is equal to $2d$.

The critical value for the 1D DTCP, with $\lambda = 1$, is estimated at $\gamma^* \approx 0.47$ [13]. What is consistent with the bounds for $PO_{n=3}$ and is reasonably close to the numerically estimated critical value of $p^*_{PO,n=3} = 0.462$ [36]. Since for $\lambda = 1$ the estimate of the lower bound in Eq. (14.2) does not depend on the number of neighbours, it can be used to improve the estimate of the lower bound for the critical value $1 - 0.819 = 0.181 < p^*$ of the POs in Eq. (14.1).

14.3 Continuous-Time Contact Process

A continuous-time contact process is defined by the transition rates ω_r^x for the state x_r on a site r; i.e. the states of sites are updated at a times given by an independent exponential random variable with a mean equal to $1/\omega_r^x$ (e.g. rate 2 means that on average an event happens twice per unit of time). The most classical transition rates ω_r^x of the CP, are given as [27],

$$\omega_r^x = \begin{cases} 1 \to 0, \text{ with rate } \gamma, \\ 0 \to 1, \text{ with rate } \lambda \sum_{s\in\eta} x_s, \end{cases} \tag{14.3}$$

here neighbourhood η does not include site r. The transition rates ω_r^x can be described as follows: particles die at rate 1 and are born at vacant sites at rate λ times the number of occupied neighbours. Since the rates ω_r^x can be interpreted as intensities of the process, it is possible to normalise them. In other words, the time of the process can be measured with respect to one of the transitions, either $1 \to 0$ or $0 \to 1$. Setting $\gamma = 1$, as in [27], defines the intensity of the process is defined in terms of the average frequency of death events.

Alternatively, the CP can be defined in terms of the following infinitesimal transition probabilities (in the sense that $P(\Delta t)/f(\Delta t) \to 1$ as $\Delta t \to 0$) [12]:

$$\begin{cases} P(1_t \to 0_{t+\Delta t}) \sim \Delta t, \\ P(0_t \to 1_{t+\Delta t}) \sim \left(\lambda \sum_{s\in\eta} x_s\right) \Delta t, \end{cases} \tag{14.4}$$

as above η does not include site r. In (14.4) the sum over the neighbourhood η can be interpreted as a sum of probabilities of independent birth events, i.e. at time $t + \Delta t$ each neighbour in state 1 can cause site r to change its state from 0 to 1 with probability $\lambda \Delta t$.

A continuous-time CP, in dimension d, is ergodic for $\lambda < \lambda^*(d)$, and non-ergodic for $\lambda > \lambda^*(d)$. The bounds for $\lambda^*(d)$ are given by:

$$\frac{1}{2d-1} \leq \lambda^*(d) \leq \frac{2}{d}. \tag{14.5}$$

For proof of the bounds see [27]; for the 1D continuous-time CP the bounds are $1.53 \leq \lambda^* \leq 2$ [28].

The continuous-time CPs, similarly to the DTCPs, have the same basic feature as the POs. Namely, the transition $0 \rightarrow 1$, on site r can occur only if there is at least one site in the state 1 in its neighbourhood; exactly as in the case of the maximum function of the POs. However, the POs and continuous-time CPs differ in terms of the probability that the site is going to be in the state 0 after updating, which for the continuous-time CPs depends on the state of neighbourhood of site r as well as on the state of the site r. For a comparison we refer the reader to cross reference with Table 14.1 using $\gamma = \Delta t$.

We gain further insight into the relation between the POs and continuous-time CPs, by comparing the parameters of continuous-time mean-field approximations of $PO_{n=2}$ and continuous-time CP in 1D with a single neighbour. For the $PO_{n=2}$, the continuous-time mean-field approximation can be described the following way, after an exponentially distributed time with mean 1:

- a site in state 1 dies with probability p,
- a site in state 0 checks the state of a randomly chosen other site and assumes state 1 with probability $1 - p$ if the chosen site is in state 1, and stays in state 0 otherwise.

If we now consider a large but finite number of sites, all that matters is the proportion θ_{PO} of sites in state 1; and in the limit of an infinite number of sites, the stochastic process for θ_{PO} converges to the deterministic Equation (14.6). Similarly, if we take the continuous-time CP and disregard lattice information (i.e. we look only at the proportion θ_{CP} of sites in state 1) then the stochastic process for θ_{CP} converges to the deterministic Equation (14.7).

$$\frac{d\theta_{PO}}{dt} = (1-p)\theta_{PO}(1-\theta_{PO}) - p\theta_{PO}, \tag{14.6}$$

$$\frac{d\theta_{CP}}{dt} = \lambda\theta_{CP}(1-\theta_{CP}) - \theta_{CP}. \tag{14.7}$$

In order to compare the two approximations it is necessary to rescale time $\tau = pt$ in Eq. (14.6). As a result, we obtain the following equation:

$$\frac{d\theta_{PO}}{d\tau} = \frac{1-p}{p}\theta_{PO}(1-\theta_{PO}) - \theta_{PO}. \tag{14.8}$$

In this way, time in Eq. (14.8) as well as in Eq. (14.7) is measured with respect to the death events, this allows us to compare the birth rate λ and the error probability p. A quick analysis reveals that the error probability p of the $PO_{n=2}$ can be considered as a mapping of the birth rate λ of the CP on the interval [0, 1]. Specifically, by taking $p = 1/(\lambda + 1)$ we observe that:

$$\lambda = 0 \rightarrow \ p = 1,$$
$$\lambda = \infty \rightarrow \ p = 0.$$

In the case of the mean-filed approximation of a continuous-time CP with $\gamma \neq 1$, to establish the relation between parameters of the two models it is necessary to rescale time in Eq. (14.7) and in Eq. (14.6). The relation between the error probability and the birth and death rates can be then expressed as $p = \gamma/(\lambda + \gamma)$.

Remark A more theoretical relation between the discrete-time and continuous-time stochastic processes can be explored through the relation of transition matrices for the states spaces of the processes:

$$\mathcal{P}^t = e^{\mathcal{Q}t}, \tag{14.9}$$

here, $\mathcal{P}$ is a matrix with probabilities of transitions between any two states of the stochastic system, and $\mathcal{Q}$ is a matrix of transition rates between states of the system that differ on a single site (*e* is matrix exponential) [35]. For the POs and CPs on the infinite lattice, their state space is $\{0, 1\}^{\mathbb{Z}^d}$, and the elements of matrix $\mathcal{Q}$ are given by the transition rates ω_r^x from Eq. (14.3).

14.3.1 Discrete Time Approximations of Continuous-Time CP

It is possible to approximate a continuous-time CP using a probabilistic cellular automaton (PCA). The simplest method of doing so, as suggested in [4], is to simultaneously update all the sites according to the infinitesimal probabilities given in Eq. (14.4), i.e.:

$$P(x_r = 0) = \begin{cases} \Delta t, & \text{if } x_r = 1 \\ 1 - \left(\lambda \sum_{s \in \eta} x_s\right) \Delta t, & \text{if } x_r = 0, \end{cases} \tag{14.10}$$

and $P(x_r = 1) = 1 - P(x_r = 0)$. Such an approximation works on a finite lattice because, as $\Delta t \rightarrow 0$ the probability of two sites being updated simultaneously tends to 0. In [4] it is also argued that the above approximation should work on an infinite lattice.

A more detailed analysis of using PCAs to approximate a continuous-time stochastic processes can be found in [29]. The PCA used in [29], called δ-approximating PCA, has transition probabilities based on the transition rates ω_r^x of the continuous-time process. More specifically, the transition probabilities of the δ-PCA, given as

$$P(x_r) = \frac{1}{2}(1 - \exp[-2\delta\omega_r^x(x_r)]),$$

are based on a cumulative distribution function of an exponential random variable and are chosen specifically to control the error between the original continuous-time process at time $N\delta$ and the Nth discrete time step of the δ-PCA. In other words, the δ-PCA approximation of the continuous-time process can be considered an analogue of Euler's method for solving ordinary differential equations [29]. The δ-PCA approximation holds if both processes are started from the same initial configuration, and if the continuous-time process is ergodic [29]. In the case of a continuous-time CP the transitions probabilities of the δ-PCA are based on the transitions rates ω_r^x given in Eq. (14.3) and have the following form:

$$P(1 \to 0) = \frac{1}{2}(1 - \exp[-2\delta\gamma]),$$
$$P(0 \to 1) = \frac{1}{2}\left(1 - \exp[-2\delta\lambda \sum_{s \in \eta} x_s]\right).$$

The continuous-time CPs can be further approximated by stochastic processes with discrete-time asynchronous updating scheme. A typical form of such approximation, introduced in [20] is also a basis of the standard asynchronous Monte Carlo scheme, and can be described as follows: at each time step of the simulation a site in state 1 is chosen at random, and with probability $p = \gamma/(\lambda + \gamma)$ dies $(1 \to 0)$ or, with probability $1 - p = \lambda/(\lambda + \gamma)$ gives birth to a new state 1 at a randomly selected neighbouring site (birth is successful only if the selected site is empty). λ and γ are the coefficients of the birth and death rates ω_r of the underlying continuous-time process. The time increment between update steps is chosen as $\Delta t = 1/N_1$, where N_1 is number of sites in state 1 before the attempted transition [7, 18]. The rates λ and γ are often scaled relative to each other, e.g. $\lambda' = \lambda/\gamma$ and $\gamma' = 1$.

The numerical estimate of the critical birth rate $\lambda^* \approx 3.29785(8)$ which is obtained by simulating a continuous-time CP (neighbourhood $\eta = \{r-1, r, r+1\}$ and $\gamma = 1$) with the above Monte Carlo scheme [7, 17], is two times bigger than the numerical estimate of the critical birth rate for the same process reported in [27] $\lambda^* \approx 1.6494$. The difference is due to the fact that, in the Monte Carlo scheme proposed in [7, 18], the rate of $0 \to 1$ transitions is equal to $\lambda(\sum_{s\in\eta} x_s)/n_s$, where n_s is a number of neighbours, while in [27] the rate of $0 \to 1$ transitions is $\lambda(\sum_{s\in\eta} x_s)$; for the 1D CP the number of neighbours is equal to $n_s = 2$.

14.3.2 Critical Values for Processes with Asynchronous and Synchronous Updating Rules

Using the transformation $p = 1/(\lambda + 1)$ we can compare the numerical estimates of the critical birth rate of the CP in 1D, with the neighbourhood $\eta = \{r - 1, r, r + 1\}$, both for synchronous and asynchronous updating rules. The critical value of the CP with asynchronous updates, reported in [27], is lower than the critical value of the CP with synchronous updates, reported in [36], $p^*_{\text{asynch}} \approx 0.3774 < p^*_{\text{synch}} \approx 0.462$. In the literature, there are several other examples of similar observations. For instance, similar behaviour of estimates for the critical values was reported for: majority voting rule with North-East-Center (NEC) neighbourhood [30], majority voting rule with von Neumann neighbourhood [4], and for Glauber dynamics with NEC neighbourhood [30]. In each of these cases, the numerical estimate for the critical value of the error rate was lower for the process with asynchronous updating rule than for the one with synchronous updating rule.

This observation can be understood intuitively by considering that in the case of asynchronous updates, the order of picking sites may influence the outcome of the process, e.g. a site in the state 1 can be turned into a 0 before it will have chance to influence its neighbours. Therefore, the effect of the asynchronous updating scheme can be interpreted as an additional source of noise, applied to the deterministic part of a PCA (updated synchronously). In other words, to account for the randomness originating from the updates, the noise of the process with asynchronous updates has to be lower than the noise of the corresponding PCA. See [19] for a comprehensive collation of references concerning the relations between models with synchronous and asynchronous updating rules.

14.4 Models of Biological Systems that are Related to POs and CPs

We now present a review of results of selected models with absorbing states that are used to study ecological phenomena. In particular, we are interested in models related to POs and CPs, which exhibit hysteresis and are reported to be bi-stable. That is, models in which the results of simulations depend on initial conditions. The existence of such dependence may indicate that, as in the case of the POs and the CPs, for some parameter range the models are non-ergodic.

14.4.1 Multi-type Percolation Operator

We start by showing an example of a simple model with an individual state space that has more than two states. In a multi-type PO each site can be in one of M states

$m \in [0, M-1]$. The $D_{\max}$ deterministic maximum operator acts in the same way as for the binary POs, and the R_p random noise operator has the following structure:

$$\begin{bmatrix} 1 & 0 & \cdots & & 0 \\ p & 1-p & \cdots & & 0 \\ \vdots & & \ddots & & \vdots \\ p/(M-2) & p/(M-2) & & 1-p & 0 \\ p/(M-1) & p/(M-1) & \cdots & p/(M-1) & 1-p \end{bmatrix}$$

For such an operator, the 'all 0s' state is absorbing. Furthermore, the multi-type PO is non-ergodic and has at least M stationary states. Specifically, the stationary states of the multi-type PO are: the 'all 0s' state and states that are homogenous mixtures of 0s and ms. Non-ergodicity of the multi-type PO follows from non-ergodicity of the binary POs, i.e. if the initial state of the multi-type PO is 'all ms', then the probability of the absorption into the 'all 0s' state is less than 1 for small enough p.

14.4.2 Models with 3 Individual States

In Fig. 14.2 we show schematic representation of different models with three individual states $\{0, 1, 2\}$: (a) a biological two-species system from [5, 6]; (b) a predator-prey model from [1]; (c) a model of mussel disturbance dynamics from [21]; (d) model of Mediterranean arid ecosystem from [24], (e) a forest fire model with immunisation from [2, 10] and (f) model of grass-bushes-trees transition from [15]. Arrows indicate possible transitions between the states. λs are proportionality constants of birth rates conditioned on the presence of the state 1 or 2 in the neighbourhood, n_1 and n_2 are the numbers of neighbours in state 1 or 2, and p and δ are the probabilities and rates of spontaneous transitions between states. Each of the models in Fig. 14.2 can be reduced to a PO or to a CP in various ways. For instance, the models in Fig. 14.2a–c can be reduced to the CPs by setting $\lambda_2 = \infty$. For $\lambda_2 = \infty$ the transition between states 1 and 2 becomes instantaneous and hence the two states can be treated as a single state. In the rest of this section we present a short description and summary of the results for each model shown in Fig. 14.2. We focus on similarities between the models, and on the existence of bi-stability.

Figure 14.2a shows the transition probabilities of a two-population (predator-prey) model, with local transition rules based on the Lotka-Voltera model. Here 0s indicate empty sites, 1s indicate sites with prey and 2s indicate sites with predators. The results of simulations of this model with a synchronous updating scheme, as well as an analysis of its mean-field approximation showed the existence of a non-homogenous stationary state with the coexistence of all three states [2, 5, 6]. For this reason, a version of a model depicted in Fig. 14.2a was used to investigate mechanisms that could be responsible for the non-homogeneity of the densities of empty sites (0s), hosts (1s) and parasitoids (2s) observed in field data [32]. Since the hosts and

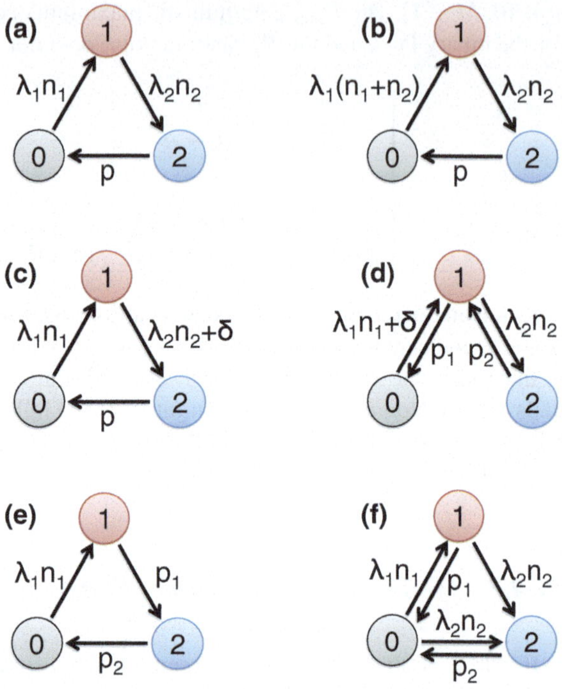

Fig. 14.2 Schematic representation of different models: **a** a biological two-species system [5, 6], **b** a predator-prey model [1], **c** a model of mussel disturbance dynamics [21], **d** model of Mediterranean arid ecosystem [24], **e** a forest fire model with immunisation [2, 10] and **f** model of grass-bushes-trees [15]

parasitoids were airborne, the model in Fig. 14.2a was modified to allow for long-range dispersal of the organisms. To this end, the neighbourhood of each site, instead of being restricted to its nearest neighbours, was expanded to the whole lattice ($n_i > 0$ if there is at least one site in state i on the lattice). The results of the analysis presented in [32] demonstrated the existence of non-homogenous states with the coexistence of all 3 predators, prays and empty sites. Furthermore, the results showed a good qualitative agreement between the simulations (performed on a grid obtained from geographical data) and the collected field data.

The model shown in Fig. 14.2b studied in [1], is another version of a predator-prey model. In the model illustrated in Fig. 14.2b, 0s indicate empty sites, 1s indicate sites with a population of prey, and 2s indicate sites with coexisting populations of predators and prey (because predators need prey to survive). Such a biologically motivated interpretation of a site in state 2 leads to a model in which the transition $0 \rightarrow 1$ is conditioned on a presence of an occupied site in the neighbourhood (it can be either in state 1 or in state 2). The results of the analysis of this model and its simulations with asynchronous updating scheme showed that the models in Fig. 14.2a, b have qualitatively the same behaviour: the states all 0s and all 1s are absorbing (in absence of predators (2s) a single surviving prey can fill the whole lattice with prey), and there exists a stationary state where all three coexist.

The model shown in Fig. 14.2c was developed, in [21], to describe the emergence of spatial patterns in mussel beds. In the model, 0s indicate an empty site, 1s indicate sites that are occupied by mussels and 2s indicate sites that are disturbed (sites from which mussels were recently removed). The only difference between the model in Fig. 14.2c and the models in panels (a) and (b), is the existence of a spontaneous transition $1 \to 2$ with rate $\delta > 0$. The results of simulations presented in [21] show that the model exhibits bi-stability. In particular, the authors of [21] demonstrated that depending on the initial state of the system, the result of the simulation is either an absorbing state of all 1s, or a mixed state of 0s, 1s, and 2s where the proportion of sites in state 1 is around 0.4. The initial states used in the simulations were either state 'all 1s' with a single site in state 2 (so called spreading analysis), or a mixed state. More interestingly, Fig. 5a in [21] shows existence of hysteresis for $\delta = 0$. Note that, for $\delta = 0$ models in panels (a) and (c) are identical.

In Fig. 14.2d we depict a desertification model presented in [24, 25], where 0s indicate degraded sites, 1s indicate empty sites, and 2s indicate vegetation. In order to include the possibility of the long-range dispersal of seeds the rate $\lambda_2 n_2$ of the transition $1 \to 2$ is conditioned on the presence of a site in state 2 anywhere on the lattice, i.e. the neighbourhood of a site in state 1 is expanded to the whole lattice. However the analysis presented in [24, 25] demonstrats emergence of non-homogenous stationary states, with characteristics of a real ecosystems, even if the neighbourhood of a site in the state 1 consists only of its nearest neighbours. In other words, analysis in [24, 25] showed that the local interactions are more important than the long-range ones. Additionally, analysis of mean-field and pair approximations (moment closure method for neighbour-site pairs) of the stochastic process from Fig. 14.2d, showed that for some parameter values the model is bi-stable [24, 25]. The two stable states are: a 'mostly 1s' state, and a mixed state of 0s, 1s, and 2s. The mechanism responsible for the bi-stability in the model shown in Fig. 14.2d [24, 25] appears to be similar to the one reported for the model shown in Fig. 14.2c [21], i.e. for some parameter values, the introduction of a single site in state 2 can lead to the emergence of the non-homogenous mixed state with 0s, 1s and 2s.

More generally, the results presented in [21, 25] demonstrate that some 3-state models with absorbing states, have multiple coexisting stationary states, and that they can exhibit hysteresis. The two coexisting (non-absorbing) states are: mostly 1s and a mixed state. The mechanism responsible for the coexistence of the two states depends on the ratio of the transitions $1 \to 2$ to the transitions $2 \to 0 \to 1$ in Fig. 14.2c [21], and on the ratio of the transitions $1 \to 2$ to transition $2 \to (1) \to 0 \to 1$ in Fig. 14.2d [25]. For example, if the system shown in Fig. 14.2c is initiated with a low density of 2s, and the transition $2 \to 0 \to 1$ is slow, then the density of 2s will be limited because the 2s need sites in state 1 to reproduce; details of the analysis can be found in [21]. Cases of bi-stability reported in [21, 25] appear very interesting, however it is known that the result of the spreading analysis (analysis simulations started with low density of one of the individual states) might depend on the implemented algorithm [23]. For that reason, they need further investigation.

The models in Fig. 14.2e, f are two examples of a 3-state models based on a continuous-time CP, i.e. they are formulated in as a continuous-time stochastic processes. Figure 14.2e depicts a forest fire model with immunisation from [10] (mentioned also in [2]), where 0s indicate green trees, 1s indicate burning trees, 2s indicate empty sites. A theorem for the existence of a non-trivial (different from 'all 0s') stationary state is proved in [14]. Additionally, the discussion presented in [14] concerns different modifications of this model. For example the authors discuss the CP limit with the rate $p_2 = \infty$, and a model in which a transition from empty site to a site occupied by a tree ($2 \to 0$) depends on the number of neighbours in the individual state 0 (the model is then a continuous-time version of the model depicted in Fig. 14.2a).

Finally, Fig. 14.2f illustrates the multi-species competition model analysed in [15]. In the model: 0s indicate grass, 1s indicate bushes and 2s indicate trees. The conditions for the existence of a non-trivial stationary state in this model are given in [15]. However, as noted in [16] this stationary state is a homogenous mixture of 0s, 1s and 2s. We note that, a discrete-time version of the model illustrated by Fig. 14.2f can be considered a multi-type PO, i.e. the transitions between the states are given by composition of the maximum operator $D_{\max}$ and a random noise operator R_p.

14.4.3 Models with Ageing

In this section we discuss two models of desertification in which the extension of individual state space is used to study processes in which transition probabilities depend on how long a site was occupied. That is, different individual states indicate age of a site.

We first review results from the analysis of a modified CP, used to model trees-grass coexistence on a savannah, and which is introduced in [40]. The modifications to a CP that are introduced and studied in [40] include: an age dependant death rate, a birth rate that can change in time, a local facilitation conditioned on the presence of two sites in state 1 being in the neighbourhood of a site r. Overall, the modification produced the following results:

- The introduction of an age-dependent death rate creates a non-homogenous stationary state.
- The addition of a time-varying birth rate to the basic CP (with $\{0, 1\}$ individual state space) results in time-varying densities of 1s. Furthermore, when changes in the birth rate were based on the paleoecological precipitation data, showed good agreement with historical tree-grass ratios on a savannah.
- The introduction of a more complicated form of local facilitation results in the appearance of a birth rate dependant hysteresis.

We now analyse in more detail the existence of the hysteresis in this model. Transition probabilities of the local facilitation process described in [40] have the following form:

$$P(1 \to 0) = d,$$
$$P(0 \to 1) = \begin{cases} b, \text{ if } \sum_{s \in \eta} x_s \geq 2 \\ 0, \text{ otherwise,} \end{cases} \tag{14.11}$$

here d is the death rate, b is the birth rate, and the summation does not include site r. The transition probabilities (14.11) show that the 'all 0s' state is an absorbing state. Furthermore, a site in the state 0 can change its state to 1, only if there are at least two sites in state 1 in its neighbourhood. Structure of (14.11) indicates that, the results of simulations will strongly depend on the initial state of the system. In particular, if the initial density of 1s is too low (sites in state 1 are too far away from each other) then the system will always end in the absorbing state 'all 0s', independent of the value of the birth rate b. More generally, hysteresis as presented in [40] might have a similar nature to the results of the spreading analysis reported in [21, 25]. This however requires further analysis.

The second system we review is the model of semi-arid vegetation presented in [3]. It is a complicated model based on a multi-type PO. Its individual state space is $M = \{0, \ldots, 60\}$, where 0s indicate empty sites and the individual states $m > 0$ correspond to different ages of vegetation. The neighbourhood is defined as 5 square shells around the site r (the first square shell is the Moore neighbourhood) and includes site r. The dynamics of the system operate as follows: at each time step, for each site, a random number h is drawn from a uniform distribution, if $h < T(x_r)$ then $x_r \to 0$, otherwise $x_r \to x_r + 1$; spontaneous transitions $0 \to 1$ occur with probability β. T is the death probability and depends on both, age (individual states x_r) and the state of the neighbourhood of a site r. Both T and β depend on the model parameters, and are chosen in such a way that $T(60) = 1$. The model can be represented as a composition of the deterministic operator D_{age} and the random operator R_T^{β}. The deterministic operator represents ageing:

$$D_{\text{age}} = ((x_r + 1) \bmod M) \cdot \min\left(1, \sum_{s \in \eta} x_s\right).$$

D_{age} grows until the value M and is equal to 0 only when all sites in the neighbourhood of site r are equal to 0 (including site r). The random operator R_T^{β} governs deaths and births, and can be represented with the following matrix:

$$\begin{bmatrix} 1-\beta & \beta & 0 & \cdots & 0 \\ T(1) & 1-T(1) & 0 & & \\ T(2) & 0 & 1-T(2) & & \\ \vdots & \vdots & \ddots & & \vdots \\ T(M-1) & 0 & & 1-T(M-1) & 0 \\ T(M) & 0 & \cdots & 0 & 1-T(M) \end{bmatrix}$$

For $\beta = 0$, the random operator R_T^β has a structure of the random operator of a multi-type PO. $\beta > 0$ means that, in contrast to the other models presented in this section, model from [3] does not have an absorbing state. Nevertheless, in simulations the model exhibited bi-stability. Specifically, simulations initiated with a high density of states $x_r > 0$ had a different stationary state to simulations initiated with the 'all 0s' initial state. The former had a high density of vegetation and for the later, the stationary state had high density of 0s. The nature of this bi-stability, with dependance on the initial state as well as on the model parameters, appears to be similar to the other cases of bi-stability discussed in this chapter.

14.5 Conclusions

In this chapter we presented a selection of results concerning, percolation operators, contact processes and related models with absorbing states. In particular:

- We explained that POs on uniform graphs, in any dimension, with a neighbourhood of size $n \geq 2$ are non-ergodic and have the eroder property.
- We discussed the relation between POs and discrete-time CPs.
- We showed that the parameter spaces of a mean-field approximations of $PO_{n=2}$ and of a continuous-time CP in 1D with a single neighbour, can be easily mapped into each other.
- We presented different discrete-time approximations of the continuous-time CPs.

Furthermore, we observed:

- That the numerical estimate of the critical value of a stochastic process, simulated with asynchronous updating rule, is lower than the numerical estimate of the critical value for the same process simulated with synchronous updating scheme.
- Many similarities between models with three or more individual states that are bi-stable and exhibit hysteresis.

Acknowledgements I am grateful to Prof Robert MacKay for introduction to the fascinating world of probabilistic cellular automata and for showing me different techniques for analysing them. I am also grateful to the editors and the reviewers for their many insightful comments, which greatly improved this chapter. Finally, I would like to thank the organisers of the workshop on "Probabilistic Cellular Automata: Theory, Applications and Future Perspectives", June 2013, Eindhoven, for the opportunity to meet and learn from the experts of the field of PCAs. The research of the author has been funded by The Alfred P. Sloan Foundation, New York.

References

1. Antal, T., Droz, M.: Phase transitions and oscillations in a lattice prey-predator model. Phys. Rev. E **63**, 056119 (2001)

2. Arashiro, E., Tome, T.: The threshold of coexistence and critical behaviour of a predator-prey cellular automaton. J. Phys. A: Math. Theor. **40**, 887–900 (2007)
3. Bailey, R.M.: Spatial and temporal signatures of fragility and threshold proximity in modelled semi-arid vegetation. Proc. R. Soc. B **278**, 1064–1071 (2011)
4. Balister, P., Bollobas, B., Kozma, B.: Large deviations for mean field models of probabilistic cellular automata. Random Struct. Algorithms **29**, 399–415 (2006)
5. de Carvalho, K.C., Tome, T.: Probabilistic cellular automata describing a biological two-specimen system. Modern Phys. Lett. B **18**(17), 873–880 (2004)
6. de Carvalho, K.C., Tome, T.: Anisotropic probabilistic cellular automaton for a predator-prey system brazilian. J. Phys. **37**(2A), 466–470 (2007)
7. de Oliveira, M.M., Dickman, R.: How to simulate the quasi-stationary state. Phys. Rev. E **71**, 016129 (2005)
8. de Santana, L.H., Ramos, A.D., Toom, A.L.: Eroders on a plane with three states at a point. Part I: Deterministic. J. Stat. Phys. **159**(5), 1175–1195 (2015). doi:10.1007/s10955-015-1226-9
9. Diakonova, M., MacKay, R.S.: Mathematical examples of space-time phases. Int. J. Bifurc. Chaos **21**, 2297–2304 (2011)
10. Drossel, B., Schwabl, F.: Forest-fire model with immune trees. Phys. A **199**(2), 183–197 (1993)
11. Durrett, R.: Stochastic growth models: bounds on criticality. J. Appl. Probab. **29**, 11–20 (1992)
12. Durrett, R.: Stochastic spatial models. SIAM Rev. **41**(4), 677–718 (1999)
13. Durrett, R., Levin, S.A.: Stochastic spatial models: a user's guide to ecological applications. Philos. Trans. R. Soc. Lond. B **343**, 329–350 (1994)
14. Durrett, R., Neuhauser, C.: Epidemics with recovery in $d = 2$. Ann. Appl. Probab. **1**, 189–206 (1991)
15. Durrett, R., Swindle, G.: Are there bushes in a forest? Stoch. Process Appl. **37**, 19–31 (1991)
16. Durrett, R., Schonmann, R.H., Tanaka, N.I.: The contact process on a finite set. III: the critical case. Ann. Probab. **17**(4), 1303–1321 (1989)
17. Enss, T., Henkel, M., Picone, A., Schollwock, U.: Ageing phenomena without detailed balance: the contact process. J. Phys. A: Math. Gen. **37**, 10479 (2004)
18. Fallert, S.V., Ludlam, J.J., Taraskin, S.N.: Simulating the contact process in heterogeneous environments. Phys. Rev. E **77**, 051125 (2008)
19. Fatès, N.: Guided tour of asynchronous cellular automata. J. Cell. Autom. **9**(5–6), 387–416 (2014)
20. Grassberger, P., de la Torre, A.: Reggeon field theory (Schlogl's First Model) on a lattice: Monte Carlo calculations of critical behaviour. Ann. Phys. **122**, 373–396 (1979)
21. Guichard, F., Halpin, P.M., Allison, G.W., Lubchenco, J., Menge, B.A.: Mussel disturbance dynamics: signatures of oceanographic forcing from local interactions. Am. Nat. **161**(6), 889–904 (2003)
22. Harris, T.E.: Contact Interactions on a lattice. Ann. Probab. **2**(6), 969–988 (1974)
23. Hinrichsen, H.: Non-equilibrium critical phenomena and phase transitions into absorbing states. Adv. Phys. **49**(7), 815–958 (2000)
24. Kefi, S., Rietkerk, M., Alados, C.L., Pueyo, Y., Papanastasis, V.P., ElAich, A., de Ruiter, P.C.: Spatial vegetation patterns and imminent desertification in Mediterranean arid ecosystems. Nature **449**, 213–217 (2007)
25. Kefi, S., Rietkerk, M., van Baalen, M., Loreau, M.: Local facilitation, bistability and transitions in arid ecosystems. Theor. Popul. Biol. **71**(3), 367–379 (2007)
26. Kinzel, W., Yeomans, J.M.: Directed percolation: a finite-size renormalisation group approach. J. Phys. A: Math. Gen. **14**, L163–L168 (1981)
27. Liggett, T.M.: Classics in Mathematics: Interacting Particle Systems. Springer, Berlin (2005)
28. Liggett, T.M.: T.E. Harris' contributions to interacting particle systems and percolation. Ann. Probab. **39**(2), 407–416 (2011)
29. Maes, Ch., Shlosman, S.B.: When is an interacting particle system ergodic? Commun. Math. Phys. **151**(3), 447–466 (1993)

30. Makowiec, D., Gnacinski, P.: Universality class of probabilistic cellular automata. In: Bandini, S., Chopard, B., Tomassini, M. (eds.) Cellular Automata. LNCS, vol. 2493, pp. 104–113. Springer, Berlin (2002)
31. Mendonza, J.R.G.: Monte Carlo investigation of the critical behavior of Stavskaya's probabilistic cellular automaton. Phys. Rev. E **83**, 012102 (2011)
32. Peltomaki, M., Rost, M., Alava, M.: Characterizing spatiotemporal patterns in three-state lattice models. J. Stat. Mech. P02042 (2009)
33. Schonmann, R.H., Shlosman, S.B.: Wulff droplets and the metastable relaxation of kinetic ising models. Commun. Math. Phys. **194**(2), 389–462 (1998)
34. Stavskaya, O., Piatetski-Shapiro, I.I.: On homogeneous nets of spontaneously active elements. Syst. Theory Res. **20**, 75–88 (1971) (Originally published in Russian in 1969)
35. Stroock, D.W.: An Introduction to Markov Processes. Graduate Texts in Mathematics, vol. 230. Springer, Berlin (2005)
36. Taggi, L.: Critical probabilities and convergence time of Stavskaya's Probabilistic Cellular Automata. J. Stat. Phys. **159**(4), 853–892 (2015)
37. Toom, A.L.: Cellular automata with errors: problems for students of probability. In: Snell L. (ed) Topics in Contemporary Probability and Its Applications, Probability and Stochastic Series. CRC Press, Boca Raton (1995)
38. Toom, A.L.: Contours, Convex Sets, and Cellular Automata - Course notes from the 23th Colloquium of Brazilian. Mathematics. UFPE Department of Statistics, Recife (2004)
39. Toom, A.L., Vasilyev, N.B., Stavskaya, O.N., Mityushin, L.G., Kurdyumov, G.L., Pirogov, S.A.: Discrete local Markov systems. In: Dobrushin, R.L., Kryukov, V.I., Toom, A.L. (eds.) Stochastic Cellular Systems, Ergodicity, Memory and Morphogenesis, pp. 1–182. Manchester University Press, Manchester (1990)
40. Vazquez, F., Lopez, C., Calabrese, J.M., Munoz, M.A.: Dynamical phase coexistence: a simple solution to the "savanna problem". J. Theor. Biol. **264**, 360–366 (2010)

Chapter 15
Phase Transitions of Cellular Automata

Franco Bagnoli and Raúl Rechtman

Abstract We explore some aspects of phase transitions in cellular automata. We start recalling the standard formulation of the Monte Carlo approach for a discrete system. We then formulate the cellular automaton problem using simple models and illustrate different types of possible phase transitions: density phase transitions of first and second order, damage spreading, dilution of deterministic rules, asynchronism-induced transitions, synchronization phenomena, chaotic phase transitions and the influence of the topology.

15.1 Introduction: Monte Carlo Simulations

The main result of statistical mechanics is that of expressing the probability distribution of a statistical ensemble in terms of its constraints. Just to be concrete, let us consider a discrete system that can be described by N Boolean variables $x_i \in \{0, 1\}$, $i = 1, \dots, N$ located in sites connected by a graph defined by an adjacency matrix $a_{ij} = 1$ if i is connected to j and zero otherwise. A configuration of the system is expressed as $\mathbf{x} = (x_1, x_2, \dots, x_N)$.

Let us denote by $E(\mathbf{x})$ is energy of such a configuration, and with $P(\mathbf{x})$ the probability of observing it.

The simplest way of deriving the equilibrium probability distribution is that following the principle of maximum entropy. One has to maximize the entropy

F. Bagnoli (✉)
Department of Physics and Astronomy and CSDC, University of Florence,
via G. Sansone 1, 50019 Florence, Italy
e-mail: franco.bagnoli@unifi.it

F. Bagnoli
INFN, sez. Firenze, Sesto Fiorentino, Italy

R. Rechtman
Instituto de Energías Renovables, Universidad Nacional Autónoma de México,
Apdo. Postal 34, 62580 Temixco, MOR, Mexico
e-mail: rrs@ier.unam.mx

P.-Y. Louis and F.R. Nardi (eds.), *Probabilistic Cellular Automata*, Emergence, Complexity and Computation 27, https://doi.org/10.1007/978-3-319-65558-1_15

$$S = -\sum_{\mathbf{x}} P(\mathbf{x}) \log(P(\mathbf{x}))$$

with the given constraints. In the so-called "canonical ensemble" (a system in contact with a heat bath that keeps the temperature constant), the average energy

$$U = \sum_{\mathbf{x}} E(\mathbf{x})P(\mathbf{x})$$

is kept constant by the heat bath. It is straightforward to derive the probability distribution $P(\mathbf{x})$,

$$P(\mathbf{x}) = \frac{1}{Z}\exp(-\beta E(\mathbf{x}))$$

where β corresponds to inverse temperature and Z the normalization constant (partition function). In principle, this solves the problem of computing the average value $\langle A \rangle$ of an observable $A(\mathbf{x})$,

$$\langle A \rangle = \sum_{\mathbf{x}} A(\mathbf{x})P(\mathbf{x}).$$

The problem is that in general the number of configurations is huge, and therefore a brute-force evaluation of this sum is not feasible.

The Monte Carlo technique allows one to compute the time-average $\overline{A}$ of the observable over a fictitious trajectory $\mathbf{x}(t)$

$$\overline{A} = \frac{1}{T}\sum_{t=0}^{T} A(\mathbf{x}(t)).$$

The trajectory is obtained by defining the conditional probability $M(\mathbf{x}|\mathbf{y})$ of getting $\mathbf{x} \equiv \mathbf{x}(t+1)$ given $\mathbf{y} \equiv \mathbf{x}(t)$ (i.e. defining a Markov chain), as a function of the difference in energy of $\mathbf{x}$ and $\mathbf{y}$. Clearly, speaking of numeric simulations, we refer to finite chains.

The main requirement is that the Markov chain has to be ergodic, i.e. each state can be reached from any other state in a finite number of steps (not being trivially periodic), and that the iteration of the procedure leads to a unique probability distribution.

Considering now the stochastic sampling, this property assures that a long enough trajectory visits all states a number of times proportional to the asymptotic distribution.

We shall now consider the temporal evolution of the probability distribution, denoted as $P(\mathbf{x}, t)$. We have

$$P(\mathbf{x}, t+1) = \sum_{\mathbf{y}} M(\mathbf{x}|\mathbf{y})P(\mathbf{y}, t),$$

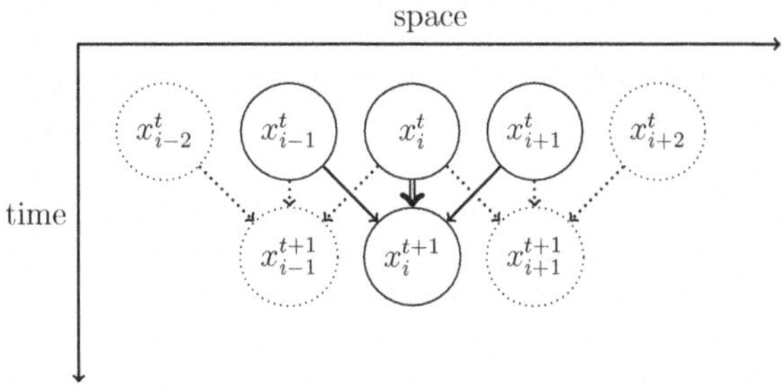

Fig. 15.1 Monte Carlo neighbourhood (2 cells plus the same-cell link)

or, in vectorial terms

$$\mathbf{P}(t+1) = \mathbf{MP}(t).$$

$M(\mathbf{x}|\mathbf{y})$ is in general decomposed into a series of N stochastic "local" moves that occur with probability $\tau(x_i|Y_i)$, which is the probability of getting x_i given its neighbourhood $Y_i = \{y_j : a_{ij} = 1\}$. One can visualize the Monte Carlo procedure as the evolution of a time–space graph, in which the connections are such that $(i, t+1)$ is connected to (j, t) if $a_{ij} = 1$, and in any case a cell at time $t+1$ is connected to the cell in the same location at time t (see Fig. 15.1). Each Monte Carlo time step is decomposed in N microscopic steps, in which just one random site is updated, and the others are copied ($x_i(t+1) = x_i(t)$).

The actual trajectory is computed by drawing, for each site i and time t, uniformly distributed random numbers $r_i(t)$ and computing Boolean quantities like $[r_i(t) < \tau(x_i|Y_i)]$, where $[\cdot] = 1$ if $\cdot$ is true and zero otherwise. If one thinks of extracting all the random numbers before the simulation, the trajectories are deterministic over the random field $r_i(t)$.

15.1.1 An Example: The Ising Model

Let us illustrate these concepts with the Ising model. Given the coupling J and the magnetic field H, the energy $E(\mathbf{s})$ of a spin configuration $\mathbf{s} = (s_1, s_2, \ldots, s_N)$ ($s_i = 2x_i - 1 = \pm 1$) is given by

$$E(\mathbf{x}) = -J \sum_{i,j} a_{ij} s_i s_j - H \sum_i s_i. \tag{15.1}$$

The Ising probability distribution is

$$P(\mathbf{s}) = \frac{1}{Z} \exp\left(-\beta \left(J \sum_{i,j} a_{ij} s_i s_j + H \sum_i s_i\right)\right), \tag{15.2}$$

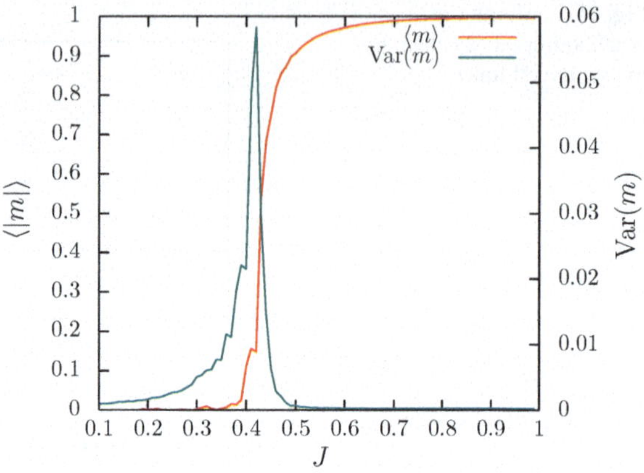

Fig. 15.2 Phase transition for the Ising model in 2D with nearest neighbour interactions. Average magnetization $|\langle m\rangle|$ and variance as a function of the rescaled coupling J for $H = 0$. Size 40×40, $T = 4000$, transient $4 \cdot 10^4$

and we can absorb the inverse temperature β in the parameters J and H (control parameters).

The magnetization m is defined as

$$m = m(J, H) = \sum_{\mathbf{s}} \left(P(\mathbf{s}) \frac{1}{N} \sum_i s_i \right).$$

It constitutes a suitable observable for this problem, as also its variance. From Onsager solution in 2D and zero magnetic field [19], we should observe a phase transition at $J_c \simeq 0.44$, with a transition from $m = 0$ to $m \neq 0$ and the divergence of its variance, see Fig. 15.2 for a numerical simulation.

There are many possible recipes for the Monte Carlo implementation, the one that we examine is the *heat bath* dynamics, for which the probability that spin i takes value s_i' is

$$\begin{aligned}\tau(s_i'|S_i) &= \frac{\exp(s_i(H + J\sum_j a_{ij}s_j)}{\exp(s_i(H + J\sum_j a_{ij}s_j)) + \exp(-s_i(H + J\sum_j a_{ij}s_j))} \\ &= \frac{1}{1 + \exp(-2s_i(H + J\sum_j a_{ij}s_j))}.\end{aligned} \tag{15.3}$$

In practice, one randomly chosen element of the Markov matrix is $M(\mathbf{s}'|\mathbf{s}) = \tau(s_i'|S_i)$, all other spins remaining the same.

15.1.2 Equilibrium Phase Transitions

There is a vast literature about phase transition in equilibrium statistical physics. We want here just recall some properties that can be useful for extending the concept to arbitrary systems, not necessarily in equilibrium, and therefore we only refer to Monte Carlo investigations.

Phase transitions are characterized by a change of the value of some observable, say the magnetization $m(J, H)$, in correspondence of a precise value of a control parameter. In practice, we can say that the dynamics of the system changes its structure in correspondence of a phase transition, for instance the phase space may effectively break in two zones that do not communicate at all. This is equivalent to say that the system is no more ergodic, and we speak of *ergodicity breaking*.

If we consider the point of view of deterministic trajectories over a random field, the phase transition can be seen as a bifurcation from a single to multiple attractors.

However, we have a kind of contradiction here: we chose the Monte Carlo dynamics to be ergodic, so how can ergodicity breaking occur? Actually, this breaking only manifests itself in a limit procedure: for a finite system (finite N), and long enough time, all the phase space is visited (it is finite), and therefore the average of observables takes a unique value. However, near the phase transition, the observables (say, the magnetization in the Ising model) maintain the same value for very long periods, with occasional switches from one extreme to another. So, while its average value has a certain value (say, zero), one never observes such value! The time that the system spends on one phase becomes longer as we approach the critical value of the control parameter and (exponentially) as we increase the system size.

If we take first the limit of infinite system size and then that of infinite time, we observe the ergodicity breaking. In practice, it is sufficient to use a large enough system. In the language of stochastic trajectories, there are two low-energy valley separated by a high (energy) and/or large (entropy) barrier. In order to connect the two valleys, a path should climb the separating saddle, and the associated probability becomes smaller and smaller with the system size, in the vicinity of the phase transition and above.

In the language of Markov processes, we always have an irreducible transition matrix (since the dynamics is ergodic), but in the previous limit the time product of matrices (denoted as $\mathbf{M}$) effectively breaks in two (or more) submatrices, that do not communicate

$$\mathbf{M} = \begin{pmatrix} \mathbf{M}_1 & \epsilon \\ \epsilon & \mathbf{M}_2 \end{pmatrix} \xrightarrow{N \to \infty} \begin{pmatrix} \mathbf{M}_1 & 0 \\ 0 & \mathbf{M}_2 \end{pmatrix},$$

where the ϵ denote the paths that connects the two valleys. The asymptotic distribution $P^{\text{eq}}(\mathbf{x})$ is proportional to the eigenvector of M with eigenvalue 1. At phase transition, this eigenvalue becomes degenerate and we have two or more asymptotic distributions, with different "basins".

We can introduce the correlation function

$$C(\rho, \tau) = \left(\sum_{i=1}^{N} \sum_{t=1}^{T} x_i(t) s_{i+\rho}(t+\tau) \right) - \left(\sum_{i=1}^{N} \sum_{t=1}^{T} s_i(t) \right) \left(\sum_{i=1}^{N} \sum_{t=1}^{T} s_{i+\rho}(t+\tau) \right)$$

The observables can be defined in terms of the correlation function.

The correlation function is expected to decrease exponentially

$$C(\rho, \tau) \sim \exp\left(-\frac{\rho}{\xi_\perp}\right) \exp\left(-\frac{\tau}{\xi_\parallel}\right).$$

defining the correlation lengths $\xi_\perp$ (with respect to space) and $\xi_\parallel$ (with respect to time).

At a phase transition (nonanalytical behaviour of some observables like discontinuities, divergence or angular points), the correlation lengths can stay finite (first-order phase transitions) or diverge (second-order phase transitions). In the latter case,

$$\xi(J, H; N) \sim N^\alpha \tilde{\xi}\left(\frac{J}{N^\gamma}, \frac{H}{N^\delta}\right),$$

where α, γ, δ are critical exponents. Also observables like the magnetization exhibit similar scaling behaviour. This phenomenology extends to systems defined directly by stochastic transition probabilities.

15.2 Probabilistic Cellular Automata

In many cases, we are looking for the asymptotic properties of a system that is just defined in terms of the local transition probabilities, of which Probabilistic Cellular Automata (PCA) are prototypical examples.

Cellular automata are defined in a way similar to the previous Monte Carlo time–space evolution, allowing for generic transition probabilities and parallel evolution of all cells at the same time. PCA are therefore Markov chains for which the matrix elements are given by the product of the local transition probabilities (generally uniform),

$$M(\mathbf{x}|\mathbf{y}) = \prod_i \tau(x_i|Y_i).$$

Again, we can define stochastic trajectories (or deterministic trajectories over a stochastic field)

$$x_i(t+1) = [r_i(t) < \tau(x_i|Y_i)].$$

Deterministic Cellular Automaton (DCA) can be considered as limit cases of PCA, where the transition probabilities τ are either zero or one.

15.2.1 *Parallel Ising Model*

For instance, we can define a parallel version of the Ising model, for which

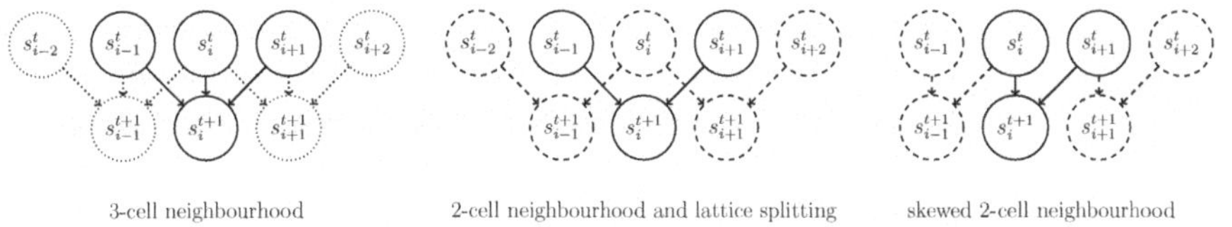

Fig. 15.3 2-cell neighbourhood

$$M(\mathbf{s}'|\mathbf{s}) = \prod_i \tau(s_i'|S_i),$$

with τ given by Eq. (15.3).

In this case, we can still have an asymptotic probability distribution if the interactions are symmetric (here they are so by definition), but the asymptotic distribution is now [9]

$$P^{\mathrm{eq}}(\mathbf{s}) = \frac{1}{Z}\prod_i e^{\beta H s_i} \cosh\left(\sum_j \beta\left(H + J\sum_j a_{ij}s_j\right)\right),$$

where Z is again the normalization constant.

Notice that the transition probabilities of Eq. (15.3) do not depend on the previous value of the site s_i. If we apply them in parallel to all sites, at least in one dimension and with nearest-neighbour interactions, the lattice decouples in two noninteracting sublattices (for even N, see Fig. 15.3), so that $s_i' = f(s_{i-1}^t + s_{i+1}(t), r_i(t))$ It is an example of a totalistic PCA, that has been studied by Kinzel [23] and shows no phase transition.

15.2.2 Domany–Kinzel Model. Absorbing States

We can extend the parallel Ising example to a general case, on the same two-neighbour network (Fig. 15.3), defining three independent totalistic transition probabilities, as shown in Table 15.1. This model has been studied by Domany and Kinzel [10, 23] and can be considered the simplest model showing a phase transition.

Table 15.1 Transition probabilities of the Domany–Kinzel model

$S = s_{-1} + s_{+1}$	$X = x_{-1} + x_{+1}$	$\tau(1\|S)$	$\tau(0\|S) = 1 - \tau(1\|S)$	Bond percolation	Site percolation
-1	0	w	$1 - w$	0	0
0	1	p	$1 - p$	p_b	p_s
1	2	q	$1 - q$	$p_b(2 - p_b)$	p_s

For generic values of w ($\tau(1|0)$), p ($\tau(1|1)$) and q ($\tau(1|2)$), this model can be mapped onto a parallel Ising model with a plaquette term [23] (we need another control parameter in addition to H and J since here we have three free probabilities),

$$E(\mathbf{S}) = -\sum_i s_i \left(H + J(s_{i-1} + s_{i+1}) + K s_{i-1} s_{i+1}\right).$$

Denoting $h = \exp(-2H)$, $j = \exp(-4J)$, $k = \exp(-2K)$, we have $w = 1/(1 + hk/j)$, $p = 1/(1 + h/k)$, $q = 1/(1 + hjk)$ and therefore

$$H = \frac{1}{6}\log\frac{wpq}{(1-w)(1-p)(1-q)}, \quad J = \frac{1}{8}\log\frac{(1-w)q}{w(1-q)}, \quad K = \frac{1}{6}\log\frac{w(1-p)q}{(1-w)p(1-q)}.$$

However, this model does not show any phase transition.

If we set $w = 0$ (by letting the coupling take infinite values with suitable limits), we leave the equilibrium condition. In this limit, the configuration $\mathbf{s} = -1$ becomes an absorbing state. We can also switch to the Boolean representation by setting $x_i = (s_i + 1)/2$. In this representation, the absorbing state is the configuration $\mathbf{x} = 0$. It is called absorbing since it cannot be left by the dynamics once entered. The order parameter is here the density of ones

$$c = \frac{1}{N}\sum_i x_i.$$

We can reformulate the phase transition in this new language: for finite N, there is always a probability $M(0|\mathbf{y})$ that brings any configuration to the absorbing state in one step. In the limit $N \to \infty$ and for a suitable value of the parameters p and q, this probability goes to zero and the Markov matrix becomes reducible. It is composed by a submatrix M_1 that maps states "near" to 0 into 0 in a few time steps, and a set of states with a nonvanishing density c.

Again, one can speak of deterministic trajectories once that the stochastic field has been laid out. The evolution equation of the system is

$$x_i' = [r_i^{(1)}(t) < p](x_{i-1}(t) \oplus x_{i+1}(t)) \oplus [r_i^{(2)}(t) < q] x_{i-1}(t) x_{i+1}(t)$$

where $\oplus$ is the XOR operation (sum modulus two). Notice that the two random numbers $r_i^{(1)}(t)$ and $r_i^{(2)}(t)$ may be the same or not, since the two conditions $(x_{i-1}(t) \oplus x_{i+1}(t)$ and $x_{i-1}(t)x_{i+1}(t)$ are never true at the same time (but this makes a difference for damage spreading, Sect. 15.2.5).

In the language of trajectories, one can say that there are two attractors, the fixed point 0 and a "chaotic" attractor with $d > 0$, each one with its own basin. More on absorbing phase transition can be found in Ref. [17].

For $w = q = 0$ and $p = 1$, we have the deterministic rule 90 in Wolfram's notation [28], so the line $q = w = 0$ corresponds to the dilution of rule 90 (see Figs. 15.4 and 15.5).

15.2.3 Mean-Field Approximation

In order not to use a heavy notation, let us apply this approximation using the DK model, assuming that a site i at time $t+1$ is connected to sites i and $i+1$ at time t (i.e. using the skewed lattice of Fig. 15.3).

The evolution equation for the probability distribution is

$$P(x_1, x_2, \ldots, x_N; t+1) = \sum_{y_1, y_2, \ldots, y_N} \left(\prod_i \tau(x_i | y_i, y_{i+1}) \right) P(y_1, y_2, \ldots, y_N; t), \tag{15.4}$$

considering appropriate boundary conditions (e.g., periodic). We can obtain the reduced probabilities $\pi_\ell(x_1, \ldots, x_\ell; t)$ by summing $P(x_1, x_2, \ldots, x_N; t+1)$ over all $i > \ell$. If the system is translation-invariant, one obtains the same result summing the elements of any set of consecutive variables. Since $\sum_{x_i} \tau(x_i | y_i, y_{i+1}) = 1$ for all x_i, we can then sum over $y_{i+2}, \ldots, y_N$, obtaining

$$\begin{aligned} \pi_1(x_1, t+1) &= \tau(x_1 | y_1, y_2) \pi_2(y_1, y_2; t), \\ \pi_2(x_1, x_2, t+1) &= \tau(x_1 | y_1, y_2) \tau(x_2 | y_2, y_3) \pi_3(y_1, y_2 y_2; t), \\ &\ldots \end{aligned}$$

i.e. a hierarchy of equations that are equivalent to Eq. (15.4).

If the correlation length ξ is less than N, two cell separated by a distance greater that ξ are practically independent. The system acts like a collection of subsystems each of length ξ (this is why ergodicity and selfaveraging holds far from the transition). Since ξ is not known a priori, one assumes a certain correlation length ℓ and computes the quantity of interest. By comparing the values of these quantities with increasing ℓ generally a clear scaling law appears, allowing to extrapolate the results to the case $\ell \to \infty$.

The very first step is to assume $\ell = 1$. In this case, we can simply factorize $\pi_2(x_1, x_2) = \pi_1(x_1)\pi_1(x_2)$. By calling $c = \pi_1(1; t)$ $(1 - c = \pi_1(0; t))$, $c' = \pi_1(1; t+1)$ and using the transition probabilities of Table 15.1 with $w = 0$, one gets (Fig. 15.6)

$$c' = 2pc(1-c) + qc^2.$$

Notice that in the mean-field approximation, the evolution of the system is given by a deterministic equation for the average value of observables. In this approximation,

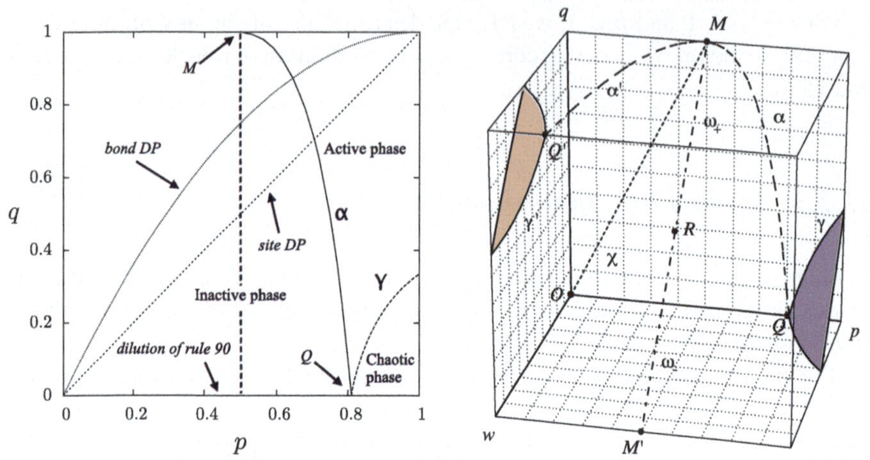

Fig. 15.4 The phase diagram of the Domany–Kinzel model, α marks the density transition and γ the damage transition. *Left* the phase diagram for $w = 0$. The *dashed line* marks the transition line for the simplest mean-field approximation. *Right* the complete phase diagram. The curves labelled α and α' belong to planes $w = 0$ and $w = 1$ resp., and correspond to the density phase transitions. The *solid curves* correspond to the intersection of the damage critical surface (*shaded*) γ and γ' with the boundaries of the cube. The *dotted-dashed lines* labelled ω_+ and ω_- correspond to the existence line for the parallel Ising model for positive and negative temperatures, resp. The points labelled M and M' to the critical points of the parallel Ising model at zero temperature (compact DP), and the point labelled R to infinite temperature. The *dotted line* labelled χ corresponds to the damage in the parallel Ising model

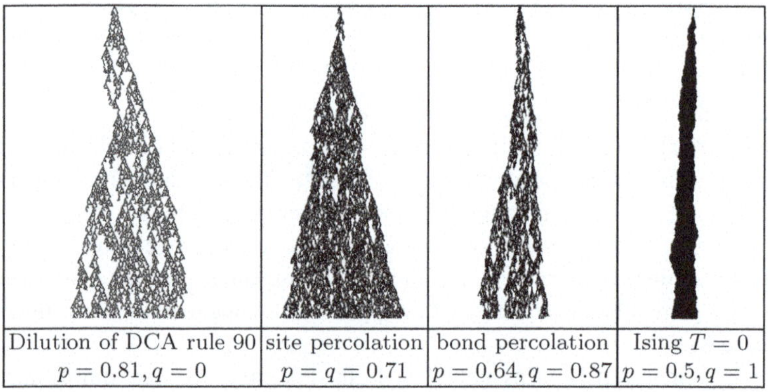

Fig. 15.5 Typical patterns of the DK model. Space runs horizontally and time vertically, from *top* to *bottom*

a phase transition corresponds to a bifurcation (change of stability of the attractors) of the map.

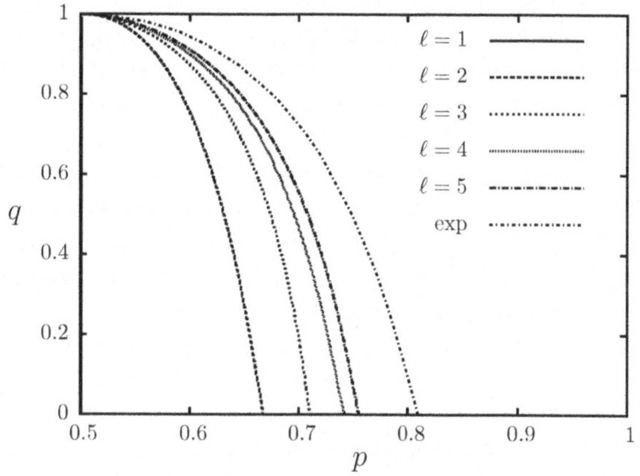

Fig. 15.6 Local structure approximation for the DK model, with several values of length ℓ. The case $\ell = 1$ is the simplest mean-field approximation and corresponds to the line $p = 1/2$. The line marked "exp" corresponds to numerical simulations as in Fig. 15.4

The fixed points ($c' = c$) are $c = 0$ and $c = 2p/(2p - q)$. There is a change of stability from $c = 0$ (the absorbing state) to $c > 0$ for $p_c = 1/2$. As shown in Fig. 15.4-left, this approximation is quite rough.

The DK model includes the Directed Percolation (DP) one [22], which can be formulated thinking to an infection process: an individual i at time t can get infected by its infected neighbours at the previous time step, with a probability that depends on the number of infected neighbours (bond percolation) or not (site percolation), see Table 15.1. In the mean-field approximation, we have for the bond percolation the line $q = p(2 - p)$, and for the site percolation the line $q = p$, as shown in Fig. 15.4-left.

There are two ways of extending the above approximation. The first one is still to factorize the cluster probabilities at single site level but to consider more time steps, for instance obtaining $\pi_1(t + 2)$ in terms of $\pi_3(t)$ and then factorizing π_3 in terms of π_1. The map is still expressed as a polynomial of the density c. The advantage of this method is that we still work with a scalar (the density), but in the vicinity of a phase transition the convergence towards the thermodynamic limit is very slow.

The second approach, sometimes called *local structure approximation* [16], is a bit more complex. Let us start from the generic ℓ cluster probabilities π_ℓ. We generate the $\ell - 1$ cluster probabilities $\pi_{\ell-1}$ from π_ℓ by summing over one variable,

$$\pi_{\ell-1}(x_1, \ldots, x_{\ell-1}) = \sum_{x_\ell} \pi_\ell(x_1, \ldots, x_{\ell-1}, x_\ell).$$

The $\ell + 1$ cluster probabilities are generated by using a Bayesian estimation

$$\pi_{\ell+1}(x_1, x_2, \ldots, x_\ell, x_{\ell+1}) = \frac{\pi_\ell(x_1, \ldots, x_\ell)\pi_\ell(x_2, \ldots, x_{\ell+1})}{\pi_{\ell-1}(x_2, \ldots, x_\ell)}.$$

Finally, one is back to the ℓ cluster probabilities by applying the transition probabilities

$$\pi'(x_1, \dots, x_\ell) = \sum_{y_1,\dots,y_{\ell+1}} \prod_{i=1}^{l} \tau(x_i | y_i, y_{i+1}).$$

This last approach has the disadvantage that the map lives in a high-dimensional (2^ℓ) space, but the results converge much better in the whole phase diagram.

This mean-field technique can be considered an application of the transfer matrix concept to the calculation of the eigenvector (asymptotic probability distribution) corresponding to the maximum eigenvalue (fundamental or ground state), by means of the iteration of the matrix.

15.2.4 Asynchronism of DCA

An unexpected phase transition occurs with an increasing level of asynchronism of some DCA rule [11, 12]. Let us denote by $f(x_{i-1}, x_i, x_{i+1})$ the deterministic rule. The evolution equation of its dilution is

$$x_i' = x_i \oplus [r_i(t) < (1-p)]\big(x_i \oplus f(x_{i-1}, x_i, x_{i+1})\big).$$

With probability $1-p$, the site follows the rule f, and with probability p it keeps its old value.

Examples of phase transitions are shown in Fig. 15.7.

An unexpected fact is that the simplest mean-field approximation completely fails for this problem. Indeed, we have

$$c' = pc + (1-p) \sum_{a,b,c=0}^{1} f(a,b,c)c^{a+b+c}(1-c)^{3-a-b-c}$$

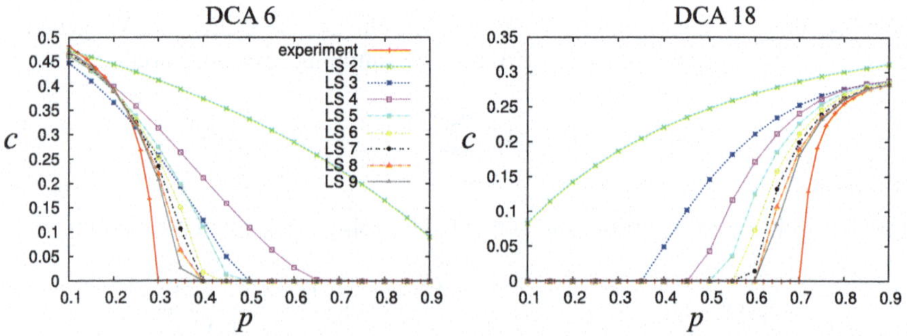

Fig. 15.7 DCA dilution phase transition for two Elementary Cellular Automaton rule 6 and 18 in Wolfram notation [28], comparisons between numerical simulations and the local structure approximation (from Ref. [13]). In the y axis, the asynchronism parameter p

and for the stationary state $c' = c$ one gets

$$c = \sum_{a,b,c=0}^{1} f(a,b,c)c^{a+b+c}(1-c)^{3-a-b-c}$$

i.e. the mean-field approximation of the deterministic rule, without any dependence on p. Increasing the order of the mean-field approximation (local structure approximation), one can approximate the actual phase transition behaviour [13], as shown in Fig. 15.7.

15.2.5 Damage Spreading

We have said that the large-time distribution $\mathbf{x}(T)$ depends in general on the random field and the initial conditions $\mathbf{x}(0)$, although, for large N, the observables like the density do not depend on them due to ergodicity and self-averaging. Actually, we can check the dependence on the initial conditions by considering the evolution of an initial difference between two replicas, evolving on the same random field and looking at the difference (or damage) $z_i = x_i \oplus y_i$,

$$\begin{aligned}
x_i' &= [r_i^{(1)}(t) < p]\big(x_{i-1}(t) \oplus x_{i+1}(t)\big) \oplus [r_i^{(2)}(t) < q]x_{i-1}(t)x_{i+1}(t),\\
y_i' &= [r_i^{(1)}(t) < p]\big(y_{i-1}(t) \oplus y_{i+1}(t)\big) \oplus [r_i^{(2)}(t) < q]y_{i-1}(t)y_{i+1}(t),\\
z_i' &= x_i' \oplus y_i' = [r_i^{(1)}(t) < p]\big(z_{i-1}(t) \oplus z_{i+1}(t)\big) \oplus [r_i^{(2)}(t) < q] \cdot\\
&\quad \big((z_{i-1}(t)z_{i+1}(t) \oplus z_{i-1}(t)x_{i+1}(t) \oplus x_{i-1}(t)z_{i+1}(t) \oplus x_{i-1}(t)x_{i+1}(t)\big).
\end{aligned}$$

Since now the two conditions can occur at the same time, there is a difference in the evolution if one uses one or two random numbers per site (or if they are otherwise correlated). Looking only at the evolution of the difference $\mathbf{z}$, the evolution of the x replica (which is not affected by z) is just another field (although it is not fully random). The quantity z shows another phase transition (Fig. 15.4) that characterizes the dependence on the initial condition: in one phase the difference goes to zero, meaning that all initial conditions will follow after a transient time the same trajectory, only depending on the stochastic field. In the other phase, the system maintains forever some memory of the initial condition.

This phase transition also belongs to the directed percolation universality class. It is possible to approximately map the density phase transition onto the damage one [2].

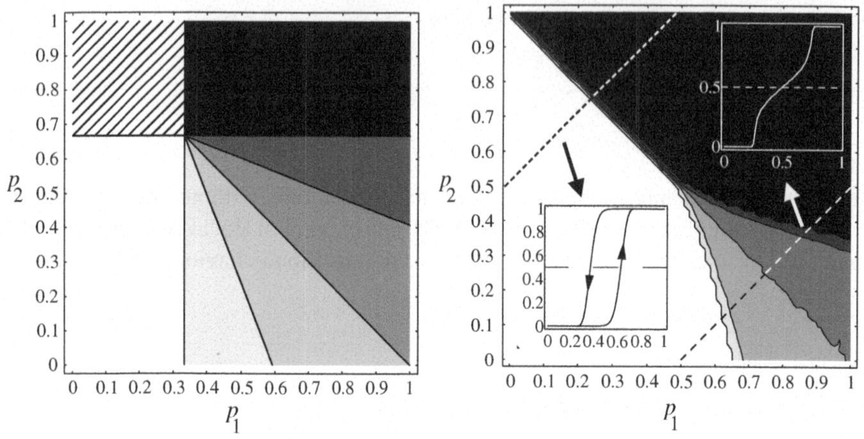

Fig. 15.8 Phase transition diagrams of the BBR model (colour code: *white*=0, *black*=1). *Left* mean-field phase diagram for the density. *Right* numerical phase diagram of the density, in the inset the variation of the density when cutting the phase diagram; the hysteresis inset at *bottom right* is obtained by setting $w = 10^{-4}$, $T = 500$. Numerical simulations with $N = T = 10^4$

15.2.6 A Richer Phase Diagram: The BBR Model

The DK model is quite useful for studying nonequilibrium phase transitions due to its simplicity. In order to explore other types of transitions beyond DP, let us introduce the BBR model [8], that is a 3-input cellular automata with two absorbing states. It is a totalistic automaton, meaning the transition probability depends on the sum S of the states in the neighbourhood, with $0 \leq S \leq 3$. The BBR transition probabilities $\tau(x'|S)$ are $\tau(1|0) = w$, $\tau(1|1) = p_1$, $\tau(1|2) = p_2$, $\tau(1|3) = 1 - w$ By setting $w = 0$, the states 0 and 1 are absorbing, and on the line $p_1 = 1 - p_2$, the system is symmetric for the inversion $1 \leftrightarrow 0$.

As can be seen in Fig. 15.8, we have here, for high-p_1 and low-p_2 value, two DP transitions reminiscent of the DK model. The two lines meet at about $p_1 = p_4 = 0.5$ ($p_1 = 1 - p_2 = 1/3$ in the mean-field approximation). In this point, the universality class changes to that of parity conservation. In the low-p_1, high-p_2 part of the diagram, we have a first-order transition: the two absorbing states are stable (as predicted by the mean-field analysis) and we can investigate the nature of an hysteresis cycle. In order to do that, we have to remove the absorbing characteristic of the states 0 and 1. We do this by imposing that $w = \tau(1|0)$ is small but different from zero, so that in principle the system does not more show a true phase transition. Indeed, that states with high or low values of the density are now metastable, so we have to tune the simulation time with the value of w. This tuning is however not critical: for a large range of values of simulations times, we obtain an hysteresis diagram similar to that of Fig. 15.8.

15.2.7 Janssen–Grassberger's Conjecture

The DP class is extremely robust with respect to the microscopic dynamic rules. The large variety and robustness of DP models led Janssen [21] and Grassberger [14] to the conjuncture that all systems with a single-order parameter and a single absorbing state will belong to the universality class of the Directed Percolation (DP) model. More precisely, the requirements are

- The model displays a continuous phase transition from a fluctuating active phase into a dominant stable absorbing state.
- The transition is characterized by a positive one-component order parameter.
- The dynamic rules involve only short-range processes.
- The system has no unconventional attributes such as additional symmetries or quenched randomness.

We have already seen that the BBR model has two absorbing states. As far as their basins are different, the phase transition belongs to the DP universality class, on the symmetry line $p_1 = 1 - p_2$ it switches to the parity conservation class.

Another way of violating this condition is that of modifying the stability of the absorbing state. This can be easily realized in the synchronization scenario [15].

15.2.8 Synchronization

The idea of a replica synchronization is the following: take two replicas of a system, either driven by a deterministic or a stochastic dynamics (in the latter case, the random field is the same for the two systems). Let one system evolve by itself, and "push" the other towards the first. If the pushing is strong enough, the system will synchronize. A simple illustration is the following. Let's consider a continuous map $x' = f(x)$ and construct the synchronization mechanism

$$\begin{aligned} x' &= f(x), \\ y' &= (1-p)f(y) + pf(x), \end{aligned}$$

for $p = 0$ the two systems are completely disconnected, and if the map f is chaotic, they stay well separated. For $p = 1$, the two system are identically the same. There is a critical value p_c such that the distance $\delta = |x - y|$ goes to zero. For small distance, δ evolves as

$$\delta' = (1-p)|f(y) - f(x)| \simeq (1-p)\left|\frac{\mathrm{d}f(x)}{\mathrm{d}x}\right|\delta$$

and thus

$$\delta(t) = (1-p)^t \delta(0) \prod_{t'} \left| \frac{\mathrm{d} f(x(t'))}{\mathrm{d}x} \right| = (1-p)^t \delta(0) \exp\left(\sum_{t'} \log \left| \frac{\mathrm{d} f(x(t'))}{\mathrm{d}x} \right| \right)$$
$$= \delta_0 \exp((\log(1-p) + \lambda) t),$$

where λ is the Lyapunov exponent of the map. Thus, when $\delta(t) = \delta(0)$ (the synchronization threshold), $p_c = 1 - \exp(-\lambda)$, and this relates the synchronization threshold to the chaotic properties of the map.

This mechanism can be applied in several ways to extended systems (coupled map lattices and cellular automata). For reference, consider the following generic coupled system

$$x_i' = f(g(x_{i-1}, x_i, x_{i+1})),$$

where g defines the coupling. One can use a homogeneous "pushing", i.e. use the same p for all sites, or, at the other extreme, a all-or-none pushing, i.e. choose a fraction p of sites to be completely synchronized and leave the other unperturbed.

Using the first mechanism, one again relates the synchronization threshold to the Maximum Lyapunov Exponent (MLE) of the system. Chaotic systems are expected to amplify the distance between replicas. For a value of p slightly below the synchronization threshold, some patches may synchronize for some time, after which they will separate. This picture resembles that of a growing interface that may stay pinned to local traps. From field theory studies, such a behaviour is denoted multiplicative noise (MN) and is equivalent to the behaviour of the "bounded" Kardar–Parisi–Zhang equation, which describes the behaviour of a growing surface that tends to pin and is pushed from below [20, 24, 25]. On the other hand, stable systems have a negative MLE. So, replicas should naturally synchronize once their distance is (locally) below the threshold of validity of linear analysis. However, when the local difference is large, nonlinear terms may maintain or amplify this distance. In this case, synchronized patches may be destabilized only at the boundaries. Again, theoretical studies associate such a behaviour to that of directed percolation (DP) [22].

However, it is questionable if the MLE exponent really captures the chaotic properties of an extended system. For instance, let us take f chaotic and $g(a, b, c) = \varepsilon(a + c) + (1 - 2\varepsilon)b$, i.e. a diffusive coupling. The Lyapunov exponent $\lambda(\varepsilon)$ in general decreases with ε, since the coupling acts like a constraint (a kind of surface tension). Thus, λ takes its maximum values for $\varepsilon = 0$, but in this case the chaos does not spread on the lattice.

On the contrary, the all-or-none ("pinching") synchronization mechanism shows that the case in which synchronization is most difficult is for $\varepsilon \simeq 1/3$, which is what one intuitively expects. Moreover, we can apply this synchronization mechanism also to cellular automata, provided that the two replicas evolve using the same random field. It is possible to show that in this case, one can develop a concept of Boolean derivative for such a discrete systems and obtain an equivalent of the maximum Lyapunov exponent, which is related to the pinching synchronization threshold [1, 4].

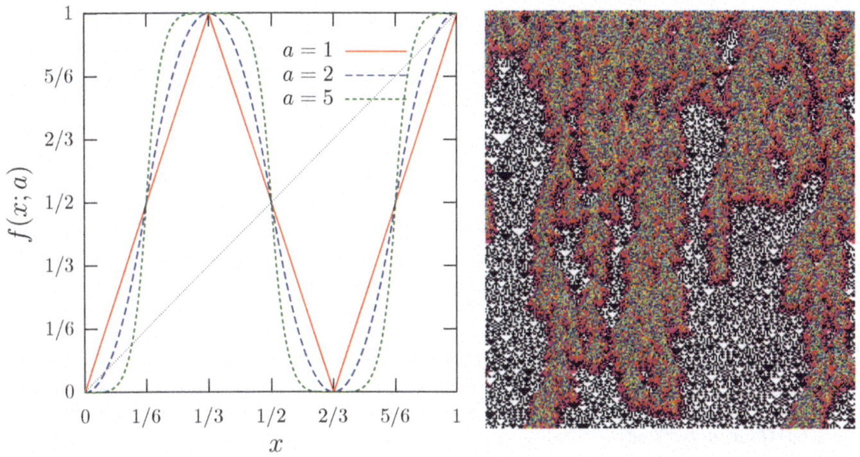

Fig. 15.9 *left* The graph of $f(x; a)$ for three values of a. *right* Space time pattern of the CML of Eq. (15.5) with $a = 1.9$ and $N = 256$ drawn horizontally for a total time of $T = 300$ time steps drawn vertically from *top* to *bottom*. The initial configuration $x(0)$ is chosen randomly. The colour code assigns *white* (*black*) whenever $x_i(t) = 0(1)$ and a *rainbow colour* scale for other values of $x_i(t)$ starting with *red* for values near zero. Patches of CA behaviour (rule 150) appear after a short transient and will eventually fill the whole pattern

The synchronized state is an example of absorbing state, but clearly in real cases one rarely expect to find a complete synchronization: the evolution may be influenced by noise, or the two replicas can be slightly different.

We can test this hypothesis using the map

$$f(x; a) = \begin{cases} (6x)^a/2 & 0 \le x < 1/6, \\ 1 - |6(1/3 - x)|^a/2 & 1/3 \le x < 1/2, \\ |6(x - 2/3)|^a/2 & 1/2 \le x < 5/6, \\ 1 - (6(1 - x))^a/2 & 5/6 \le x < 1, \end{cases} \tag{15.5}$$

where $1 \le a < \infty$ (see Fig. 15.9-left), see Ref. [5]. This map that has the advantage of reducing to the DCA rule 150 for a large, and to a chaotic map from a small. For $a \gtrsim 1.81$ (stable chaos), one observes a transient chaos, with positive Lyapunov exponent, followed by a cellular automata pattern. One may wonder about the unpredictability of such map: in the chaotic phase, an infinitesimal damage will amply, while in stable chaos phase infinitesimal damages are absorbed (and thus the word "stable") but finite ones spread (and thus the word "chaos"). The synchronization procedure applied to a lattice of such maps indeed shows that a certain effort is needed even in the "stable" phase to get the synchronization. In agreement with the Janssen–Grassberger conjecture, one finds the synchronization phase transition for $a < 1$ do belongs to the MN universality class, while for $a \gtrsim 1.81$.

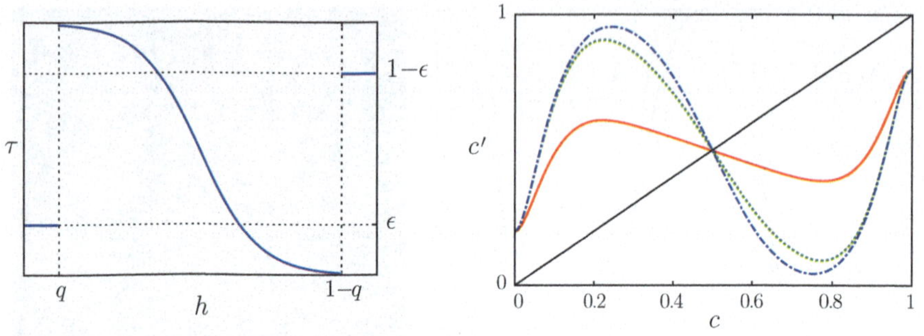

Fig. 15.10 *left* the transition probability $\tau(h)$ given by Eq. (15.6) with $J = -3$, $k = 20$, $q = 0.1$ and $\varepsilon = 0.2$. *right* graphs of the mean-field map, Eq. (15.7) for different values of J and $k = 20$. From *bottom* to *top* for $c < 1/2$, $J = -0.5$ (*red*, *lower line*), $J = -3.0$ (*green*, *middle line*) and $J = -6.0$ (*blue*, *upper line*)

Such a behaviour is not limited to systems that reduce to DCA, see Ref. [3] for an example.

15.2.9 Topology and Chaotic Phase Transitions

Up to now, we have not investigated the influence of the topology, i.e. of the connections defined by the adjacency matrix a_{ij}. It is well known that if we replace a regular lattice with a random network of the same connectivity, the global behaviour becomes that of the mean-field, since in this way correlations are disrupted.

We can study the influence of the topology by adopting the Watts–Strogatz rewiring mechanism [27]: start with a regular lattice of connectivity k in 1D and, for each site, rewire at random a fraction p of incoming links.

In order to show the effects of the mechanism and also to present a new type of phase transition, let us consider a cellular automaton whose mean-field approximation is chaotic. This model has been developed originally as an opinion formation model [6].

The average local opinion or social pressure h_i, is defined by

$$h_i = \frac{\sum_j a_{ij} s_j}{k}.$$

The opinion of agent i changes in time according to the transition probability $\tau(s_i|h_i)$ that agent i will hold the opinion s_i at time $t + 1$ given the local opinion h_i at time t. This transition probability, shown in Fig. 15.10-left, is given by

$$\tau(h) = \begin{cases} \varepsilon & \text{if } h < q, \\ \dfrac{1}{1+\exp(-2J(2h-1))} & \text{if } q \le h \le 1-q, \\ 1-\varepsilon & \text{if } h > 1-q, \end{cases} \tag{15.6}$$

with $\tau(h) = \tau(1|h)$.

The simplest mean-field description of the model is given by

$$c' = f(c) = \sum_{w=0}^{k} \binom{k}{w} c^w (1-c)^{k-w} \tau\left(\frac{w}{k}\right), \tag{15.7}$$

with $c' = c(t+1)$ and $c = c(t)$. The term in parenthesis on the *r.h.s* of this expression denotes the w-combinations from a set of k elements. In Fig. 15.10-right, we show some graphs of f. The bifurcation diagram of this map after varying J is shown in Fig. 15.11-left. The doubling bifurcation route to chaos ends at $J = J_c$. For $0 > J \ge J_2$ and $J_3 > J \ge 6$, there is only one attractor (blue, darker dots). For $J_2 > J \ge J_c$, there are two, one corresponding to the lower branches that bifurcate up to J_c (red, lighter dots), and the other one to the upper branches (blue, darker dots). For $J_c > J \ge J_3$, there are two chaotic attractors, one corresponding to the lower branches (blue, darker dots), the other to the top branches (red, lighter dots). For every value of J, the dots are 64 iterates of the map after a transient of 10^3 time steps. For values of J with only one basin of attraction, the orbits do not depend on the initial average opinion $c(t = 0)$. For values of J that correspond to two attractors, one of them was found with $c(0) = 0.1$, the other one with $c(0) = 0.9$.

By varying the long-range probability p, we observe the transition towards the mean-field behaviour, as reported in Fig. 15.12. This induces a stochastic bifurcation diagram by varying p, Fig. 15.11-right that is quite similar to that obtained in the mean-field approximation by varying J, Fig. 15.11-left. For $p \lesssim p_0$, there are almost periodic orbits of period one and for $p_0 \lesssim p \lesssim p_1$ of period two. For $p_1 \lesssim p \lesssim p_2$, we find two attractors, one (in red, lighter) in the lower branches, the other one (in blue, darker) in the top ones.

Notice that up to now we have met phase transition that, in the mean-field description, implies the change of stability of fixed points, while here we observe a real bifurcation diagram with coexistence of basins, period-doubling and chaos.

15.3 Conclusions

The main aim of this presentation was that of discussing some characteristics of nonequilibrium phase transitions.

We have illustrated some aspects of phase transitions in probabilistic cellular automata, trying to show how such a problem arises in different contexts and some of the method used for its study.

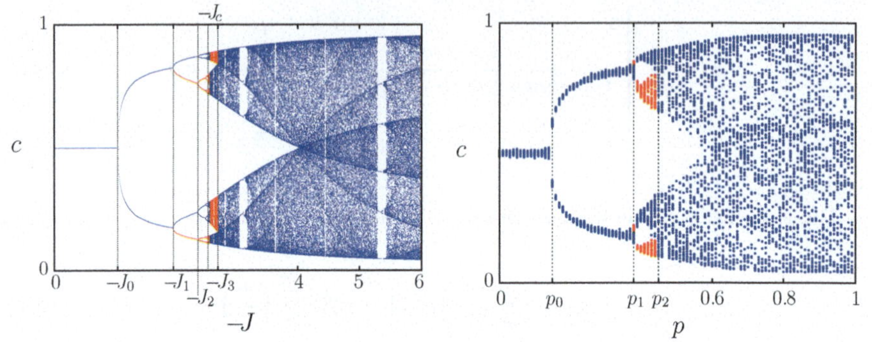

Fig. 15.11 *left* bifurcation diagram of the mean-field map, Eq. (15.7), by varying J. *right* small-world probabilistic bifurcation diagrams as functions of the long-range probability p. The colours mark different attractors

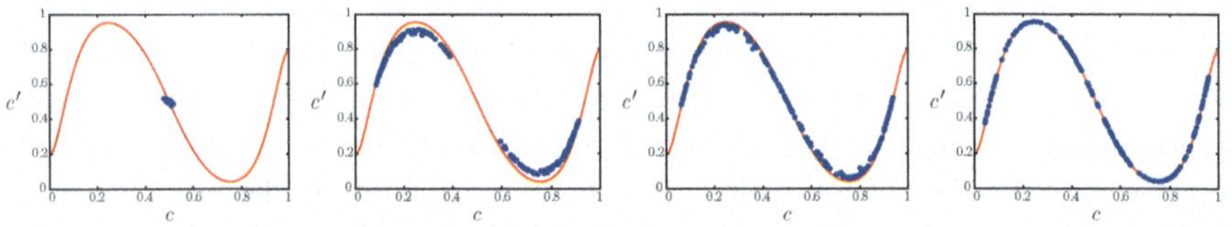

Fig. 15.12 (Colour online) Return map of the average opinion c on small-world networks for several values of the long-range connection probability p with $J = -6$, $k = 20$, $N = 10^3$ and a transient of 10^3 time steps. The following 200 iterations are shown as (*blue*, *darker*) *dots*. The (*red*, *lighter*) continuous curve is Eq. (15.7). From *left* to *right* $p = 0.0$, $p = 0.5$, $p = 0.6$ and $p = 1.0$

The real-life problems are usually more complex than those faced here. However, a phase transition separates qualitatively different states, and since continuous phase transitions are related to the divergence of the correlation function, the general scenario is quite independent of the details of the model, so that the investigation of simplified models is justified.

The numerical simulations of phase transitions constitute also a challenge by itself: the need of approaching the limit of infinite space and time requires particular techniques and an efficient implementation.

This study can be complemented by an analysis based on the principles of the renormalization group (see for instance Ref. [26]), that in principle allows to group several models into a few universality classes.

The study of phase transitions in stochastic systems (and the related one of bifurcations in dynamical systems) can constitute a good training ground for both theoretical, computational and experimental students.

Acknowledgements This work was partially supported by EU projects 288021 (EINS – Network of Excellence in Internet Science) and project PAPIIT-DGAPA-UNAM IN109213.

References

1. Bagnoli, F.: Boolean derivatives and computation of cellular automata. Int. J. Mod. Phys. C **3**, 307–320 (1992). https://doi.org/10.1142/S0129183192000257
2. Bagnoli, F.: On damage spreading transitions. J. Stat. Mech. **85**, 151–164 (1996). https://doi.org/10.1007/BF02175559
3. Bagnoli, F., Cecconi, F.: Synchronization of non-chaotic dynamical systems. Phys. Lett. A **282**, 9–17 (2001). https://doi.org/10.1016/S0375-9601(01)00154-2
4. Bagnoli, F., Rechtman, R.: Synchronization and maximum Lyapunov exponents of cellular automata. Phys. Rev. E **59**, R107–R1310 (1999). https://doi.org/10.1103/PhysRevE.59.R1307
5. Bagnoli, F., Rechtman, R.: Synchronization universality classes and stability of smooth coupled map lattices. Phys. Rev. E **73**, 026202 (2006). https://doi.org/10.1103/PhysRevE.73.026202
6. Bagnoli, F., Rechtman, R.: Topological bifurcations in a model society of reasonable contrarians. Phys. Rev. E **88**, 062914 (2013). https://doi.org/10.1103/PhysRevE.88.062914
7. Bagnoli, F., Baroni, L., Palmerini, P.: Synchronization and directed percolation in coupled map lattices. Phys. Rev. E. **59**, 409–416 (1999). https://doi.org/10.1103/PhysRevE.59.409
8. Bagnoli, F., Boccara, N., Rechtman, R.: Nature of phase transitions in a probabilistic cellular automaton with two absorbing states. Phys. Rev. E **63**, 046116 (2001). https://doi.org/10.1103/PhysRevE.63.046116
9. Derrida, B.: Dynamical phase transitions in spin models and automata. In: Van Beijeren, H. (ed.) Fundamental Problems in Statistical Mechanics VII, pp. 273–309. Elsevier Science Publisher, Amsterdam (1990)
10. Domany, E., Kinzel, W.: Equivalence of cellular automata to Ising models and directed percolation. Phys. Rev. Lett. **53**, 311–314 (1984). https://doi.org/10.1103/PhysRevLett.53.311
11. Fatès, N.: Asynchronism induces second order phase transitions in elementary cellular automata. J. Cell. Autom. **4**, 21–38 (2009)
12. Fatès, N.: A guided tour of asynchronous cellular automata. J. Cell. Autom. **9**, 387–416 (2014)
13. Fukś, H., Fatès, N.: Local structure approximation as a predictor of second order phase transitions in asynchronous cellular automata. Nat. Comput. **14**(4), 507–522 (2015)
14. Grassberger, P.: On phase transitions in Schlögl's second model. Z. Phys. B **47**, 365 (1982)
15. Grassberger, P.: Synchronization of coupled systems with spatiotemporal chaos. Phys. Rev. E **59**, R2520–R2524 (1999). https://doi.org/10.1103/PhysRevE.59.R2520
16. Gutowitz, H.A., Victor, J.D., Knight, B.K.: Local structure theory for cellular automata. Physica **28D**, 18–48 (1987)
17. Henkel, M., Hinrichsen, H., Lübeck, S.: Non-Equilibrium Phase Transitions Volume 1: Absorbing Phase Transitions. Springer Science, Dordrecht (2008)
18. Hinrichsen, H., Weitz, J.S., Domany, E.: An algorithm-independent definition of damage spreading, application to directed percolation. J. Stat. Phys. **88**, 617–636 (1997)
19. Huang, K.: Statistical Mechanics. Wiley, New York (1963)
20. Kardar, G.P.M., Zhang, Y.-C.: Dynamic scaling of growing interfaces. Phys. Rev. Lett. **56**, 889–892 (1986)
21. Janssen, H.K.: On the nonequilibrium phase transition in reaction-diffusion system with an absorbing stationary state. Z. Phys. B **42**, 151 (1981)
22. Kinzel, W.: Directed percolation. In: Adler, J., Zallen, R. Deutscher, G. (eds.) Percolation Structures and Processes. Annals of the Israel Physical Society, vol. 5, p. 425 (AIP, New York, 1983)
23. Kinzel, W.: Phase transition of cellular automata. Z. Phys. B **58**, 229–244 (1985). https://doi.org/10.1007/BF01309255
24. Muñoz, M.A., Hwa, T.: On nonlinear diffusion with multiplicative noise. Europhys. Lett. **41**, 147–152 (1998)
25. Tu, Y., Grinstein, G., Muñoz, M.A.: Systems with multiplicative noise: critical behavior from KPZ equation and numerics. Phys. Rev. Lett. **78**, 274–277 (1997)
26. Tomé, T., de Oliveira, M.J.: Renormalization group of the Domany-Kinzel cellular automaton. Phys. Rev. E **55**, 4000–4004 (1997). https://doi.org/10.1103/PhysRevE.55.4000

27. Watts, D.J., Strogatz, S.H.: Collective dynamics of 'small-world' networks. Nature **393**, 440–442 (1998). https://doi.org/10.1038/30918
28. Wolfram, S.: Statistical mechanics of cellular automata. Rev. Mod. Phys. **55**, 601–644 (1983). https://doi.org/10.1103/RevModPhys.55.601

Part III
Applications to Natural Sciences and Computational (Cell) Biology

Chapter 16
A Trade-Off Between Simplicity and Robustness? Illustration on a Lattice-Gas Model of Swarming

Nazim Fatès, Vincent Chevrier and Olivier Bouré

Abstract We re-examine a cellular automaton model of swarm formation. The local rule is stochastic and defined simply as a force that aligns particles with their neighbours. This lattice-gas cellular automaton was proposed by Deutsch to mimic the self-organisation process observed in various natural systems (birds, fishes, bacteria, etc.). We explore the various patterns the self-organisation process may adopt. We observe that, according to the values of the two parameters that define the model, the alignment sensitivity and density of particles, the system may display a great variety of patterns. We analyse this surprising diversity of patterns with numerical simulations. We ask where this richness comes from. Is it an intrinsic characteristic of the model or a mere effect of the modelling simplifications?

"Everything should be made as simple as possible, but not simpler". The aim of this contribution is to examine a model of collective motion in the light of this semi-apocryphal word of Einstein.[1]

We investigate in detail the behaviour of a cellular model that was proposed to mimic the formation of swarms: under some conditions, self-propelled particles that are initially randomly scattered on a lattice spontaneously form coherent groups. These groups may stay stable for a very long time or progressively merge in a single coherent entity that takes the form of a diagonal stripe.

Our study focuses on the model introduced by Deutsch [4, 5] some twenty years ago: in this particular kind of cellular automata, called lattice-gas cellular automata, the cells contain particles which can move along the lattice directions.

[1] See http://quoteinvestigator.com/2011/05/13/einstein-simple/ for more details (consulted Jan. 2015).

N. Fatès (✉)
inria, Université de Lorraine, CNRS, LORIA, F-54000 Nancy, France
e-mail: nazim.fates@inria.fr

V. Chevrier · O. Bouré
Université de Lorraine, CNRS, LORIA, F-54000 Nancy, France
e-mail: vincent.chevrier@loria.fr

O. Bouré
e-mail: olivier.boure.pro@gmail.com

P.-Y. Louis and F.R. Nardi (eds.), *Probabilistic Cellular Automata*, Emergence, Complexity and Computation 27, https://doi.org/10.1007/978-3-319-65558-1_16

The interactions between particles are local and stochastic: each particle tends to align with the neighbouring particles according to a single parameter, called the alignment sensitivity. Observations show the presence of a "minimal cohesion force": the particles self-organise into patterns only when the alignment sensitivity is set higher than a threshold, which depends on the number of particles in the grid. If this sensitivity is too low, no pattern emerges. When it is higher than a given threshold, a spontaneous symmetry breaking occurs: the particles form a stripe pattern that moves diagonally on the lattice; the direction the stripe takes results from a distributed consensus between particles.

The organisation in stripes is only one of the various patterns that can be observed [3]. Depending on the number of particles and on the value of the sensitivity, particles may also organise into clusters or, more surprisingly, they may "anti-align" and form what we call "chequerboard patterns". The purpose of this chapter is to explore the various behaviours of the model and to examine how they depend on the two main parameters of the model, the sensitivity and the density. We then examine whether this behaviour is robust to various modifications of the simulation conditions. We then discuss the question as to how the changes of behaviour we observe show a limit in the interpretation of the simulation results.

16.1 Formal Description of the System

16.1.1 First Approach

Before we start describing formally the model, we recommend to the readers to observe visually the phenomenon of self-organisation as shown in Fig. 16.3. We should take this figure as a metaphor: in the same way that, sitting on the sea shore, we can observe hundreds of birds group and form a flock, let us watch how particles that are randomly spread on the lattice progressively unite their movements and form a group that travels diagonally.

This simple experiment illustrates how randomness is a means to attain a consensus in a decentralised system. Indeed, the rules that govern the particles are stochastic, and despite this randomness, or more exactly *thanks to* the randomness, an agreement can be reached. To understand this phenomenon, many questions need to be raised. First, the role of the local rule has to be clarified. In particular, what level of "noise" must be introduced to allow the particles to form patterns? Beyond the local rule, how does the self-organisation phenomenon depend on the form of the grid? How does the system react to perturbations of its updating? etc.

In the lines that follow, we will first define formally our system and then examine with numerical simulations the behaviour of the model.

16.1.2 The LGCA Swarming Model

The bricks of the model are the *cells*. The cells are arranged as a regular lattice, denoted by $\mathcal{L}$. Here, since our approach mainly relies on simulations, we will use a finite grid with periodic boundary conditions; we thus define the grid as $\mathcal{L} = (\mathbb{Z}/X.\mathbb{Z}) \times (\mathbb{Z}/Y.\mathbb{Z})$. For the sake of simplicity, when the grid has square dimensions we use $L = X = Y$. The generalisation to the infinite grid $\mathbb{Z}^2$ is straightforward.

Each cell of the lattice consists of four *channels*, denoted by $\mathtt{n}, \mathtt{e}, \mathtt{s}, \mathtt{w}$, and respectively associated with North, East, South and West directions. A channel can be *occupied* or *empty*, depending on whether it contains or not a *particle*. We thus represent the state of a cell as a quadruple $(q_\mathtt{n}, q_\mathtt{e}, q_\mathtt{s}, q_\mathtt{w}) \in Q = \{0, 1\}^4$, called a *tile*, where the 0 or 1 value, respectively, represents the absence or presence of a particle in a given channel.

The state of the system at a given time is called a *configuration*, and we here choose to represent a configuration as a *mapping* from $\mathcal{L}$ to Q. The space of configurations is denoted by $Q^\mathcal{L}$.

The evolution of the system is described by a discrete dynamical system $F : Q^\mathcal{L} \to Q^\mathcal{L}$. The *global function* F is decomposed into two steps: (a) the *interaction* step, where particles of a cell are "rearranged" in this cell, that is, each particle is affected to a channel, and (b) the *propagation* step, where particles are moved from a cell to its neighbours. Formally, we write that: $x^{t+1} = F(x^t) = F_{\text{prop}} \circ F_{\text{int}}(x^t)$.

Let us first define the **interaction step**. Let $q = (q_\mathtt{n}, q_\mathtt{e}, q_\mathtt{s}, q_\mathtt{w})$ be a *tile* (the state of a given cell). The step consists of permuting the positions of the particles; this amounts to replacing this tile by a tile which contains the same number of particles. To define how we choose a new tile, let us denote by $N(q) = q_\mathtt{n} + q_\mathtt{e} + q_\mathtt{s} + q_\mathtt{w}$ the number of particles of a given tile q and by $\pi(q)$ the set of its "permuted tiles": $\pi(q) = \{q' \in Q, N(q') = N(q)\}$. Now, since the interaction step is stochastic, we assign to each permuted tile a *weight* and select this tile with a probability that is proportional to this weight.

We now explain how to calculate the weights; we need to introduce intermediary definitions. For a configuration x and a given cell c, two quantities will be used:

- The *local flux* $\varphi : Q \to \{-1, 0, 1\}^2$ is the resulting "momentum" of the particles contained in this cell. Noting $x(c) = q = (q_\mathtt{n}, q_\mathtt{e}, q_\mathtt{s}, q_\mathtt{w})$, we have:

$$\varphi(q) = q_\mathtt{n} \cdot v_\mathtt{n} + q_\mathtt{e} \cdot v_\mathtt{e} + q_\mathtt{s} \cdot v_\mathtt{s} + q_\mathtt{w} \cdot v_\mathtt{w}.$$

  Each local flux is assigned a color according to the convention shown on Fig. 16.1.
- The *director field* $D : Q^\mathcal{L} \times \mathcal{L} \to \mathbb{Z}^2$ is the sum of the local fluxes of the four immediate neighbours of c:

$$D(x, c) = \varphi(x(c + v_\mathtt{n})) + \varphi(x(c + v_\mathtt{e})) + \varphi(x(c + v_\mathtt{s})) + \varphi(x(c + v_\mathtt{w})),$$

  where $v_\mathtt{n} = (0, 1)$, $v_\mathtt{e} = (1, 0)$, $v_\mathtt{s} = (0, -1)$, $v_\mathtt{w} = (-1, 0)$ denote the four elementary vectors that follow the cardinal directions.

Fig. 16.1 Colour code associated with each local flux in $\{-1, 0, 1\} \times \{-1, 0, 1\}$

For a configuration x, the probability to update a cell c with a tile $q = x(c)$ into a tile $q_{\mathrm{I}} \in \pi(q)$ is then given by:

$$P(q \to q_{\mathrm{I}}) = \frac{\exp[\,\sigma \cdot D(x, c) \odot q_{\mathrm{I}}\,]}{Z(x, c)}$$

where:

- σ is a positive constant called the *alignment sensitivity*,
- $\odot$ denotes the scalar product of two vectors of $\mathbb{Z}^2$, and
- $Z(x, c) = \sum_{q \in \pi(x(c))} \exp[\,\sigma \cdot D(x, c) \odot \varphi(q)\,]$ is the renormalisation factor.

In words, given that σ is a positive factor, the tiles which are "in agreement" with the local field have the greatest weight while the tiles that are "in opposition" (negative scalar product) are assigned a little weight.

By noting $x_I = F_{\mathrm{int}}(x)$ the result of the interaction step, our system can now be defined with the probability $P(x \to x_{\mathrm{I}})$ to update x into x_I:

$$P(x \to x_{\mathrm{I}}) = \prod_{c \in \mathcal{L}} P\big(x(c) \to x_{\mathrm{I}}(c)\big).$$

Let us now define the **propagation step**. The idea is simple: each particle moves in the direction of its channel (there is thus no possibility of collisions).

Noting $y = F_{\mathrm{prop}}(x)$ and: $\forall c \in \mathcal{L}$, $y(c) = \big(y_{\mathrm{n}}(c), y_{\mathrm{e}}(c), y_{\mathrm{s}}(c), y_{\mathrm{w}}(c)\big)$, we simply have:

$$y_{\mathrm{n}}(c) = x_{\mathrm{n}}(c - v_{\mathrm{n}}) \qquad y_{\mathrm{e}}(c) = x_{\mathrm{e}}(c - v_{\mathrm{e}})$$
$$y_{\mathrm{s}}(c) = x_{\mathrm{s}}(c - v_{\mathrm{s}}) \qquad y_{\mathrm{w}}(c) = x_{\mathrm{w}}(c - v_{\mathrm{w}})$$

where v_δ represents the movement by one cell in direction δ. In words, the North channel of a cell c *after* propagation is equal to the value of the North channel of the cell that is at the *South* of c *before* propagation (see Fig. 16.2).

16.2 Exploration of the Stable Patterns

In the following experiments, we use Bernoulli random initial conditions, i.e., each channel is set to be occupied independently of the other channels with a probability d,

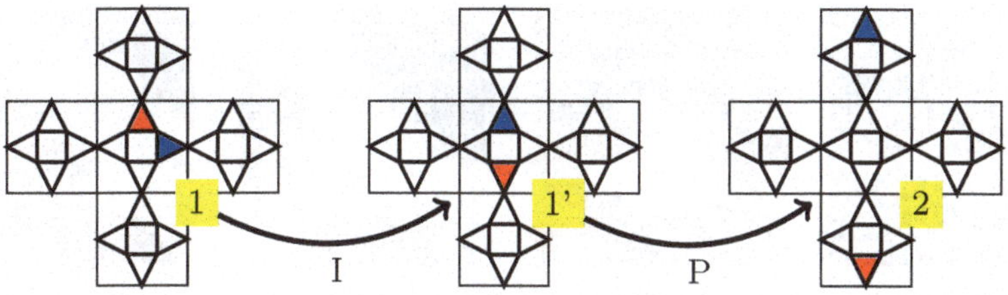

Fig. 16.2 Example of an evolution of five cells on one time step. Cells are represented with their four channels (*triangles*); each channel can be empty (*white*) of filled with a particle (*red* or *blue*, *color* online). The time step is decomposed into two steps: the interaction step (I) and the propagation step (P)

called the *initial density*, or simply the *density*, as this quantity is conserved during the evolution of the system. Note that we do not consider densities higher than $1/2$ since the local rule behaves symmetrically by exchanging particles and "holes", that is, empty channels.

The experiments that we will present are in part a re-examination of the experiments presented in previous articles [2, 3]. They are completed by new experiments which aim at showing the variety of behaviours this model can generate. Videos showing the temporal evolution of the systems which appear in the figures can be found on the preprint Web page of this chapter.[2]

16.2.1 Quantification of the Order

We denote by $\gamma(x)$ the *mean alignment* of a configuration x: this parameter evaluates the tendency of a particle to be aligned with the particles of its neighbouring cells. It is defined with:

$$\gamma(x) = \frac{1}{4N(x)} \sum_{c \in \mathcal{L}} \varphi(x, c) \odot D(x, c)$$

Note that since each particle counts four times in the calculation of $D(x)$ (one per direction), we need to divide by 4 in order to assign to γ a value in $[-1, 1]$. A value close to 1 indicates that in general particles are aligned with their neighbours, a value close to 0 indicates an absence of alignment and a value close to -1, the fact that particles have a direction which is opposed to that of their neighbours. In an abuse of language, we will call this phenomenon *anti-alignment*.

[2]See HAL preprint 01230145 (https://hal.inria.fr/hal-01230145) and open "annex files". The simulations were obtained with the **`FiatLux`** software [7].

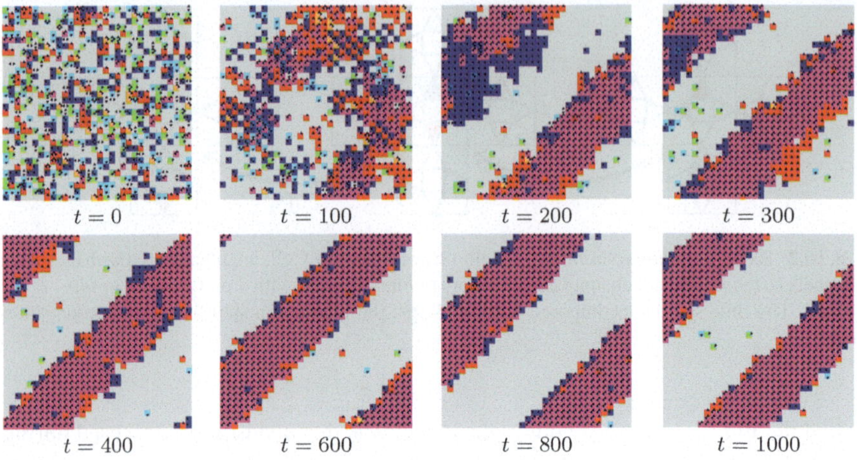

Fig. 16.3 Stipe formation: evolution of the system from a Bernoulli random initial configuration. The grid size is $L = 32$, the initial density is $d = 0.2$, the sensitivity is $\sigma = 2.0$

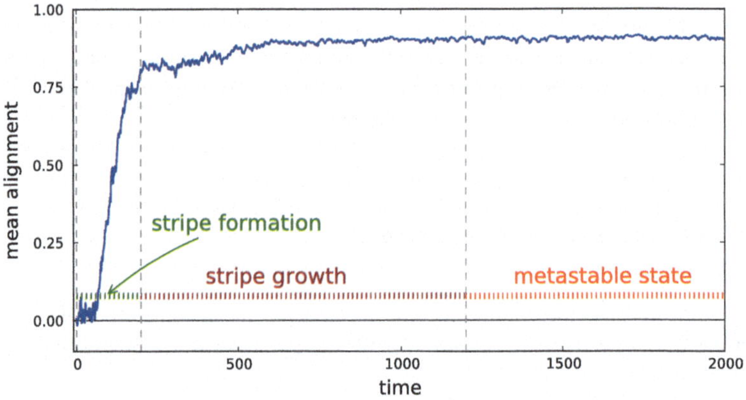

Fig. 16.4 Stripe formation: evolution of the mean alignment for the sampled trajectory observed in Fig. 16.3: $L = 32$, $d = 0.2$, $\sigma = 2.0$. The *vertical bars* show an (informal) separation of the evolution in three phases

16.2.2 Stripe Patterns

To observe our first pattern, we fix the alignment sensitivity to $\sigma = 0.2$, and take a density of $d = 0.2$.

Observing a Trajectory

Figure 16.4 shows the evolution of the mean alignment parameter for the trajectory observed in Fig. 16.3. This evolution follows three phases: first, the stripe forms in a relatively short time. Then, the stripe strengthens progressively and its shape gets

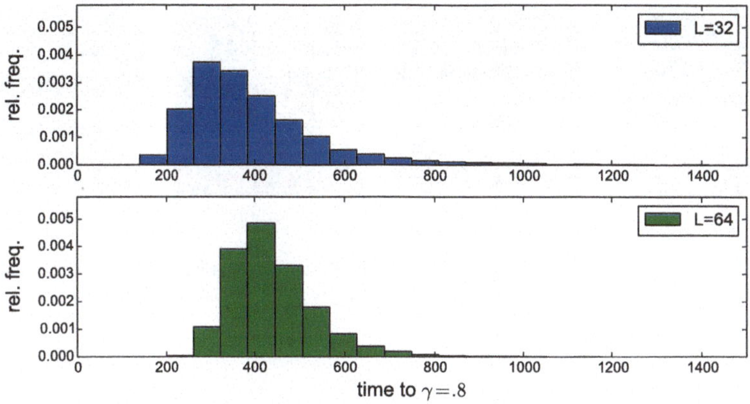

Fig. 16.5 Stripe formation: distribution of the time to reach $\gamma = 0.8$ for $L = 32$ and $L = 64$, with $d = 0.2$, $\sigma = 2.0$ and 10000 samples

more and more regular. After a thousand steps, the system has reached a metastable state and its characteristics do not seem to change any more.

Statistical Distributions

Since we are working on a stochastic system, this phenomenon varies from one experiment to the other. For example, the trajectory shown in Fig. 16.4 shows that the stripe appears in approximately 200 time steps, but for other simulations with $L = 32$, the stripe appeared only after 1200 steps.

In order to assess the statistical variation of the time needed to form a stripe, we repeated this experiment, and for each trajectory, we measured the time needed to attain a high level of order. This level of order was arbitrarily set to the threshold of $\gamma = 0.8$. Empirically, this value corresponds to a point where the stripe is almost formed and, in practice, has only an infinitesimal chance to disappear. The order in the pattern is too large to be destroyed by the random fluctuations of the system. However, remark that since the system is recurrent (see Sect. 16.4), the stripe is bound to disappear after a very long time.

Figure 16.5 shows the distribution of the time needed to reach the threshold $\gamma = 0.8$ for $L = 32$. The statistics are obtained with 10,000 random samples. It can be asked whether the distribution of this time obeys a Poisson law; we observed informally that the "tail" of the distribution is heavier than expected but larger statistical samples are needed to confirm the observations.

When we double the size of the lattice to $L = 64$, we notice a paradoxical effect: the average time to reach the threshold increases slightly, but the "tail" of the distribution becomes *less* developed. One possible explanation of this paradox is that for small lattices there are locally stable patterns that form and destroy, whereas for large lattices, such patterns are less present. We refer to our previous work [3] for more details on this issue. It is an open problem to infer the behaviour of the system for lattice sizes that tend to infinity.

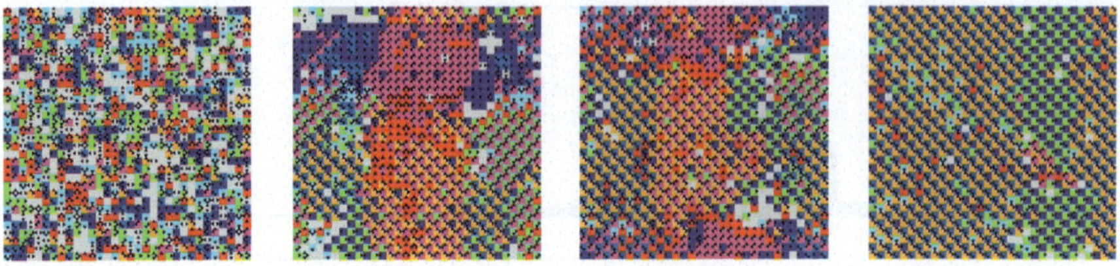

$t = 0$ $t = 250$ $t = 500$ $t = 1000$

Fig. 16.6 Chequerboard formation: evolution of the system from a Bernoulli random initial configuration. The grid size is $L = 32$, the initial density is $d = 0.4$, the sensitivity is $\sigma = 2.0$

16.2.3 Chequerboard Patterns

To observe our second pattern, we keep the alignment sensitivity constant $\sigma = 0.2$, but we double the density, by taking $d = 0.4$.

Observing a Trajectory

Figure 16.6 shows one trajectory with these settings: contrarily to what was observed before, the particles now tend to *anti-align* with their neighbours! We observe that rapidly, homogeneous and stable zones appear; in these zones, the colours, which represent local fluxes, are alternating as in a chequerboard.

The stability of this pattern can be explained by looking at the effects of the interaction step and the propagation step. Indeed, in the interaction step, particles in a cell tend to align with the particles of the neighbouring cells. However, as the same movement happens in the neighbouring cells, all the particles reverse their direction. After the propagation step, the situation is unchanged: we are back to the initial condition.

The argument above shows that chequerboard patterns are potentially stable, but it does not explain how the pattern forms. Empirically, we observe that chequerboards merge when they touch each other. Their frontiers move randomly, which makes two neighbouring chequerboards grow or shrink. These movements make them merge and create a tendency to form a single chequerboard that spans all over the lattice. This phenomenon is qualitatively similar to what we observed with the two-dimensional minority rule under fully asynchronous dynamics [11] or with an Ising spin system with ferromagnetic alignment.

The process of chequerboard formation can be (partially) quantified by monitoring the evolution of the alignment. Figure 16.7 shows how the alignment evolves for the same trajectory as observed on Fig. 16.6. We see that it decreases until it reaches a metastable state characterised by $\gamma = -0.8$ (approximately). This value corresponds to the existence of a single chequerboard "zone" that covers a large part of the lattice, while in some parts of the lattice, we find no particle or only one particle per cell.

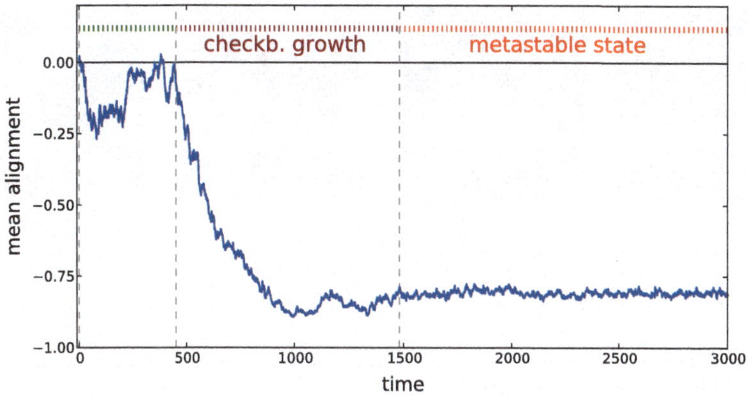

Fig. 16.7 Chequerboard formation: evolution of the mean alignment for the sampled trajectory observed in Fig. 16.6: $L = 32$, $d = 0.4$, $\sigma = 2.0$

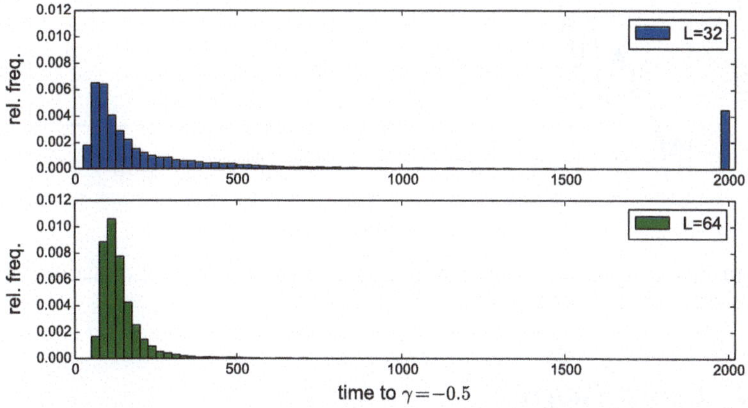

Fig. 16.8 Chequerboard formation: distribution of the time to reach $\gamma = -0.5$ for $L = 32$ and $L = 64$, with $d = 0.4$, $\sigma = 2.0$ and 10000 samples

Statistical Distributions

As for the stripe, a statistical estimation of the time needed to form a chequerboard was performed. For 10000 initial conditions, we measured the time needed to attain the arbitrary threshold of $\gamma = -0.5$. The distribution of this time is shown on Fig. 16.8. We observe that for $L = 32$, some trajectories take more than 2000 time steps to stabilise. However, for $L = 64$, such outliers are no longer observed and the average time to attain the threshold is 140 steps.

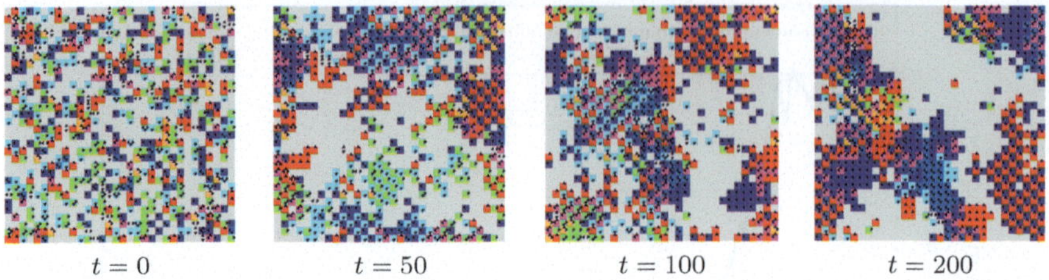

Fig. 16.9 Clusters formation: evolution of the system from a Bernoulli random initial configuration. The grid size is $L = 32$, the initial density is $d = 0.2$, the sensitivity is $\sigma = 4.0$

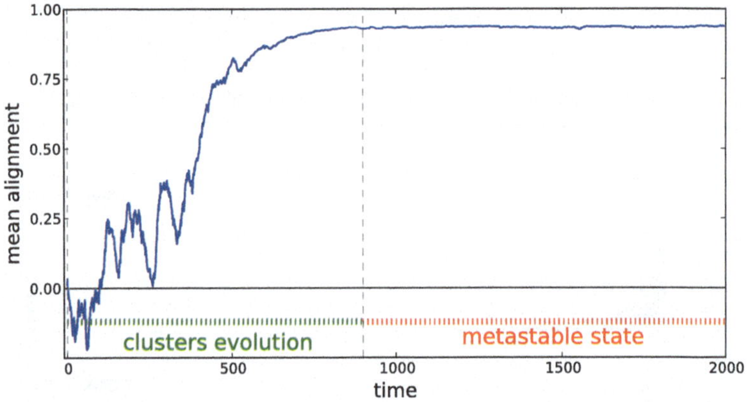

Fig. 16.10 Cluster formation: evolution of the mean alignment for the sampled trajectory observed in Fig. 16.9: $L = 32$, $d = 0.2$, $\sigma = 4.0$

16.2.4 *Cluster Patterns*

We now observe the third pattern: we take a density of $d = 0.2$ and a "high" alignment sensitivity: $\sigma = 4$. (This value corresponds to the double of the one used in the first experiment.)

Observing a Trajectory

Figure 16.9 shows a trajectory of the system from a Bernoulli random initial condition. We see that the system stabilises on a stationary state formed of groups of particles, the *clusters*, which move in the same directions. These group of particles are coherent and give the impression to traverse each other without interfering much. However, the shape of the groups evolve slowly by loosing or gaining particles.

Figure 16.10 shows the mean alignment γ as a function of time for the same trajectory as the one observed in Fig. 16.9. For these initial conditions, the phenomenon of clusters formation is in general very rapid.

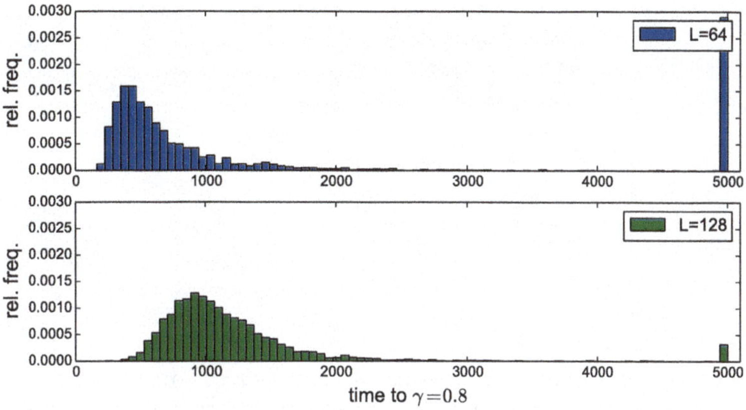

Fig. 16.11 Cluster formation: distribution of the time to reach $\gamma = 0.8$ for $L = 64$ and $L = 128$, with $d = 0.2$, $\sigma = 4.0$ and 3000 samples

Statistical Distributions

We measured the distribution of the time needed to attain $\gamma = 0.8$. We were surprised to observe that for $L < 128$, it is relatively frequent to observe trajectories that do *not* form clusters. (See in particular Ref. [2].) It is thus necessary to increase the size to values much higher than $L = 128$ in order to obtain stable statistical results that do not suffer from finite-size effects (see Fig. 16.11).

To conclude this part, we identified three main types of organisations of the system, or, to use the vocabulary of statistical physics, three phases. In fact only clusters correspond to what we first expected: coherent groups of particles that move in the same direction. Stripes are surprising because they imply a symmetry breaking (one direction is chosen) and chequerboards are even more surprising because their particles are anti-aligned.

16.3 Robustness of the Patterns

We now examine whether the patterns are due to the simple and discrete nature of cellular automata. Do they come from the regularity of the lattice? Are they a "perverse effect" of the perfect synchronous updating of the lattice?

In the following, we will examine what happens to the three main patterns when they are submitted to the following perturbations:

- change in the lattice size,
- change in the type of boundary conditions in the lattice,
- change in the type of updating.

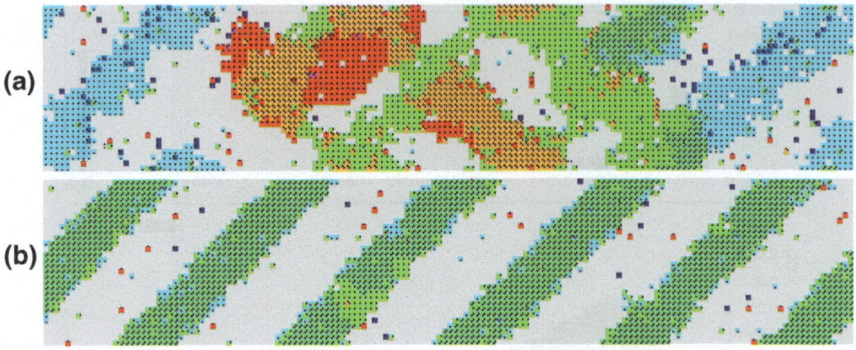

Fig. 16.12 Stripe formation for a rectangular grid: $(X, Y) = (150, 30)$ and $d = 0.2$, $\sigma = 2.0$; (**a**) $t = 2000$, the process is still under evolution and (**b**) $t = 4000$, the metastable state is reached (a multi-loop stripe)

We will give a succinct view on our previous observations and indicate in which directions the research could be continued.

16.3.1 The Size Ratio Is a Key Parameter

Changing the ratio of the lattice reveals interesting facts on stripe formation. We refer to our previous studies for more information on this phenomenon [3]. The main observation is that if the lattice ratio is an integer value k (for a small k), the stripe "loops" in order to form a closed pattern (Fig. 16.12).

It is interesting to note that for some particular values of the grid ratio, a new "bifurcation" appears. In some cases, the stripe does not totally succeed in forming a "closed" pattern, and in some other cases, the stripe forms a multi-loop closed pattern. For instance, for $(X, Y) = (100, 75)$, it was observed that the system may stabilise on *three* different patterns depending on the value of σ. As shown in Fig. 16.13, for a sensitivity $\sigma \in [1, 1.5]$, an irregular stripe appears; for $\sigma \in [1.5, 1.8]$, a regular multi-loop stripe appears; and for higher values of σ, we recover the clusters pattern. This situation contrasts with what can be observed for a square grid of size (100, 100) in which only simple loops are observed with the same settings of σ and d.

These observations let us think that the stripe results from a "resonance effect": particles cross the grid several times and interact regularly with the particles that come in perpendicular directions. In the case where a good "harmony" in these interactions is found, an amplification effect exists, which stabilises the diagonal pattern and strengthens it. It is an open question to analyse more precisely this phenomenon.

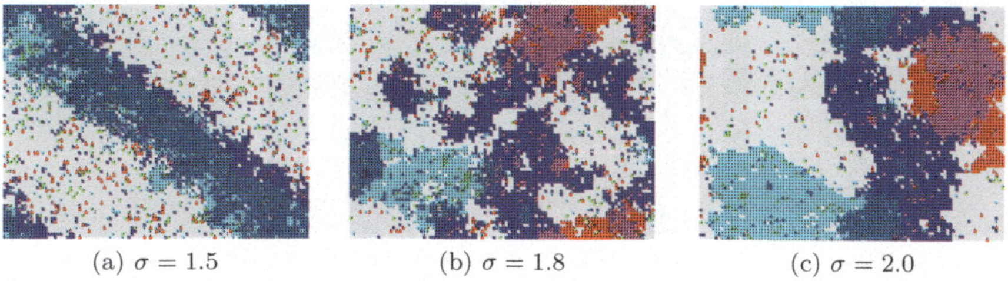

Fig. 16.13 Patterns observed for rectangular grid dimensions $(X, Y) = (100, 75)$ at $t = 5000$. Depending on the value of σ, we observe: (**a**) a non-perfect stripe, (**b**) a regular stripe with multiple loops on the grid, (**c**) clusters

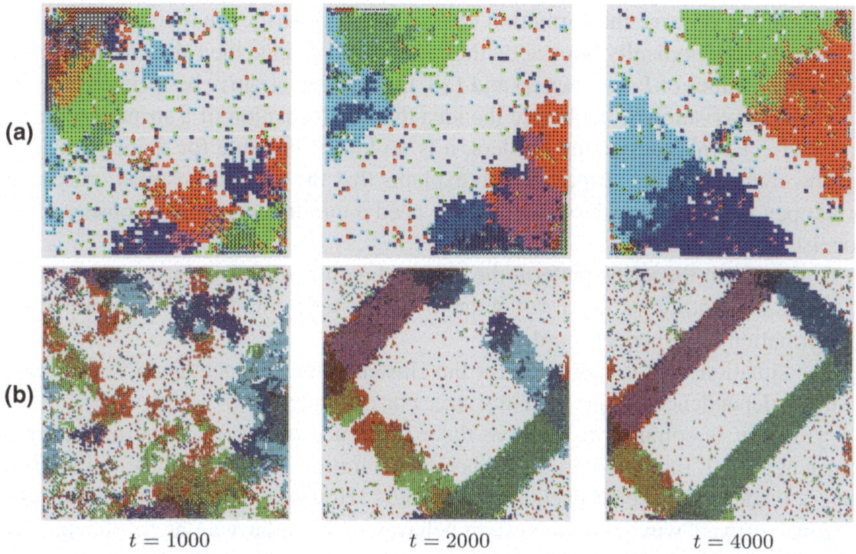

Fig. 16.14 Patterns observed with reflecting borders with two square lattices: (**a**) $L = 64$, (**b**) $L = 128$. Settings are: $d = 0.2$ and $\sigma = 2.0$

16.3.2 Reflecting Borders Suppress the Single Stripe

In order to get a better understanding of the mechanisms involved in the stripe formation, we propose to observe a new experiment where we use *reflecting borders*. For the sake of brevity, we put the technical descriptions of these particular boundary conditions in the Appendix.

Figure 16.14 shows a comparison of two trajectories of the system for the simulation conditions that normally lead to the formation of a stripe: $d = 0.2$ and $\sigma = 2.0$. It is surprising to observe that in both cases we no longer observe the formation of a single stripe: (a) For $L = 64$, we observe the formation of "big" clusters that travel horizontally and vertically and periodically "bounce" on the corners. (b) For

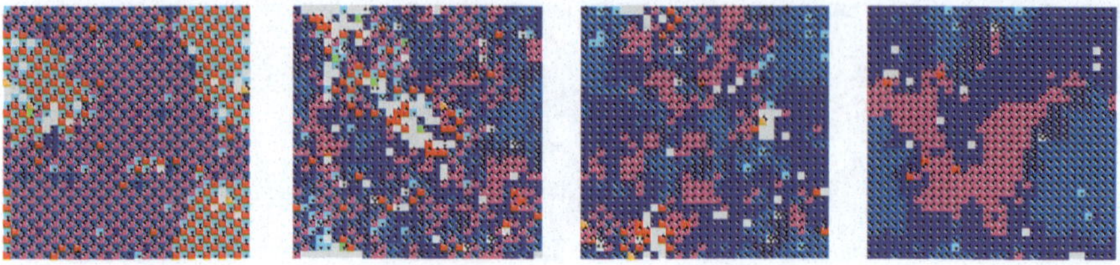

Fig. 16.15 Application of an asynchronous interaction at time $t = 1000$ after a chequerboard was formed. Simulation conditions: $L = 32$, $d = 0.4$, $\sigma = 2.0$

$L = 128$, we observe the formation of *four* diagonal stripes that coexist and with particles that periodically bounce on the "walls" and change their directions. For intermediary values of L between 64 and 128, these two qualitative behaviours coexist.

These observations corroborate the idea that the formation of a stripe is a "resonance effect". With reflecting borders, it is no longer possible for stripes to merge by absorbing the other stripes, which explains that the four stripes may coexist. It is an open problem to understand why we need a minimal grid size to observe four coexisting stripes.

Contrarily to the stripe pattern, the chequerboard pattern is robust to the change of boundary conditions: for $d = 0.4$ and $\sigma = 2.0$, reflecting borders do not perturb the formation of a chequerboard.

The situation is more complex for the clusters pattern. Once again, we observe a strong dependence on the system's size. For example, for $d = 0.2$, $\sigma = 4.0$ and $L = 128$, it is common to observe the coexistence of various patterns during a transient period. In particular, the higher concentration of particles near the borders have a tendency to create chequerboards there. However, after a long period, the diagonal stripes "win" and the system behaves in a similar way as for the smaller values of σ. Reflecting borders thus have a tendency to suppress the frontier between the diagonal stripe phase and the cluster phase.

Such observations are preliminary and need to be consolidated by a more extensive study. It is important to note that they all reveal a strong and unexpected dependence on the lattice size. These emergent patterns are thus as much dependent on the local rule as on the topology.

16.3.3 Asynchronism Creates a New Pattern

Asynchronism in cellular automata is a rich source of questions. It has given rise to numerous studies, with various motivations and approaches. We refer to a recent survey paper for readers who would like to have an overview on the topic [8].

We proposed a first model of an asynchronous lattice gas in Ref. [2]. Applying an asynchronous updating in this model is less straightforward than for "classical"

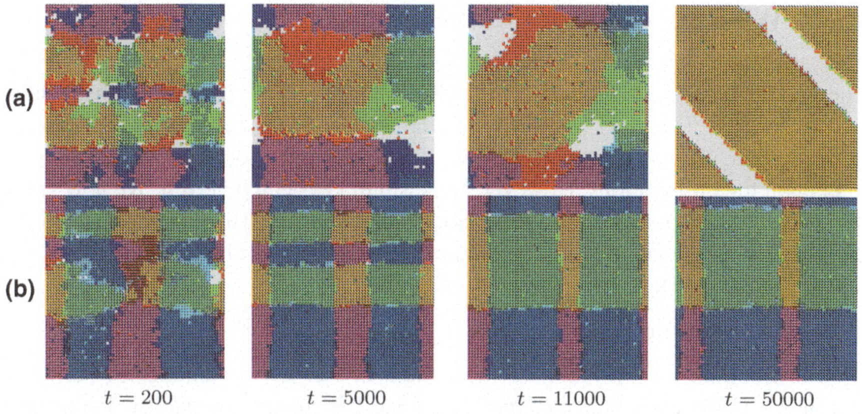

Fig. 16.16 Observation of an asynchronous interaction of $\alpha_I = 0.5$ and $L = 64$, $\sigma = 2.0$. **a** A tartan pattern that transforms into a stripe: $d = 0.4$. **b** A tartan pattern that remains stable for a very long time: $d = 0.5$

cellular automata. The updating of the lattice gas is done in two steps, and we may apply the asynchrony to one step or the other, or to both steps. We will here focus on the asynchronism of *interaction*: we apply the interaction rule with probability α_I and leave the cells' state unchanged with probability $1 - \alpha_I$. This provides a way to progressively "deviate" from the regular synchronous case. The question we ask is: What happens if we apply a small amount of asynchronism?

We empirically observed that if we start from a stripe or from a "clustered" state, no qualitative change is visible. However, if we start from a chequerboard, then we have a quick destruction of the pattern. This dissolution is shown on Fig. 16.15.

In some cases, a new pattern can be formed: horizontal and vertical lines appear, which span throughout the lattice and "loop". This pattern was named tartan, with a reference to the Scottish textile. This pattern is not always stable: in most cases, the tartan is destroyed after a few thousand steps and is replaced by a diagonal stripe (see Fig. 16.16).

The dissolution of the tartan pattern greatly depends on how close to 1/2 the density is. For instance, trajectory shown on Fig. 16.16b shows a tartan pattern that remains stable for more than 5.10^5 steps. It is another open problem to understand the dynamics of the tartan pattern. We believe that, contrarily to the other "major" patterns whose stability time would be exponential in the grid size, the stability of the tartan is "only" quadratic.

16.4 What do We Know Mathematically?

The phenomena discussed previously call for an immediate question. What can be said from a mathematical basis? In fact, the answer is: so far, very little. As mentioned earlier, the authors are not aware of any analytical tool that would allow us to explain the observations. However, some small keys can be given. We now study two properties: (a) the existence of a "hidden" invariant and (b) the recurrence of the system, that is, the fact that it almost surely returns to its initial condition (possibly in a very long time).

A Hidden Invariant

By construction, the evolution of the system conserves the number of particles. But the LGCA construction also introduces another conserved quantity. Let us take a grid with even dimensions, and name "even" and "odd" particles, the particles which are respectively located on an even or odd cell, that is, a white or black cell of the corresponding chequerboard. Then, the total number of even and odd particles is exchanged at each time step. Formally, let (x^t) represent a given evolution of the system, and let us denote by $\mathcal{L}_0 = \{c = (c_x, c_y) \in \mathcal{L}, c_x + c_y \in 2\mathbb{Z}\}$ and by $\mathcal{L}_1 = \mathcal{L} \setminus \mathcal{L}_0$ the set of even and odd cells, respectively. We also denote by: $N_0(t) = \sum_{c \in \mathcal{L}_0} N(x^t(c))$ and by $N_1(t) = \sum_{c \in \mathcal{L}_1} N(x^t(c))$ the number of even and odd particles, respectively. Remark that a cell that belongs to $\mathcal{L}_0$ has its four neighbours in $\mathcal{L}_1$ and reciprocally. Also, remark that the interaction step does not modify N_0 and N_1. Then, by the composition of the interaction and propagation step: $(N_o, N_1)(t+1) = (N_1, N_0)(t)$.

The presence of this "chequerboard invariant" is often seen as a drawback of the model: indeed, it was not included in it deliberately. This is a well-known problem of LGCA on a square grid, and this is why some authors prefer to use hexagonal grids, for which this invariant disappears. Note that using grids with odd dimensions is a dangerous expedient: indeed, even if the invariant does not exist at the global scale, it is still locally conserved on a short-time evolution. We refer to the work of Barberousse and Imbert for an in-depth discussion on how the existence of invariant quantities may affect the interpretation of the numerical simulations obtained on LGCA [1].

Defining Patterns

In what follows, the term "pattern" will be used informally to designate a given subset of the configuration space $Q^{\mathcal{L}}$. These configurations have the same visual appearance and their evolution shows a form of visual stability during long periods (typically a few thousand steps). Note that the difficulty is that the patterns that we visually observe are not perfect; their formal definition would thus need to "tolerate" a given degree of noise.

Ideally, we would like to study the system by calculating the stationary distributions of the Markov chain which defines the dynamics of the system. Unfortunately, even though this Markov chain depends only on two parameters, coming up with a closed formula that would help us predict the properties of the system is out of

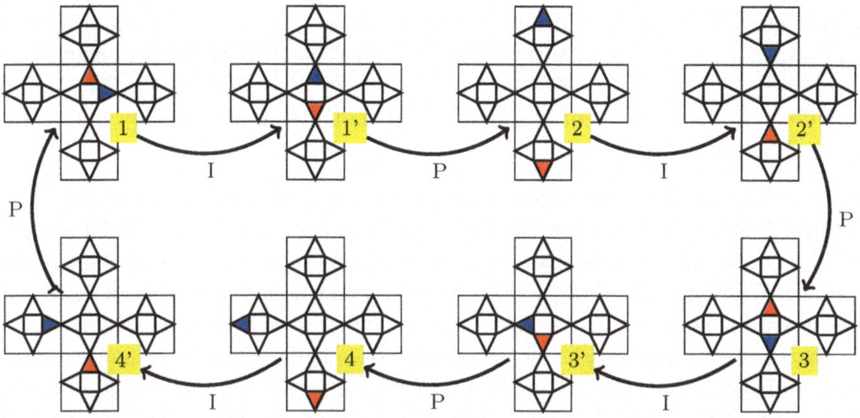

Fig. 16.17 Example of a recurrent behaviour. The first time step (*arrows* 1 and 2) shows an arbitrary move of the particles. The three following time steps shows a sequence that reverses the system to the initial state

reach. (This is already difficult for one-dimensional binary systems, see e.g. Ref. [9, 12].) Note that even if we use numerical analysis, the state space on which we need to compute the stationary distribution grows exponentially with the size of the grid and the number of particles.

To our knowledge, only few analytic results exist on this model. A mean-field approximation was used to predict the phase transition between non-ordered and ordered states (see Ref. [4] and Ref. [6, Chap. 8]). However, this predicts the existence of a phase transition but does not reflect the various types of stable patterns observed in the system. Partial results have also been given by Marcovici who studied the invariant measures of an infinite-size system for some specific cases [10, Chap. 2.6].

Recurrence of the Markov Chain

The second important property to remark is the *recurrence* of the Markov chain that represents the evolution of the system. Various proofs of this property can be given; we will here give an informal argument by showing that if a configuration y is reachable from x in one step, then x is reachable from y in three steps. As an example, consider the case illustrated in Fig. 16.17.

Let us consider a particular cell c and a given particle, e.g. the one which points to the East on step (1) of the figure (in blue or dark grey). After the first time step, composed of an interaction (1') and a propagation (2), the particle p is now in the cell at the North of c.

Let us now show that there is a sequence of updates that reverses this move.

- The interaction step makes p reverse its direction (2'); the propagation step (3) will thus make p come back to c, but with a direction not necessarily equal to the original direction.

- The interaction orients p in the inverse direction of its initial direction. In our example, it now points to West (3'), and it is propagated to the Western neighbour cell (4).
- In this cell, the direction of p is again inverted (4'). After propagation, p is again in c; it points East, its initial direction (4'=1).

It can be observed that the same argument also applies to the second particle of Fig. 16.17. Since there is no possibility of collision between particles in lattice-gas cellular automata, this argument can be generalised to any cell with any number of particles. Consequently, if a configuration y can be reached from x, then there is a nonzero probability that x is reached from y; on other words, the system is recurrent.

This recurrence property has an important consequence: when the system "moves" from a disordered state to an ordered state (a pattern) that seems stable, there is always a nonzero probability to destroy the order and to return to the disordered initial condition. As a consequence, for *finite* grids, the patterns that we will observe are all subject to *metastability*: they may be stable for an average time that is very long, typically exponential in the grid size, and therefore unreachable by simulation, but they are always bound to be destroyed.

Patterns can thus be characterised by their *attractivity*, which quantifies the probability to reach a given pattern when starting from a random initial condition and their *stability*, which quantifies the average time that the system will remain on a given pattern before leaving it. In the experiments described above, a minimal stability of the patterns was implicitly demanded: we required that the statistical properties remained stable in large intervals of simulation time (typically a few thousand steps). It is an open problem to determine how to estimate the stability of the patterns with mathematical analysis.

16.5 Discussion

Let us review the main elements of our study, pattern by pattern. In particular, we come back to our initial question: Can we use a simple LGCA to model the swarming phenomenon or does this tool make our model too simplistic?

About the Diagonal Stripe Pattern

This pattern is an example of spontaneous symmetry breaking: from a state that is initially disordered, the system "cools down" and "chooses" one of the four possibilities of diagonal orientations for the stripe. Experiments tend to show the presence of a "resonance effect": the pattern is created by an amplification of the waves that spontaneously emerge and travel across the lattice. Waves enter into collision and merge until they form one single stripe. The size ratio of the lattice plays a key role: if the grid is a square, a stripe forms easily; otherwise, the system may stabilise on various shapes depending on the value of the sensitivity and the size ratio. If we replace periodic boundary conditions by reflexive boundary conditions, then the symmetry breaking is no longer observed and the system shows four coexisting stripes that periodically "bounce" on the borders.

About the Chequerboard Pattern

The very existence of this pattern is surprising because the local evolution rule is *stochastic*. Indeed, the transition from diagonal stripes to the chequerboards occurs when we keep the sensitivity constant and only increasing the density of particles on the lattice. All happens as if the presence of more particles remove degrees of freedom and force the system to oscillate. This pattern shows that in some cases the "noise" produced by the local rule is not sufficient to smooth out the discrete nature of the system.

We saw that the existence of this pattern is not a direct consequence of chequerboard invariant (see Sect. 16.4) and that by adding a small amount of asynchrony, we destroy the pattern. It is not yet clear whether there exists a minimal amount of asynchrony to introduce to make the chequerboard patterns unstable.

About the Clusters Pattern

Clusters are groups of particles that move coherently. They have the strange property of traversing each other without interacting much. This property results both from the local rule *and* from the fact that the LGCA model allows particles with different directions to be simultaneously present on a single cell. It is an open problem to measure precisely the transition between the diagonal stripe pattern and the clusters pattern. Note that a simple quantitative parameter to discriminate between the two behaviours would be the average "occupancy" per non-empty cell: in the case of diagonal stripe, this parameter should be close to 2, while it should be observed close to 1 for the clusters pattern.

Finite Versus Infinite Lattices

As mentioned earlier, the central problem to understand the behaviour of the model is its *metastability*. The patterns we encountered may remain an exponential time in the lattice size, but not an *infinite* time. We thus aim at studying *infinite lattices*, but we have to remember that the behaviour of an infinite lattices may not always reflect the limit behaviour for finite lattices whose size tend to infinity. For example, the mechanism that produces diagonal stripes is probably different on an infinite lattice and will require a different approach than the one we had here.

To sum up, this model has an extraordinary richness of behaviour although it involves only a minimal set of definitions. Obviously, compared to more classical models such as the well-studied Ising model, it is the existence of a *propagation step* which here plays a key role. This step implies that the density of particles changes, which produces phenomena of amplification of noise from a local to a global scale. We find it puzzling that the three main patterns react so differently to small changes in the simulation conditions (for instance to an asynchronous updating or to the use of a non-square lattice). Would it be that the existence of non-robust patterns is a signature that the model is too simple? But at the same time, if we are unable to correctly analyse it mathematically, can't we also say that the model is already too complex?

Chapter's Appendix: Simulation of Reflecting Borders in a LGCA

The simulation of the *reflecting borders* boundary conditions is applied as follows.

We initialise the system by letting each channel contain a particle with probability d. There are exceptions: (a) The border cells are all empty. (b) For the cells situated immediately next to the northern, eastern, southern and western borders cells, we respectively empty the North, East, South and West channels. Formally, we use $\mathcal{L} = \{0, \ldots, X\} \times \{0, \ldots, Y\}$ and:

- $\mathcal{B} = \{(i, j) \in \mathcal{L}, i = 0 \text{ or } i = X - 1 \text{ or } j = 0 \text{ or } j = Y - 1\}$,
- $\mathcal{B}_N = \{(i, j) \in \mathcal{L}, j = Y - 1\}$,
- $\mathcal{B}_E = \{(i, j) \in \mathcal{L}, i = X - 1\}$,
- $\mathcal{B}_S = \{(i, j) \in \mathcal{L}, j = 0\}$,
- $\mathcal{B}_W = \{(i, j) \in \mathcal{L}, i = 0\}$.

By noting the initial condition $x \in Q^{\mathcal{L}}$ for each $(i, j) \in \mathcal{L}$, taking $x(i, j) = (q_n, q_e, q_s, q_w)$, we have $q_n = 0$ if $(i, j) \in \mathcal{B} \cup \mathcal{B}_N$, $q_e = 0$ if $(i, j) \in \mathcal{B} \cup \mathcal{B}_E$, etc. We call this last set of 4 conditions, the *integrity condition.*

The integrity condition guarantees that no particle will travel to a border cell (in $\mathcal{B}$). It is easy to see that the propagation step preserves the integrity condition, but not the interaction step. Our method thus consists in checking if the north channel of a cell of $\mathcal{B}_N$ is occupied. In this case, the particle is re-affected among the free channels of the cell, with a uniform probability. All happens as if the particle has "bounced" on a northern wall. Clearly, such a rearrangement is always possible as this cell cannot contain four particles. The same procedures is applied for the three other directions.

References

1. Barberousse, A., Imbert, C.: New mathematics for old physics: the case of lattice fluids. Stud. Hist. Philos. Sci. Part B: Stud. Hist. Philos. Mod. Phys. **44**(3), 231–241 (2013). https://doi.org/10.1016/j.shpsb.2013.03.003
2. Bouré, O., Fatès, N., Chevrier, V.: First steps on asynchronous lattice-gas models with an application to a swarming rule. Nat. Comput. **12**(4), 551–560 (2013). https://doi.org/10.1007/s11047-013-9389-2
3. Bouré, O., Fatès, N., Chevrier, V.: A robustness approach to study metastable behaviours in a lattice-gas model of swarming. In: Kari, J., Kutrib, M., Malcher, A. (eds.) Proceedings of Automata'13. Lecture Notes in Computer Science, vol. 8155, pp. 84–97. Springer, Berlin (2013). https://doi.org/10.1007/978-3-642-40867-06 (Extended version available as a tech. report at https://hal.inria.fr/hal-00768831)
4. Bussemaker, H.J., Deutsch, A., Geigant, E.: Mean-field analysis of a dynamical phase transition in a cellular automaton model for collective motion. Phys. Rev. Lett. **78**(26), 5018–5021 (1997). https://doi.org/10.1103/PhysRevLett.78.5018
5. Deutsch, A.: Orientation-induced pattern formation: swarm dynamics in a lattice-gas automaton model. Int. J. Bifurc. Chaos **06**(09), 1735–1752 (1996). https://doi.org/10.1142/S0218127496001077

6. Deutsch, A., Dormann, S.: Cellular Automaton Modeling of Biological Pattern Formation - Characterization, Applications, and Analysis. Modeling and Simulation in Science, Engineering and Technology. Birkhäuser, Basel (2005)
7. Fatès, N.: FiatLux: a simulation program in Java for cellular automata and discrete dynamical systems available, http://fiatlux.loria.fr (Cecill licence) APP IDDN.FR.001.300004.000.S.P.2013.000.10000
8. Fatès, N.: A guided tour of asynchronous cellular automata. J. Cell. Autom. **9**(5–6), 387–416 (2014)
9. Mairesse, J., Marcovici, I.: Around probabilistic cellular automata. Theor. Comput. Sci. **559**, 42–72 (2014). https://doi.org/10.1016/j.tcs.2014.09.009
10. Marcovici, I.: Automates cellulaires probabilistes et mesures spécifiques sur des espaces symboliques. Ph.D. thesis, Université Paris 7 (2013). https://tel.archives-ouvertes.fr/tel-00933977 (Text in English)
11. Regnault, D., Schabanel, N., Thierry, E.: Progresses in the analysis of stochastic 2D cellular automata: a study of asynchronous 2D minority. Theor. Comput. Sci. **410**(47–49), 4844–4855 (2009). https://doi.org/10.1016/j.tcs.2009.06.024
12. Taggi, L.: Critical probabilities and convergence time of percolation probabilistic cellular automata. J. Stat. Phys. **159**(4), 853–892 (2015). https://doi.org/10.1007/s10955-015-1199-8

Chapter 17
PCA Modelling of Multi-species Cell Clusters: Ganglion Development in the Gastrointestinal Nervous System

Kerry A. Landman and Donald F. Newgreen

Abstract A defining characteristic of the enteric nervous system (ENS) is mesoscale patterned entities called ganglia. Ganglia are clusters of neurons with associated enteric neural crest (ENC) cells, which form in the simultaneously growing gut wall. At first, the precursor ENC cells proliferate and gradually differentiate to produce the enteric neurons; these neurons form clusters with ENC scattered around and later lying on the periphery of neuronal clusters. By immunolabelling neural cell–cell adhesion molecules, the adhesive capacity of neurons is determined to be greater than that of ENC cells. Using a probabilistic cellular automata (PCA) model, we test the hypothesis that local rules governing differential adhesion of neuronal agents and ENC agents will produce clusters that emulate ganglia. The clusters are relatively stable, relatively uniform and small in size, of fairly uniform spacing, with a balance between the number of neuronal and ENC agents. These features are attained in both fixed and growing domains, reproducing, respectively, organotypic in vitro and in vivo observations. Various threshold criteria governing ENC agent proliferation and differentiation and neuronal agent inhibition of differentiation are important for sustaining these characteristics. This investigation suggests possible explanations for observations in normal and abnormal ENS development.

17.1 Introduction

Probabilistic cellular automata are useful tools for modelling many biological processes such as cell motility, growth, adhesion and differentiation, since they capture individual-level properties and interactions of a biological system [5, 20, 23,

K.A. Landman (✉)
School of Mathematics and Statistics, University of Melbourne, Parkville, VIC 3010, Australia
e-mail: kerryl@unimelb.edu.au

D.F. Newgreen
Murdoch Children's Research Institute, Royal Children's Hospital, Parkville, VIC 3052, Australia
e-mail: don.newgreen@mcri.edu.au

P.-Y. Louis and F.R. Nardi (eds.), *Probabilistic Cellular Automata*, Emergence, Complexity and Computation 27, https://doi.org/10.1007/978-3-319-65558-1_17

26, 27]. PCA models in conjunction with a discrete model for probabilistic domain growth are needed for developmental biology, where the underlying tissue is also expanding through cell proliferation [2]. This adds an additional mechanism for the transport of cellular agents [1].

When different cell populations are dissociated and mixed, they re-associate [36] and "sort out" into mutually exclusive domains [34]. In classic papers, Steinberg and co-workers analysed such phenomena theoretically and experimentally, based on the physics of immiscible liquids, resulting in the differential adhesion hypothesis as the mechanism for cell sorting [6, 7, 30–32]. Various PCA models have also reproduced features of the Steinberg experiments [8, 9, 33].

Here, we consider differential adhesion in the development of ganglia in the enteric nervous system (ENS), present experimental biological observations and develop a novel PCA model on a non-growing and growing domain which provides biological insight.

The gastrointestinal nervous system, known as the enteric nervous system (ENS), is a large complex network of neurons lining the wall of our gastrointestinal tract, controlling normal gut function and peristaltic contraction. ENS development in vertebrate embryos starts with a relatively small number of neuronal precursor cells (enteric neural crest or ENC cells) entering the stomach and then progressively colonising the whole gut to the anal end, as a highly time-tabled invasion wave. These ENC cells eventually differentiate into ENS neurons that coalesce into the ganglionated network typical of the mature ENS [12]. This invasion occurs within the dense environment of the intestinal wall made up of densely packed gut mesenchymal cells which go on to form the smooth muscle and connective tissues of the gut.

A relatively common birth defect, called Hirschsprung's disease, results when ENC cells fail to colonise the anal end. This means that this region of the gut cannot generate peristaltic waves. This causes chronic constipation which is potentially fatal if not surgically treated [24]. The colonisation process, involving ENC cell motility and a vast expansion of ENC cell numbers by cell division [40], has been extensively modelled using probabilistic cellular automata (PCA) models [18, 19, 41]. However, ENS abnormalities also occur in the size, density and organisation of the ganglia [3, 14, 22, 35, 37].

ENC cells progressively differentiate into neurons [38], which become non-proliferative and have a lower migratory speed [11, 38]. The ENC cells also differentiate into glia (support cells). We will refer to glia and true ENC precursors as ENC cells since at the early stage here they are difficult to distinguish. Early in ENC invasion, the neuron density is low, while later (that is, further behind the wavefront), the neuron density rises [11]. Simultaneously, the underlying mesenchymal cells which make up the gut tissue are rapidly dividing, leading to a dramatic elongation of the tissue [2], impacting gut colonisation by ENC cells [25]. However, in gut organ cultures, overall growth is negligible, but interestingly, the important features of ENC cell migration and differentiation remain the same [13].

Far behind the wavefront, the ratio of neurons to ENC cells stabilises to approximately 1.2:1. A mesoscale pattern becomes evident as neurons and ENC cells begin to arrange themselves into (essentially) two-dimensional cell clusters, as shown in

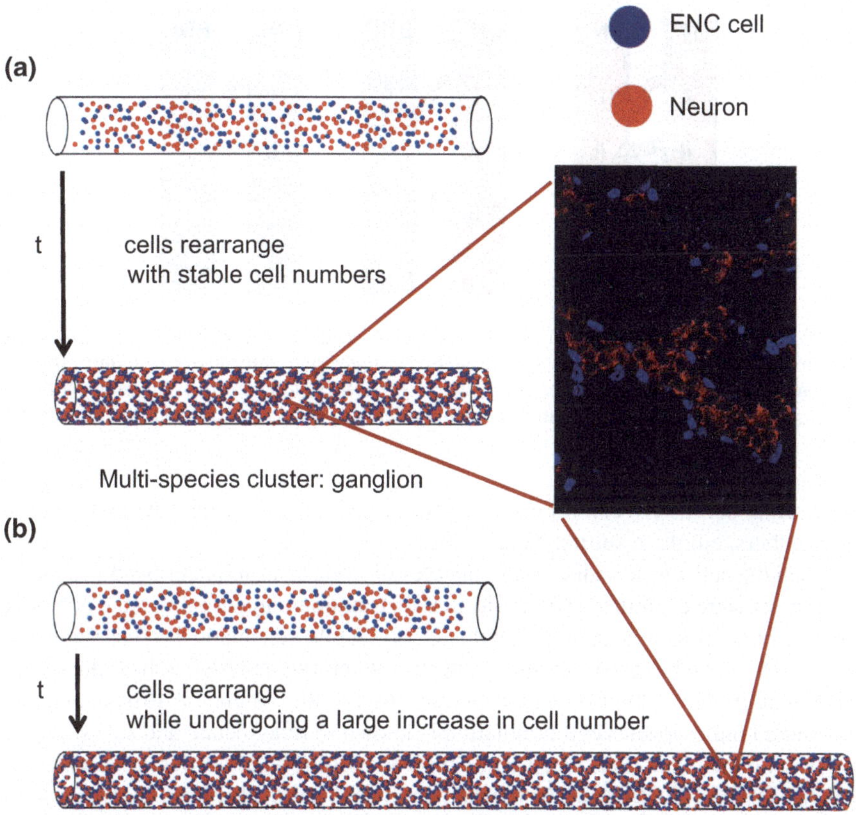

Fig. 17.1 Schematic diagram of a section of fully colonised (far from the wavefront) gut tube. **a** An organ culture in vitro experiment (fixed gut length). Initially, the ENS cell population is uniformly mixed. The two cell types move and rearrange into clusters called ganglia. The clusters consist of neurons loosely surrounded by ENC cells. Quail embryonic age 8 days midgut: neurons form large coherent clusters with ENC cells around the edges. Neuronal cytoplasm labelled for HuC/D (*red*), ENC cell nuclei labelled for SoxE (*blue*). **b** An in vivo experiment. Initially, the system is uniformly mixed. The ENC cells move, proliferate and differentiate at the same time as the gut grows. They rearrange into clusters called ganglia. The clusters consist of neurons loosely surrounded by ENC cells as in (**a**)

Fig. 17.1. The ENS forms on the surface of gut muscle layers and hence forms arrays spaced out as on the surface of a cylinder. Typically, there are two of these cylindrical ENS arrays, and each ganglion is also flattened to approximate a two-dimensional group. The ganglia are relatively regularly spaced and of a similar size. Initially, aggregation can only be discerned between neurons, while the ENC cells remain scattered. Later, the ENC cells congregate around the edges of the tight neuron clusters, and isolated neurons are no longer seen, as illustrated in Fig. 17.1a. This is the typical structure of the young ganglia in organ culture in vitro and in vivo conditions.

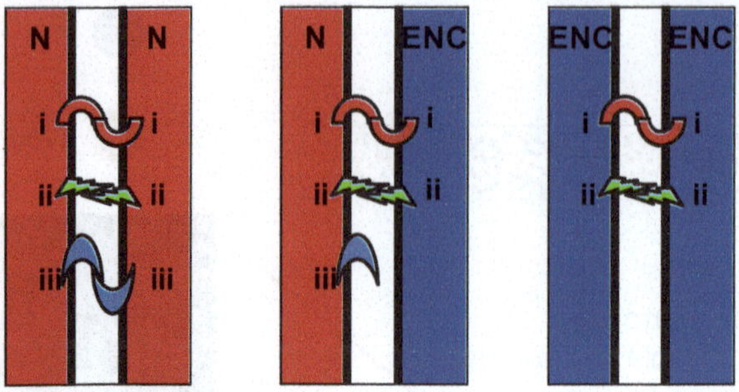

Fig. 17.2 Schematic representation of adhesion molecules associated with neurons N (*red*) and ENC cells (*blue*). Segment of cell surface with *red*, *green* and *blue* shapes denoting different homophilic surface adhesion molecules (types i, ii, iii). (Different numbers of the same molecule are not shown.) Total adhesive potential is the sum of the adhesive potentials of all adhesive modes

However, under in vivo conditions, there is a huge increase in both ENC and neuron cell numbers, but the resulting ganglia structure is unchanged.

When the cell surface adhesion molecules of ENC cells and neurons during gangliogenesis were examined [10], molecular distributions suggested that the adhesive capacity of neurons was greater than that of ENC cells, schematically represented in Fig. 17.2. Could it be that differential adhesion of neurons and ENC cells is the mechanism that produces the distinct self-organisation of the ganglia? A dissociated and then mixed multi-species cell population is known to re-associate and self-sort [34, 36]. Such phenomenon has been well-studied resulting in the differential adhesion hypothesis being identified as the mechanism for cell sorting [6, 7, 30–32]. Our work here uses the ideas of differential adhesion to build a mesoscale pattern of partially self-sorted clusters called ganglia.

17.2 PCA Model

The aim of this work is to use a multi-species PCA model to test the hypothesis that differential adhesion is the mechanism for ganglion formation in both a non-growing and growing gut. Moreover, we examine whether the less adhesive agents congregate around the edges of the more strongly adhesive agents in any clusters that evolve, thereby having the characteristics of those that develop in the ENS.

A discrete-time agent-based PCA model on a square lattice with unit lattice spacing is used. Since the ENS is restricted to a cylindrical surface within the intestinal tissue, a two-dimensional lattice of length L (the length of the cylindrical surface) and height Y (the circumference of the cylindrical surface) is appropriate, with periodic boundary conditions along the horizontal boundaries.

We introduce two species of agents, namely ENC cell agents and neuron agents, denoted ENC and N, respectively. At most a single agent occupies a single lattice site

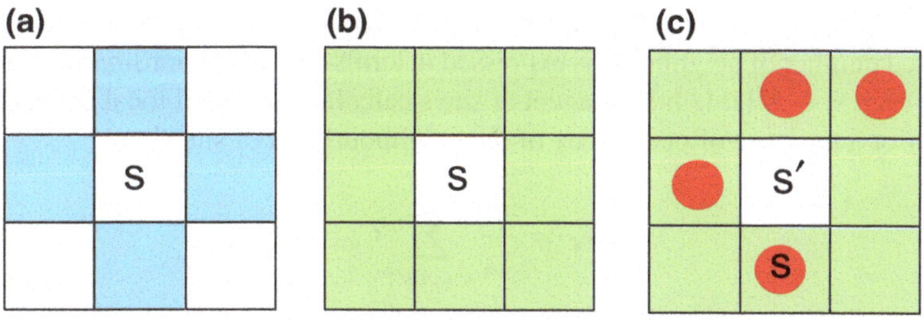

Fig. 17.3 Definitions. **a** von Neumann neighbourhood $\mathcal{N}\{\mathbf{s}\}$ (*blue*). **b** Moore neighbourhood $\mathcal{A}\{\mathbf{s}\}$ (*green*). **c** Potential agent–agent bonds for an agent at site **s** moving to site **s′**. Agents (*red circles*), the Moore neighbourhood $\mathcal{A}\{\mathbf{s}'\}$ (*green*). A motile agent at **s** contributes to $K_{\mathbf{s}'}$, so $K_{\mathbf{s}'} = 4/8$

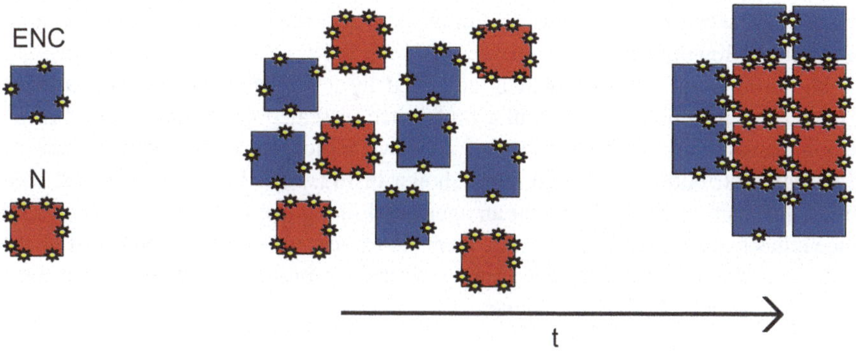

Fig. 17.4 Total adhesive potential, represented by number of *yellow shapes*, is greater on neuron N agents (*red*) than on ENC agents (*blue*). The N:ENC adhesive bias ratio is 8:4 here. The well-mixed population evolves in time, such that N agents form clusters surrounded by ENC agents

at any time t. An ENC agent can move, proliferate and differentiate into an N agent, while an N agent can only move. These functions are governed by probabilities [1, 28]. (No agent death is included here, because there is little evidence of ENC cell death at this stage of development [4].)

The PCA rules and their implementation are now described.

Agent Movement Taking into Account Adhesion Molecules

For any site **s** on the square lattice, we define the von Neumann neighbourhood (of nearest neighbour sites) $\mathcal{N}\{\mathbf{s}\}$ and the Moore neighbourhood $\mathcal{A}\{\mathbf{s}\}$ as illustrated in Fig. 17.3a, b. The numbers of sites in these neighbourhoods are $z_{\mathcal{N}} = 4$ and $z_{\mathcal{A}} = 8$, respectively. The occupancy of site **s** is $C_{\mathbf{s}}$, with $C_{\mathbf{s}} = 1$ for an occupied site and $C_{\mathbf{s}} = 0$ for an unoccupied site.

Each agent type has a certain number of adhesion molecules, represented schematically in Fig. 17.4. These molecules determine the ability of an agent to form bonds with other agents. Since N agents have more adhesion molecules than ENC agents, N agents can potentially form more bonds than the ENC agents. We model this bond number ability as a preferred number of neighbours in a local neighbourhood. This preferred number of neighbours is expressed in terms of a local coordination number.

For each $\mathbf{s}' \in \mathcal{N}\{\mathbf{s}\} \cup \{\mathbf{s}\}$, an agent at site $\mathbf{s}$ calculates a scaled local coordination number or a measure of occupancy of the neighbourhood of site $\mathbf{s}'$, as

$$K_{\mathbf{s}'} = \frac{1}{z_{\mathcal{A}}} \sum_{\mathbf{s}^* \in \mathcal{A}\{\mathbf{s}'\}} C_{\mathbf{s}^*}, \tag{17.1}$$

so that $K_{\mathbf{s}'} \in [0, 1]$ is the scaled coordination number (Fig. 17.3c). It determines the potential number of agent–agent attachments that can be made by an agent placed at each target site. In determining the coordination number, we do not distinguish between the two agent types; that is, our adhesion molecules are not agent type-specific. We weight the probability of movement according to the coordination number in the following way.

The probability of movement is governed by a non-negative *binding function* $f(K)$, chosen to reflect a particular preference in scaled coordination number K, which corresponds to the number of adhesive molecules on the cell surface. A Gaussian distribution is chosen here, shown in Fig. 17.5. Let $T(\mathbf{s}'|\mathbf{s})$ be the conditional transition probability that an agent will move from site $\mathbf{s}$ to site $\mathbf{s}' \in \mathcal{N}\{\mathbf{s}\}$ once it has been chosen to move. An agent also assesses its current site based on the scaled coordination number, determining a transition probability of remaining at the same site, namely $T(\mathbf{s}|\mathbf{s})$. We define

$$T(\mathbf{s}'|\mathbf{s}) = \frac{f(K_{\mathbf{s}'})(1 - C_{\mathbf{s}'})}{f(K_{\mathbf{s}}) + \sum_{\mathbf{s}'' \in \mathcal{N}\{\mathbf{s}\}} f(K_{\mathbf{s}''})(1 - C_{\mathbf{s}''})}, \quad \mathbf{s}' \in \mathcal{N}\{\mathbf{s}\}, \tag{17.2}$$

and

$$T(\mathbf{s}|\mathbf{s}) = \frac{f(K_{\mathbf{s}})}{f(K_{\mathbf{s}}) + \sum_{\mathbf{s}'' \in \mathcal{N}\{\mathbf{s}\}} f(K_{\mathbf{s}''})(1 - C_{\mathbf{s}''})}. \tag{17.3}$$

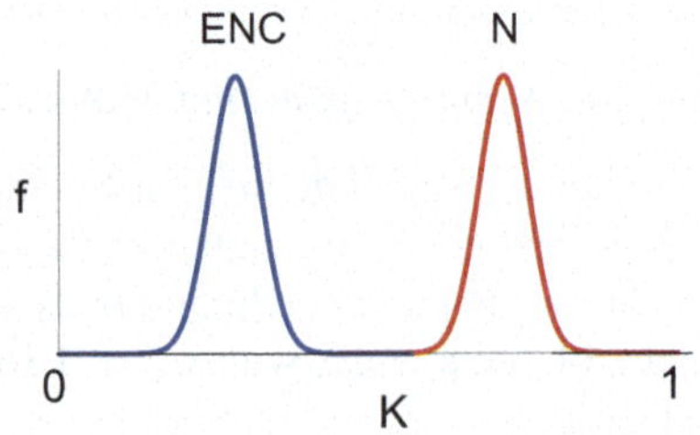

Fig. 17.5 The binding function $f(K)$ representing the adhesion capacity. Here, $f(K) = e^{-100(K - K_*^{\mathrm{S}})^2}$, where S stands for species ENC or N. The constant K_* with superscripts denoting the agent type is the preferred scaled coordination number. Here, $K_*^{\mathrm{N}} > K_*^{\mathrm{ENC}}$ for the differential adhesion hypothesis

The factors like $1 - C_{\mathbf{s}'}$ ensure that the target site is empty. These are included since cells sense their local neighbourhood and are aware of neighbouring cells; we therefore specify that agents will only attempt to move to those sites that are unoccupied. The denominator is a normalisation factor. These rules ensure that the probability of attempting to move to an occupied site is zero and that the sum of all the probabilities is unity.

Agent Proliferation and Differentiation

When an ENC agent is chosen to proliferate (with probability P_{p}), it inspects its immediate neighbourhood and proceeds to proliferate if there is a vacant site in $\mathcal{N}\{\mathbf{s}\}$. One daughter remains in the original site **s**, and the other daughter is randomly placed into one of the vacant sites in $\mathcal{N}\{\mathbf{s}\}$. Since ENC cell density increases through proliferation to reach a preferred density [29], we restrict the agent density by introducing a proliferation threshold $\lambda_{\mathrm{p}} \in [0, 1]$. An agent at **s** will only proliferate if its scaled coordination number satisfies $K_{\mathbf{s}} < \lambda_{\mathrm{p}}$.

When an ENC agent is chosen to differentiate (with probability P_{d}), the ENC agent converts to an N agent. Motivated by ENS observations, differentiation is restricted in two ways. We introduce (i) a threshold λ_{d} such that a chosen ENC agent at **s** will only differentiate if its local scaled coordination number satisfies $K_{\mathbf{s}} > \lambda_{\mathrm{d}}$; (ii) a neuronal inhibition threshold λ_{NI} such that a chosen ENC agent at **s** will only differentiate if its local scaled coordination number satisfies $K_{\mathbf{s}}^{\mathrm{N}} < \lambda_{\mathrm{NI}}$. Here, $K_{\mathbf{s}}^{\mathrm{N}}$ is the scaled N agent coordination number at **s**, that is the number of neurons in $\mathcal{A}\{\mathbf{s}\}$ divided by $z_{\mathcal{A}}$ and therefore is just the fraction of $K_{\mathbf{s}}$ occupied by N agents.

Domain Growth

We incorporate this growth into our lattice model on $L(t) \times Y$ (where Y is fixed). For the case of an elongating gut tissue in vivo, the length of the lattice increases exponentially in time t with growth rate α, as $L(t) = L_0 e^{\alpha t}$ [2], while the height remains constant. This is implemented through a random insertion of new lattice sites [1, 2] in each row, where the number of insertions at each time step τ is given by "round" of $[L(t + \tau) - L(t)]$. However, each row is independent of any other row. This technique ensures the insertions occur uniformly throughout the domain, emulating gut cell division.

On top of the underlying lattice, there may be an ENC or N agent occupying the site. If a cellular agent is occupying the same site that moves to a new position in the lattice due to a random lattice site insertion in the domain growth mechanism, then the ENC or N agent is transported or carried to this new position in the lattice.

Implementation

A sequential independent random choice of agents is made for each operation as described above. For a growing domain, the domain growth operation precedes cel-

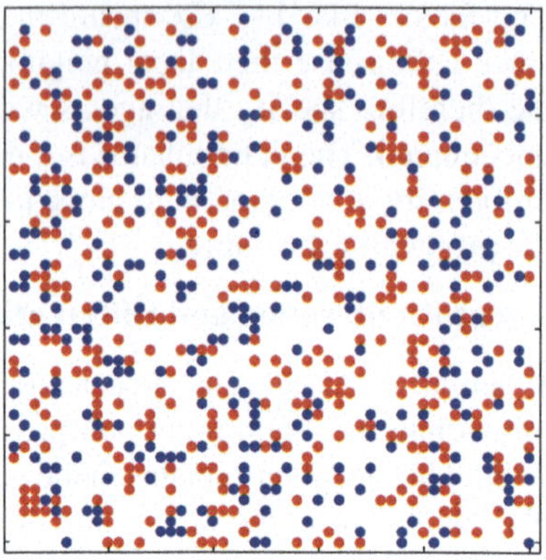

Fig. 17.6 Example of initial conditions. $\rho = 0.3$. The N to ENC agent number ratio is $R = 1.2$

lular agent operations. The cellular operations of agent movement and then proliferation and differentiation (if these are implemented) occur in this order.

The parameters are estimated from the following information: (a) the speed observed for an ENC cell is approximately $\sim$ 100μm/h [39]. For an idealised cell with diameter 5–10 μm, this speed corresponds to $\sim$ 10 lattice spacings/h. (b) A realistic midgut elongation exponent is 0.41/day [2]. Therefore, the doubling time for midgut length is $\ln(2)/0.41 \approx 1.69$ days ≈ 40 h.

For the growing domain case, simulations have $\alpha = 0.05$, giving a doubling time of the order of 10 time steps, and hence, (b) yields 1 h $\approx$ 0.25 time steps. In addition, agents move on average G times per time step, where G is estimated from (a) to be $G \in [1, 40]$. In the fixed domain case, we assume agents moves occur on average once per time step, giving $G = 1$, and hence, (a) gives 1 h $\approx$ 10 time steps. These timescales allow us to fix the cell proliferation rate P_{p} in terms of the cell cycle time, T, measured in hours. In the non-growing case, $P_{\mathrm{p}} = 0.1/T$ per time step, while in the growing case $P_{\mathrm{p}} = 4/T$.

The differentiation rate, P_{d}, is taken to be half the proliferation rate in each case.

17.3 Testing the Hypothesis: Simulations on a Fixed Domain

We present simulations on a 50×50 lattice with periodic boundary conditions, where initially the lattice is randomly seeded with both N and ENC agents at a prescribed overall density ρ and N:ENC agent number ratio R, illustrated in Fig. 17.6. We permit agent movement but no proliferation or differentiation. Therefore, the number of agents in each species remains constant.

Different values of N:ENC adhesive bias rules (where for simplicity a:b represents $8K_*^{\mathrm{N}} = \mathrm{a}$ and $8K_*^{\mathrm{ENC}} = \mathrm{b}$, and so represents the preferred coordination numbers) lead to very different mesoscale patterns (Fig. 17.7a). For instance, an agent bias

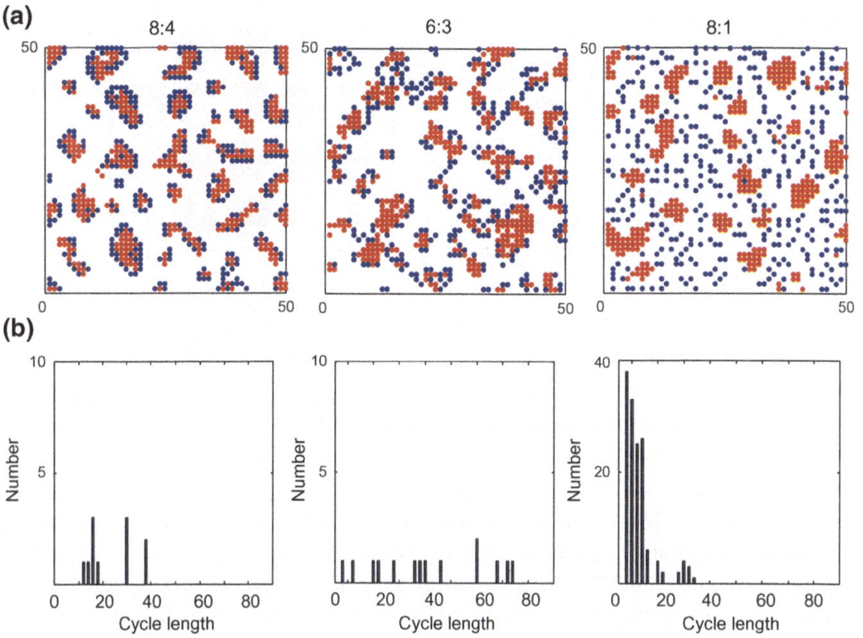

Fig. 17.7 Simulations on a fixed domain with agent density $\rho = 0.3$ and N to ENC agent number ratio $R = 1.2$ with N:ENC adhesive bias rules of 8:4, 6:3, 8:1 (*left*, *centre* and *right* columns, respectively). **a** Distribution of agents at time $t = 200$, N agents (*red*), ENC agents (*blue*). **b** Minimum cycle basis, showing number of cycles versus cycle length at $t = 200$

rule of 8:4 leads to tightly bound N clusters coated by ENC agents, while the 6:3 rule leads to looser clusters with ENC agent chains connecting different clusters. A bias rule of 8:1 also gives tight N clusters, but the ENC agents are not located on the clusters, but rather appear in groups of singlets or doublets in the gaps between clusters. The 6:3 pattern is realistic for early ganglia development, whereas the 8:4 pattern is realistic in later stages of development and is representative of the final ganglia organisation. It is desirable to have quantitative measures to characterise the patterns observed in Fig. 17.7a, so that the differences can be explained.

We define a cluster as all agents which form a connected component on a Moore neighbourhood. Useful quantitative measures are the cluster size distribution, cluster perimeter distribution, number of clusters (all defined by the Hoshen–Kopelman algorithm [16]), percentage of clusters close to neurons, the density ρ and number ratio R. Here, we report on the cluster perimeter which can be determined from the minimum cycle basis. At a prescribed time ($t = 200$ here), we calculate all closed loops within the empty space, the complement of the agent occupied sites. We construct a graph and corresponding adjacency matrix, where vertices are the centres of each unoccupied site and edges join any two vertices that are in neighbouring sites (in a Moore neighbourhood). The minimum cycle basis is determined by the standard

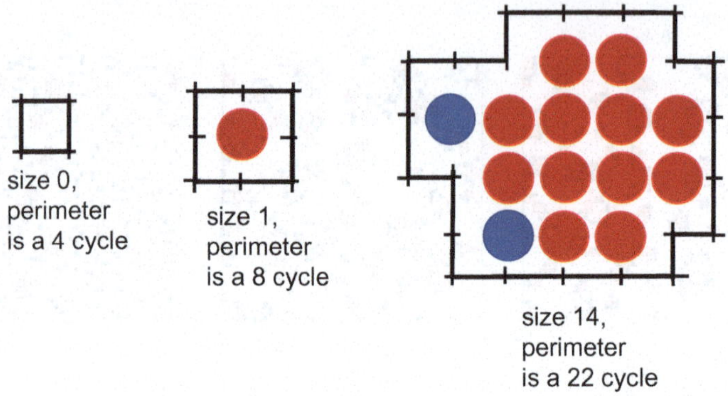

Fig. 17.8 Examples of different cluster sizes and cycles surrounding agent clusters (determining the perimeter). All agents are considered equally for this calculation

methods [15, 21]. There is a large number of 4-cycles, corresponding to adjacent unoccupied sites not enclosing any agents, while all other n-cycles ($n > 4$) enclose clusters of agents (Fig. 17.8). This measure (Fig. 17.7b) gives a good indication of how agents are distributed: a large number of small cycles indicate many isolated agents in singlets or doublets (8:1 case), while a very large cycle indicates a large connected cluster (6:3 case). We observe that the three rules give quite different frequency distributions which allows us to distinguish between them. In particular, the 8:4 rule has no small cycles and has several large cycles. The 6:3 rule has a few small cycles, and agents are well connected, while the 8:1 rule has a very large number of small cycles (corresponding to ENC singletons and small groups) together with large aggregates which correspond to the tight clusters of N agents.

17.4 Criteria for Ganglion-Like Clusters

We have demonstrated that our adhesion bias rules produce clusters for a given agent density $\rho = 0.3$ and ratio $R = 1.2$ on a non-growing domain with no proliferation and differentiation. These are biologically relevant values for the avian ENS in the region behind the wavefront where ganglia begin to form. Since precise experimental values are not available, instead we determined a convex area of acceptable (ρ, R) parameter space (e.g. Fig. 17.9 for the 6:3 rule). In this region, clusters reproduce ganglion-like characteristics when the following three properties hold:

(i) total number of clusters lies in the range [5, 35], for a 50×50 lattice,
(ii) average size of N agent clusters in the range [4.5, 17],
(iii) percentage of ENC agents having an N agent within $2\sqrt{2}$ lattice spaces is greater than 97%.

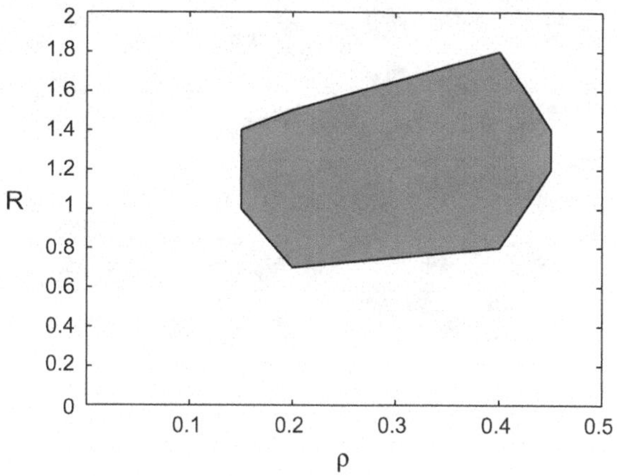

Fig. 17.9 Region of (ρ, R) parameter space (*shaded*) producing clusters with ganglion-like characteristics for 6:3 bias rule

We tested the inclusion of ENC agent proliferation and differentiation in the non-growing case. In all simulations, ρ and R evolve in time and the patterns are destabilised. The resulting patterns always move outside the allowable (ρ, R) parameter space to form clusters which are unacceptably large.

Domain growth must be included in order to stabilise both ρ and R within the range of acceptable values. In the next section, domain growth is included. The aim is to find a set of parameters which gives rise to clusters with the following properties:

1. number of N agent clusters increases with time; experimental observations indicate that clusters remain a similar size but increase in number,
2. ρ and R remain in the convex region in (ρ, R) space, giving ganglion-like clusters,
3. clusters do not elongate too much,
4. clusters have N agents surrounded by ENC agents.

17.5 Simulations on a Growing Domain

We present simulations on a lattice, initially size 50×50, which elongates in the x-direction. Since this surface is emulating a growing cylindrical surface, it is appropriate to have periodic boundary conditions on the horizontal boundaries and with no flux boundary conditions on the vertical boundaries. The initial condition is a random arrangement of N and ENC agents with initial density $\rho_0 = 0.5$ and initial number ratio $R_0 = 1.2$.

Figure 17.10 presents an example where the four properties listed in Sect. 17.4 are satisfied for the time period studied. By investigating a broad spectrum of parameter values [10], we determined that the first property (1) above is generally satisfied, while the other three properties are more difficult to satisfy.

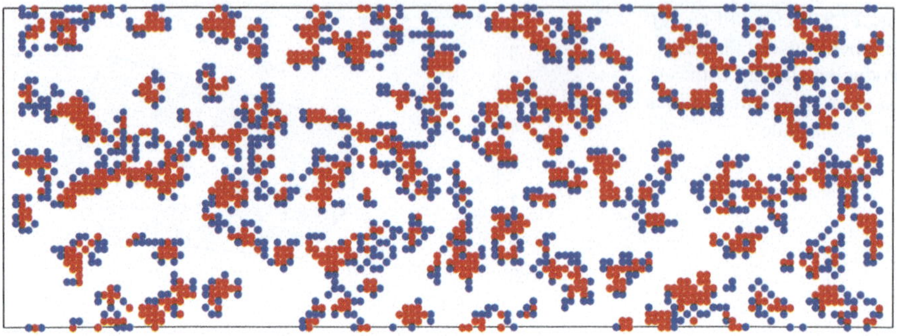

Fig. 17.10 Distribution of agents with 6:3 bias rule at $t = 20$; $\alpha = 0.05$, $\lambda_p = 0.4$, $\lambda_{NI} = 0.25$, $\rho_0 = 0.5$, $R_0 = 1.2$, $G = 10$. $P_p = 0.1$, $P_d = 0.05$, $\lambda_d = 0.25$; final values $\rho_f = 0.31$, $R_f = 1.05$

It is important to note that exponential growth will eventually overtake any constant G value. This will lead to elongating clusters and/or isolated N agents (breakdown of property 3 and/or 4). Agents will not have enough motility steps to keep up with the domain growth. Our simulations run for 20 time steps, corresponding to 80h. By this time, other processes are occurring in the development of avian ENS, such as the development of oriented muscles within the maturing gut wall. The rate of movement of ENS cells in this older tissue environment is decreased [17], and the orientation of the muscle is likely to constrain directionality of ENC cell movement. For these later stages, the issues highlighted here may no longer be relevant.

17.6 Any Differences Between the Growing and Fixed Domain Results?

Since little difference is observed in the features of ganglia formed in non-growing and growing environments (Fig. 17.1), it is interesting to investigate whether there are any quantifiable differences between the resulting distribution of agents for the non-growing and growing scenarios.

First consider the uniaxially growing domain scenario, starting with a lattice size $L_0 \times Y$ with a certain bias rule. At some later time t_f, the lattice is size $L_f \times Y$, where $L_f = L(t_f)$, with final density and number ratio ρ_f and R_f, respectively. Next consider a fixed domain, size $L_f \times Y$, starting with ρ_f and R_f with the same bias rule. We evolve this system in time, with no agent proliferation and differentiation. The two cluster structures are compared using our statistical measures. By way of example, Fig. 17.11b shows the resulting cluster structure starting from an initial random distribution of agents with L_f, ρ_f and R_f values from Fig. 17.10, and the growing and non-growing examples are presented together in Fig. 17.11. Table 17.1 and Fig. 17.12 compare statistical quantities.

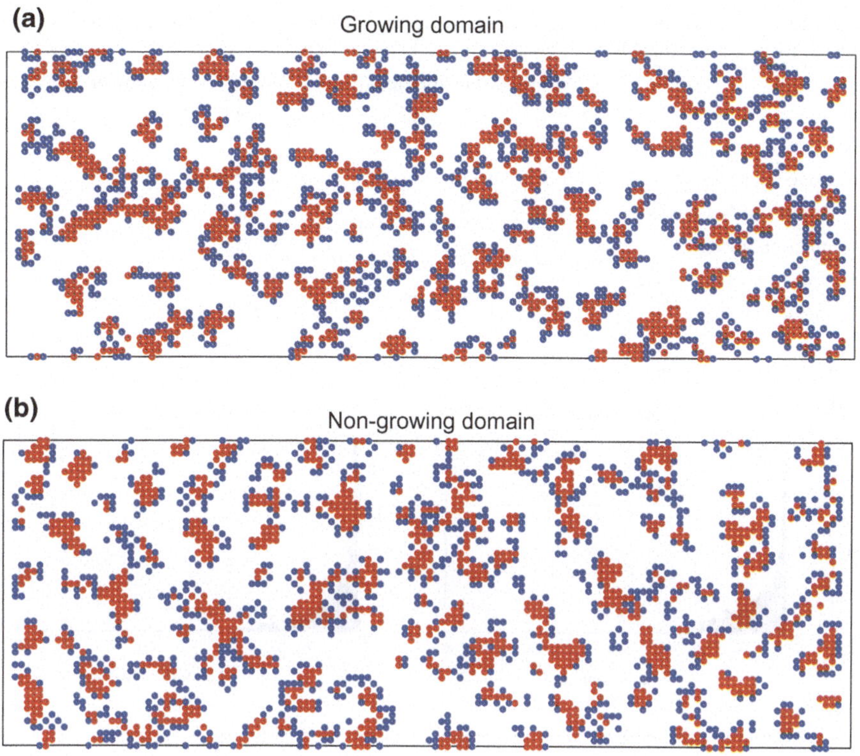

Fig. 17.11 Comparison of growing and fixed domain results. **a** Distribution of agents from a growing domain—this is just Fig. 17.10. Final values $\rho_f = 0.31$, $R_f = 1.05$. **b** Distribution of agents at $t = 100$ evolved using the bias rules 6:3 on a fixed 136×50 lattice from randomly scattered agents at $\rho_f = 0.31$ and $R_f = 1.05$. ENC agents do not proliferate or differentiate

When the two agent types are considered together, the number of clusters, average size and number of singlet clusters show no measurable difference between the growing and non-growing cases (Table 17.1). However, if we only consider the N agents, both the number of clusters and the number of singlet N agents are slightly higher, while the average cluster size is slightly lower for the growing domain. (note that singlet N agents are not true isolates—they are generally surrounded by ENC agents). These differences are quite minimal, since the error bars for the two cases overlap. Normalised frequency distributions of N agent cluster size are given in Fig. 17.12a.

Normalised frequency distributions of the minimum cycle basis, reflecting the cluster perimeter distribution, show that larger cycles occur for the growing domain case (Fig. 17.12b). However, high perimeter clusters are rare: 0.3% versus 4.4% of cycles larger than size 100 occur for non-growing and growing domain cases, respectively. Overall this does not indicate a significant difference in cluster distribution.

Table 17.1 Comparison of statistical quantities. Mean and standard deviation calculated over 20 identically prepared simulations

	Growing domain Fig. 17.11a	Non-growing domain Fig. 17.11b
Number of N clusters	135 ± 10	125 ± 5
Average N cluster size	7.9 ± 0.6	8.8 ± 0.5
Frequency of N singlets	32 ± 7	21 ± 4
Number of clusters	41 ± 7	47 ± 5
Average total cluster size	54 ± 10	46 ± 6
Frequency of singlet	2.7 ± 1.1	2.7 ± 1.8

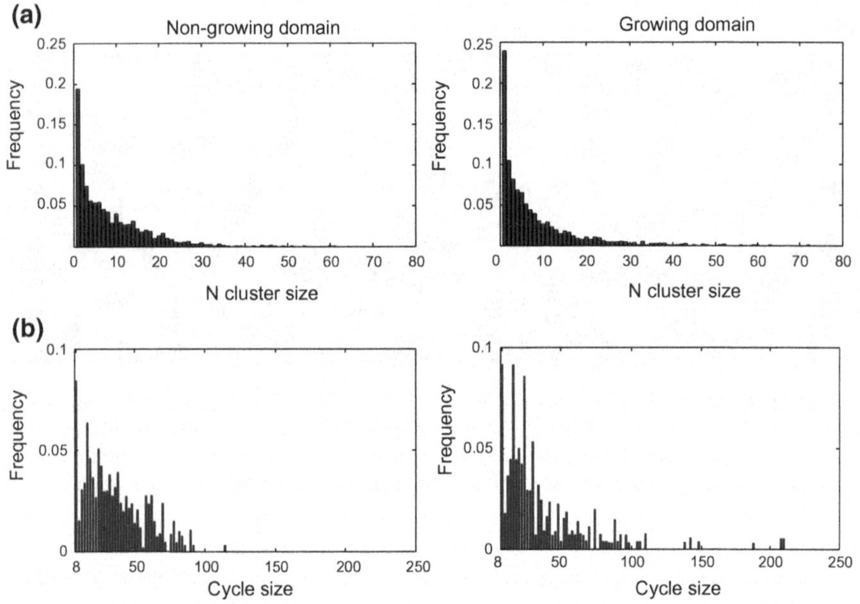

Fig. 17.12 Comparison of non-growing and growing cases of Fig. 17.11. Normalised frequency distributions of **a** N cluster size and **b** minimum cycle bases, averaged over ten identically prepared simulations

17.7 Conclusions

We have developed a model of ganglion formation which takes into account differential adhesion of neurons and ENC cells in a PCA formulation, resulting in clusters of cellular agents which emulate ganglia. We have shown how differential adhesion bias rules produce stable clusters, with relatively uniform size and spacing with a balance between N agents surrounded by ENC agents. This only occurs for a non-growing domain if the ENC agents do not proliferate or differentiate, that is when the density and N to ENC agent number ratio are fixed. In contrast, for a growing domain, the ENC agents must proliferate and also differentiate to replenish the ENC and N agent numbers as the domain increases in size. Furthermore, there is little difference in the cluster structure that evolves from the non-growing and growing domain models with the same adhesion bias rule.

These results match biological observations, where little difference is observed in the features of ganglia formed in organ cultures and in vivo. In the non-growing environment (organ culture), ENC cells have no opportunity to proliferate in the region well behind the wavefront where ganglia form [29]. In contrast, in the growing environment (in vivo), proliferation and differentiation of ENC cells is evident well behind the wavefront [40], due to the elongation of the tissue substrate [2, 25]. Therefore, the PCA model provides insight into why stable ganglia form under these different experimental circumstances. Furthermore, the models reveal how various threshold criteria which govern ENC agent proliferation and differentiation and N agent inhibition of differentiation are important for sustaining the characteristic features of ganglia formation. This work suggests that hyperganglionosis or hypoganglionosis occurs when these thresholds are relinquished or are too severe. Further experiments are needed to test whether up- or downregulation of adhesion molecules on the neuron and ENC cell populations disturbs the gangliogenesis patterns.

In the early stages of gangliogenesis in avian embryonic gut, neurons are grouped, but the ENC cells seem randomly scattered, while a day later regions appear which contain no neurons or ENC cells, as shown in Fig. 17.1. Our differential adhesion PCA model produces similar transitions when the N:ENC adhesive bias ratio is tuned from 8:1 to 6:3 to 8:4 as illustrated in Fig. 17.7. Therefore, this work suggests that the transition in mesoscale patterns may be due to either an increase in the adhesive capacity of ENC cells or an increase in the adhesive capacity of both cell types with time.

Acknowledgements This work was supported by Australian Research Council and National Health and Medical Research Council grants. Thanks are given to Emily Hackett-Jones and Dongcheng Zhang. MCRI facilities are supported by the Victorian Government's Operational Infrastructure Support Program.

References

1. Binder, B.J., Landman, K.A.: Exclusion processes on a growing domain. J. Theor. Biol. **259**, 541–551 (2009)
2. Binder, B.J., Landman, K.A., Simpson, M.J., Mariani, M., Newgreen, D.F.: Modeling proliferative tissue growth: a general approach and an avian case study. Phys. Rev. E **78**, 031912 (2008)
3. Breau, M.A., Pietri, T., Eder, O., Blanche, M., Brakebusch, C., Fassler, R., Thiery, J.P., Dufour, S.: Lack of $\beta 1$ integrins in enteric neural crest cells leads to a Hirschsprung-like phenotype. Development **33**, 1725–1734 (2006)
4. Chalazonitis, A., Gershon, M.D., Green, L.A.: Cell death and the developing enteric nervous system. Neurochem. Int. **61**, 839–847 (2012)

5. Daub, J.T., Merks, R.M.H.: A cell-based model of extracellular-matrix-guided endothelial cell migration during angiogenesis. Bull. Math. Biol. **75**, 1377–1399 (2013)
6. Foty, R.A., Steinberg, M.S.: Cadherin-mediated cell-cell adhesion and tissue segregation in relation to malignancy. Int. J. Dev. Biol. **48**, 397–409 (2004)
7. Foty, R.A., Steinberg, M.S.: The differential adhesion hypothesis: a direct evaluation. Dev. Biol. **278**, 255–263 (2005)
8. Glazier, J.A., Graner, F.: Simulation of the differential adhesion driven rearrangement of biological cells. Phys. Rev. E **47**, 2128–2154 (1993)
9. Graner, F., Glazier, J.A.: Simulation of biological cell sorting using a two-dimensional extended Potts model. Phys. Rev. Lett. **69**, 2013–2016 (1992)
10. Hackett-Jones, E.J., Landman, K.A., Newgreen, D.F., Zhang, D.: On the role of differential adhesion in gangliogenesis in the enteric nervous system. J. Theor. Biol. **287**, 148–159 (2011)
11. Hao, M.M., Anderson, R.B., Kobayashi, K., Whitington, P.M., Young, H.M.: The migratory behaviour of immature enteric neurons. Dev. Neurobiol. **69**, 22–35 (2009)
12. Hao, M.M., Anderson, R.B., Young, H.M.: Development of enteric neuron diversity. J. Cell. Mol. Med. **13**, 1193–1210 (2009)
13. Hearn, C.J., Young, H.M., Ciampoli, D., Lomax, A.E., Newgreen, D.F.: Catenary cultures of embryonic gastrointestinal tract support organ morphogenesis, motility, neural crest cell migration, and cell differentiation. Dev. Dyn. **214**, 239–247 (1999)
14. Hendershot, T.J., Liu, H., Sarkar, A.A., Giovannucci, D.R., Clouthier, D.E., Abe, M., Howard, M.J.: Expression of Hand2 is sufficient for neurogenesis and cell type-specific gene expression in the enteric nervous system. Dev. Dyn. **236**, 93–105 (2007)
15. Horton, J.D.: A polynomial-time algorithm to find a shortest cycle basis of a graph. SIAM J. Comput. **16**, 359–366 (1987)
16. Hoshen, J., Kopelman, R.: Percolation and cluster distribution. I. Cluster multiple labeling technique and critical density algorithm. Phys. Rev. B **14**, 3438–3445 (1976)
17. Hotta, R., Anderson, R.B., Kobayashi, K., Newgreen, D.F., Young, H.M.: Effects of tissue age, presence of neurones and endothelin-3 on the ability of enteric neurone precursors to colonize recipient gut: implications for cell-based therapies. Neurogastroenterol. Motil. **22**, 331–e86 (2010)
18. Landman, K.A., Fernando, A.E., Zhang, D., Newgreen, D.F.: Building stable chains with motile agents: Insights into the morphology of enteric neural crest cell migration. J. Theor. Biol. **276**, 250–268 (2011)
19. Landman, K.A., Binder, B.J., Newgreen, D.F.: Modeling development and disease in our "second" brain. Lect. Notes Comput. Sci. **7495**, 405 (2012)
20. Longo, D., Peirce, S.M., Skalak, T.C., Davidson, L., Marsden, M., Dzamba, B., DeSimone, D.W.: Multicellular computer simulation of morphogenisis: blastocoel roof thinning and matrix assembly in Xenopus laevis. Dev. Biol. **271**, 210–222 (2004)
21. Mehlhorn, K., Michail, D.: Implementing minimum cycle basis algorithms. J. Exp. Algorithmics **11**, 1–14 (2006)
22. Meier-Ruge, W.A., Bruder, E., Kapur, R.P.: Intestinal neuronal dysplasia type B: one giant ganglion is not good enough. Pediatr. Dev. Pathol. **9**, 444–452 (2006)
23. Merks, R.M.H., Glazier, J.A.: A cell-centered approach to developmental biology. Physica A **352**, 113–130 (2005)
24. Newgreen, D.F., Southwell, B., Hartley, L., Allan, I.J.: Migration of enteric neural crest cells in relation to growth of the gut in avian embryos. Acta Anat. **157**, 105–115 (1996)
25. Newgreen, D., Young, H.M.: Enteric nervous system: development and developmental disturbances-part 1. Pediatr. Dev. Pathol. **5**, 224–247 (2002)
26. Peirce, S.M., Van Gieson, E.J., Skalak, T.C.: Multicellular simulation predicts microvascular patterning and in silico tissue assembly. FASEB J. **18**(6), 731–733 (2004). https://doi.org/10.1096/fj.03-0933fje
27. Savill, N.J., Sherratt, J.A.: Control of epidermal stem cell clusters by notch-mediated lateral induction. Dev. Biol. **258**, 141–153 (2003)

28. Simpson, M.J., Merrifield, A., Landman, K.A., Hughes, B.D.: Simulating invasion with cellular automata. Phys. Rev. E **76**, 021918 (2007)
29. Simpson, M.J., Zhang, D.C., Mariani, M., Landman, K.A., Newgreen, D.F.: Cell proliferation drives neural crest cell invasion of the intestine. Dev. Biol. **302**, 553–568 (2007)
30. Steinberg, M.S.: Mechanism of tissue reconstruction by dissociated cells.II. Time course of events. Science **137**, 762–763 (1962)
31. Steinberg, M.S.: On the mechanism of tissue reconstruction by dissociated cells, III. Free energy relations and the reorganisation of fused, heteronomic tissue fragments. Proc. Natl. Acad. Sci. USA **48**, 1769–1776 (1962)
32. Steinberg, M.S.: On the mechanism of tissue reconstruction by dissociated cells, I. population kinetics, differential adhesiveness, and the absence of directed migration. Proc. Natl. Acad. Sci. USA **48**, 1577–1582 (1962)
33. Sulsky, D.: A model of cell sorting. J. Theor. Biol. **106**, 275–301 (1984)
34. Townes, P.L., Holtfreter, J.: Directed movements and selective adhesion of embryonic amphibian cells. J. Exp. Zool. **128**, 53–120 (1955)
35. Wedel, T., Roblick, U.J., Ott, V., Eggers, R., Schiedeck, T.H.K., Krammer, H.J., Bruch, H.P.: Oligoneuronal hypoganglionosis in patients with idiopathic slow-transit constipation. Dis. Colon Rectum **45**, 54–62 (2002)
36. Wilson, H.V.: On some phenomena of coalescence and regeneration in sponges. J. Exp. Zool. **5**, 245–258 (1907)
37. Yin, M., King, S.K., Hutson, J.M., Chow, C.W.: Multiple endocrine neoplasia type 2B diagnosed on suction rectal biopsy in infancy: a report of 2 cases. Pediatr. Dev. Pathol. **9**, 56–60 (2006)
38. Young, H.M., Bergner, A.J., Muller, T.: Acquisition of neuronal and glial markers by neural crest-derived cells in the mouse intestine. J. Comp. Neurol. **456**, 1–11 (2003)
39. Young, H.M., Bergner, A.J., Anderson, R.B., Enomoto, H., Milbrandt, J., Newgreen, D.F., Whitington, P.M.: Dynamics of neural crest-derived cell migration in the embryonic mouse gut. Dev. Biol. **270**, 455–473 (2004)
40. Young, H.M., Turner, K.N., Bergner, A.J.: The location and phenotype of proliferating neural-crest-derived cells in the developing mouse gut. Cell Tissue Res. **320**, 1–9 (2005)
41. Zhang, D., Brinas, I.M., Binder, B.J., Landman, K.A., Newgreen, D.F.: Neural crest regionalisation for enteric nervous system formation: implications for Hirschsprung's Disease and stem cell therapy. Dev. Biol. **339**, 280–294 (2010)

Chapter 18
Cellular Potts Model: Applications to Vasculogenesis and Angiogenesis

Sonja E.M. Boas, Yi Jiang, Roeland M.H. Merks, Sotiris A. Prokopiou and Elisabeth G. Rens

Abstract The cellular Potts model (CPM, a.k.a. Glazier–Graner–Hogeweg or GGH model) is a somewhat liberal extension of probabilistic cellular automata. The model is derived from the Ising and Potts models and represents biological cells as domains of CA-sites of the same state. A Hamiltonian energy is used to describe the balance of forces that the biological cells apply onto one another and their local environment. A Metropolis algorithm iteratively copies the state from one site into one of the adjacent sites, thus shifting the domain interfaces and moving the biological cells along the lattice. The approach is commonly used in applications of developmental biology, where the CPM often interacts with systems of ordinary-differential equations that model the intracellular chemical kinetics and partial-differential equations that model the extracellular chemical signal dynamics to constitute a hybrid and multiscale description of the biological system. In this chapter we will introduce the cellular Potts model and discuss its use in developmental biology, focusing on the development of blood vessels, a process called vascular morphogenesis. We will start by introducing a range of models focusing on uncovering the basic mechanisms of vascular morphogenesis: network formation and sprouting and then show how these models are extended with models of intracellular regulation and with interactions with the extracellular micro-environment. We then briefly review the integration of models of vascular morphogenesis in several examples of organ development in

---

S.E.M. Boas · R.M.H. Merks (✉) · E.G. Rens
Centrum Wiskunde & Informatica, Science Park 123,
1098 XG Amsterdam, The Netherlands
e-mail: merks@cwi.nl

S.E.M. Boas · R.M.H. Merks · E.G. Rens
Leiden University, Mathematical Institute, Niels Bohrweg 1,
2333 CA Leiden, The Netherlands

S.A. Prokopiou
Department of Integrated Mathematical Oncology, H. Lee Moffitt Cancer Center
and Research Institute, 12902 Magnolia Drive, Tampa, FL 33612, USA

Y. Jiang
Department of Mathematics and Statistics, Georgia State University, Atlanta,
GA 30303, USA
e-mail: yjiang12@gsu.edu

P.-Y. Louis and F.R. Nardi (eds.), *Probabilistic Cellular Automata*, Emergence, Complexity and Computation 27, https://doi.org/10.1007/978-3-319-65558-1_18

health and disease, including development, cancer, and age-related macular degeneration. We end by discussing the computational efficiency of the CPM and the available strategies for the validation of CPM-based simulation models.

18.1 Introduction

Probabilistic cellular automata (PCA) are widely applied as a modeling framework for biological and biomedical research. In particular, they are used in the study of biological pattern formation, to help to understand how biological structures can form from biological elements that follow simple rules. In this way, PCAs have been used in diverse applications, ranging from spatial structuring of ecosystems [37] to a range of biomedical problems, including the self-organization of the autonomous nervous system in the gut (Chap. 17 of this book), the growth and plasticity of tumors [3, 27] or the formation of blood vessels [17]. The central question in these applications of PCA is how cells can self-organize into tissues and organs. The states of the PCA represent the types of biological entities, while the probabilistic nature of the PCA reflects the "noisiness" inherent to most physical and biological systems. The PCA here helps unravel how biological patterns can persist in the presence of homogenizing noise, or, perhaps more interestingly, PCAs demonstrated that noise can become a driving force of pattern formation, i.e., no patterns would form in a deterministic model [50, 81]. The "PCAs" in the above applications typically deviate from the strict definition of PCA, as a system of locally coupled, homogenous system of Markov chains with synchronous updates. The updates can be asynchronous (one by one in random order), additional rules are applied (e.g., rules for mass-conserved random walks or diffusion), or the systems are hybridized with systems of partial-differential equations, e..g., to model diffusing molecular signals.

18.2 Cellular Potts Model

A generalization of PCA, which is particularly widely applied to biomedical problems, is the cellular Potts model (CPM), also known as the Glazier–Graner–Hogeweg or GGH model [33]. The CPM is used to model structures of biological cells and extracellular materials. It is a generalization of the large q-Potts model, which derives from the Potts model. Glazier et al. [32] reviewed the derivation of the cellular Potts model from its predecessors in detail; a brief recap is useful in the present context to better understand the structure and notation of the CPM. The Potts model studies the interactions between domains on a lattice, e.g., during the solidification of a fluid. It is defined on a regular lattice $\Lambda \subset \mathbb{Z}^2$ or $\Lambda \subset \mathbb{Z}^3$, with $\mathbf{x} \in \Lambda$ the coordinates in the lattice. The clusters of like spins, $\sigma(\mathbf{x}) \in \{0, \ldots, q\}$, represent individual domains, where the same spin can identify multiple domains if they are well separated spatially. Assuming that (without external fields) the spins follow Boltzmann statistics, the relative probability of each configuration of spins $\{\sigma(\mathbf{x})\}$ is,

$$P(H) = e^{\frac{-H(\{\sigma(\mathbf{x})\})}{kT}}, \tag{18.1}$$

where the Hamiltonian, $H(\{\sigma(\mathbf{x})\})$, describes the energy of the configuration, k is the Boltzmann constant, and T is the absolute temperature. In the Potts model, the interfaces between two domains contribute to the configuration energy:

$$H(\{\sigma(\mathbf{x})\}) = J \sum_{(\mathbf{x},\mathbf{x}')} (1 - \delta(\sigma(\mathbf{x}), \sigma(\mathbf{x}')). \tag{18.2}$$

Here J (typically $J \geq 0$, but see also Ref. [61]) is the energy associated with a unit length of the domain interfaces, $(\mathbf{x}, \mathbf{x}')$ is a pair of adjacent lattice sites, and the Kronecker delta function ($\delta(a, b) = 1$, if $a = b$; 0, if $a \neq b, a \wedge b \in \mathbb{Z}$) selects adjacent lattice sites of unequal spin. Minimizing the Hamitonian energy using Monte Carlo methods tends to minimize the number of interfaces in the system by forming domains of identical spin, which would coarsen to form fewer and fewer domains.

The key innovation of the CPM was to associate each of the domains formed in the Potts model with a biological cell. To ensure that the volume of the cells (or area in 2D) is approximately conserved, the CPM adds a volume energy to the Hamiltonian, $H_{\text{volume}} = \lambda_{\text{volume}}(v(\sigma) - V(\sigma))^2$, where λ_{volume} is a Lagrange multiplier to the volume constraint; similar terms with Lagrange multipliers can be added to represent additional optimization conditions. The volume energy of a domain with spin s is zero if its actual volume, $v(s) = |\{\mathbf{x}|\mathbf{x} \in \Lambda \wedge \sigma(\mathbf{x}) = s\}|$ (i.e., the number of sites in the lattice of spin equal to s), is equal to a target, or resting volume $V(s)$. Any deviation of the actual volume to the target volume contributes elastically to the total energy. For the medium that surrounds the cells (locations $\mathbf{x}$ with $\sigma(\mathbf{x}) = 0$), no volume constraint is applied. Note that the volume constraint adds a non-local, cellular scale to the system. The energy of a configuration in the CPM thus depends both on local, nearest-neighbor interactions, as well as on non-local properties of all sites in the lattice of equal spin value.

A further innovation of the CPM is a differentiation between interfacial energies, such that one type of interface may be favored over another. Each cell, $\sigma(\mathbf{x})$, also has a type, $\tau(\sigma(\mathbf{x})) \in \mathbb{N}$, with each value of τ classifying the domain as a particular biological cell type (e.g., neuron, muscle cell), or a cell state (e.g., proliferating vs quiescent), or non-cellular material (e.g., fluid, substrate, and so forth). The interfacial energy, J, then becomes a function of the pair of types at the interface, $(\tau(\sigma(\mathbf{x})), \tau(\sigma(\mathbf{x}'))$.

The full Hamiltonian of the CPM becomes,

$$H = \sum_{(\mathbf{x},\mathbf{x}')} J\left(\tau(\sigma(\mathbf{x})), \tau(\sigma(\mathbf{x}'))\right)\left(1 - \delta\left(\sigma(\mathbf{x}), \sigma(\mathbf{x}')\right)\right) + \lambda_{\text{volume}} \sum_{\{\sigma>0\}} (v(\sigma) - V(\sigma))^2 + H'. \tag{18.3}$$

The term H' identifies any additional constraints on the cell behavior, particularly those involving interactions with external fields [73, 91] or additional constraints at cell level (e.g., constraints on the length of the cellular interfaces [40], or constraints on cell shape [56, 99]).

Minimization of the Hamiltonian energy function corresponds to solving the balance of forces applied to the cell. The Hamiltonian in the CPM is usually minimized using Metropolis dynamics, which compares configurations differing by one spin at a time. In the CPM, the Metropolis algorithm is modified such that it mimics natural fluctuations of the membranes driven by the activity of the cytoskeleton. This feature introduces a temporal ordering into the energy minimization procedure, such that both the equilibrium configuration and the transition to equilibrium are of physical and biological interest.

More specifically, the CPM-version of the Metropolis algorithm selects a pair of adjacent lattice sites, $(\mathbf{x}, \mathbf{x}')$, at random from the lattice; i.e., first a target site $\mathbf{x} \in \Lambda$ is selected at random, then a lattice site $\mathbf{x}'$ is selected at random from $\mathrm{NB}(\mathbf{x})$, the set of neighbors of $\mathbf{x}$. On a square lattice, typical choices for $\mathrm{NB}(\mathbf{x})$ include the Moore neighborhood (the eight nearest neighbors); larger neighborhoods, e.g., the twenty neighbors of order 1–3 are also used to reduce lattice effects [52]. Next the algorithm attempts to change the spin of $\mathbf{x}$ (the *target* site), into the spin of its neighbor $\sigma(\mathbf{x}')$ (the *source* site). If the attempted update will reduce the energy, i.e., $\Delta H = H_{\text{after}} - H_{\text{before}} < 0$, the change occurs with probability 1. If the attempt increases the energy ($\Delta H > 0$), the change will be accepted with Boltzmann probability:

$$P(\Delta H) = \begin{cases} 1 & \text{if } \Delta H < 0 \\ \exp(-\Delta H/T) & \text{if } \Delta H \geq 0. \end{cases} \tag{18.4}$$

In contrast to the Potts model, in the CPM the temperature T is a *cellular* temperature, reflecting the amplitude of active cell membrane fluctuations. For lack of measurements of the distribution of the energy that is mechanically dissipated during these fluctuations, the CPM follows the Potts model by assuming Boltzmann probability. Time in the cellular Potts model is measured in Monte Carlo Steps (MCS), where one MCS corresponds with $|\Lambda|$ copy attempts, i.e., the number of sites in the lattice. To identify the real time corresponding with one MCS, the kinetics of the CPM itself [33, 95] or the kinetics of coupled models [59, 92] is matched with the kinetics of experiments. Similar approaches are used to parameterize the real volume or area corresponding to one lattice site.

18.2.1 Generic Behavior of the CPM

The cellular Potts model (CPM) is widely applied to biomedical problems involving cell shape changes and cell–cell adhesion. It was introduced [33, 34] in the early 1990s as a model for differential-adhesion-driven cell rearrangement: a proposed mechanism for spontaneous rearrangement of cell types in mixtures of embryonic cells [38]. The differential adhesion hypothesis suggests that these cellular rearrangements, also known as cell sorting, are driven by relative adhesion surface energies of different cell types [84]. This hypothesis has been tested with the CPM, using

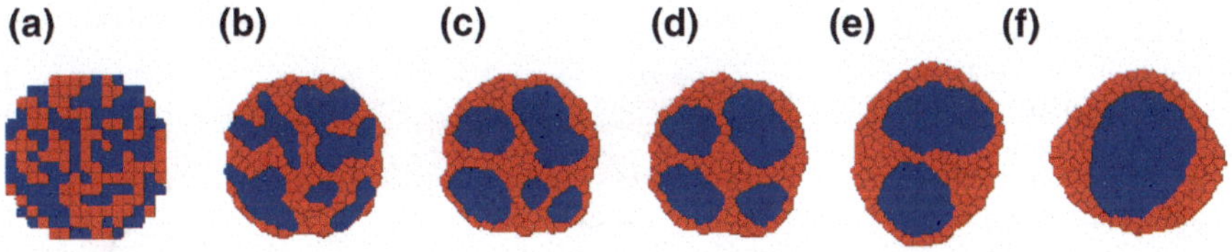

Fig. 18.1 Typical time course of a cellular Potts simulation of binary cell sorting through differential adhesion. Cell type r (*red*) engulfs cell type b (*blue*) due to differences in adhesion energies; in this example, $J_{dd} < J_{ld} < J_{ll} < \{J_{lM}, J_{dM}\}$ **a** 0 MCS; **b** 100 MCS; **c** 500 MCS; **d** 1000 MCS; **e** 5000 MCS; **f** 10000 MCS

only a volume constraint and adhesion energies [33, 34]. The model was initialized with two cell types, light (l) and pigmented, dark (d) colored cells, mixed in a random aggregate surrounded by medium (M). If the adhesion energies are set as $J_{dd} < J_{ld} < J_{ll} < \{J_{lM}, J_{dM}\}$, the two cell types would segregate into clusters of l and d, and the light aggregate would eventually engulf dark ones d (Fig. 18.1). By changing the relative adhesion energies, other patterns can be generated, including checkerboard-patterns. The differential adhesion hypothesis has been tested experimentally by Krieg et al. [45]. They measured cell adhesion at the single cell level by using an atomic force microscopy and set the relative values of J in the CPM accordingly. As a result, the CPM sorted the cells the wrong side out: consistently the so called *ectodermal* cells ended up surrounded by mesodermal cells, whereas based on the measured cell adhesion values the CPM predicted the mesodermal cells should end up in the middle. Based on experimental observations they predicted that interfacial tensions generated by muscle-like actomyosin structures near the cell surface must also be considered to correctly predict the outcome of cell sorting. Krieg et al. [45] have incorporated the additional source of interfacial energy in the values of J. An alternative approach for modeling cortical tensions is to constrain the perimeter or surface area of the cells.

18.2.2 Hybrid Modeling

Although explicit modeling of shapes and adhesion is essential in many research problems dealing with cells, other components in biomedical systems benefit from a continuum modeling approach. For example, the diffusion of a chemical is typically modeled with a partial-differential equation (PDE). Many research problems lie at the interface of these two modeling approaches, in which case we can use hybrid modeling. Hybrid models combine multiple types of modeling techniques, such as discrete and continuum modeling.

A widely used approach is to couple a field or a set of fields representing, e.g., the distribution of a chemical signal, to the CPM. The CPM is then modified such that cells respond to the chemical field by moving to higher or lower concentrations,

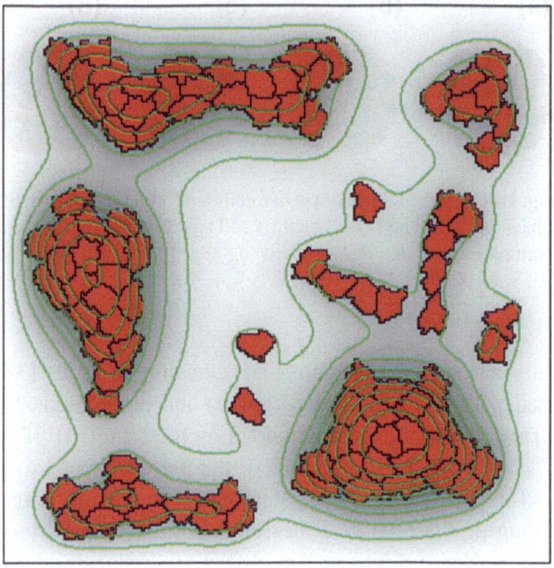

Fig. 18.2 Chemotactic cell aggregation. Chemotaxis toward an chemoattractant secreted by the cells themselves results in clustering of initially dispersed cells. Isolines (*green lines*) indicate ten chemoattractant levels relative to the maximum concentration in the simulation

a mechanism called *chemotaxis*. The typical way to model chemotaxis is to modify ΔH during an attempted update, such that moves up the chemical gradient occur with higher (or lower) probability proportional to the gradient [73],

$$\Delta H_{chem} = \lambda_{\text{chem}} \left(c(\mathbf{x}) - c(\mathbf{x}') \right), \tag{18.5}$$

with c the chemical field. The resulting bias lets cells gradually move up (or down) the gradient of the chemoattractant with a sensitivity λ_{chem}.

Figure 18.2 shows an example simulation of a hybrid CPM proposed by Merks et al. [56, 59]), where the cells (colored in red) in medium (white background) aggregate because of a chemical signal. In this model, the cells secrete a chemical to attract surrounding cells. This *chemoattractant* diffuses and slowly decays in the medium, following the PDE,

$$\frac{\partial c}{\partial t} = \alpha(1 - \delta(\sigma(\mathbf{x}), 0)) - \epsilon\delta(\sigma(\mathbf{x}), 0)c + D\nabla^2 c, \tag{18.6}$$

with α the secretion rate by cells, D the diffusion constant of VEGF and ϵ the decay of VEGF in the medium. After each MCS, the PDE (Eq. (18.6)) is solved numerically using a discretization matching the grid of the CPM, allowing the CPM to read out the chemoatractant concentration on each lattice site. The mutual attraction results in cell aggregation.

18.2.3 Implementations of the CPM

The cellular Potts model has been coded into a number of simulation environments for multicell modeling of tissues, organs, and organisms, most notably the open source package CompuCell3D [86]; see http://www.compucell3d.org. CompuCell3D provides an accessible simulation environment for the standard CPM and includes a large number of extensions, including CPM-PDE hybrid models, compartmental CPMs (see Sect. 18.6.2) and spring-like connections between cells. CompuCell3D allows end users to set up biological simulation models with little to no programming experience or in-depth knowledge of the CPM, whereas it multitiered and extensible architecture provides access to the underlying CPM algorithms or allows users to add additional terms to the Hamiltonian. It is currently growing into the standard platform of the CPM community, allowing end users to share new extensions and applications. The simulations reported in Sect. 18.6 used CompuCell3D.

An alternative open source implementation of the CPM is Tissue Simulation Toolkit (TST) [57]; it is available at http://sourceforge.net/projects/tst. TST is a C++ library providing implementations of the CPM on two-dimensional, square lattices and functionality for hybrid CPM-PDE models. While limited in functionality relative to CompuCell3D, the simplicity of the TST allows more straightforward access to the CPM implementation. This makes it particularly well suited as a test bed for new CPM algorithms and extensions of the Hamiltonian. A recent tutorial [24] provides detailed instructions for how to adopt TST for one's own needs. Most simulations reported in Sects. 18.4 and 18.5 used the TST. The hybrid CPM and finite-element model reviewed in Sect. 18.4.4 was implemented in an independent C-code, released as supplements to its publication [92]; it may soon be merged with the TST.

Other implementations of the CPM are part of the free simulations environments Morpheus [83] and Chaste [67]; both these environments provide a range of biological modeling formalisms, including cellular automata, ordinary-differential equations and partial-differential equations, and off-lattice cell-based modeling techniques. Morpheus provides a high-level, XML-based declarative programming language to describe model rules and has an attractive user interface. The first releases of Morpheus were closed-source, limiting its applicability, but an open source release has been announced as of this writing (April 2015). Morpheus is available from http://imc.zih.tu-dresden.de/wiki/morpheus/doku.php. Chaste is a large C++ software library focusing on cardiac electrophysiology and cell-based modeling. The latter component contains cellular Potts functionality. Chaste is available from http://www.cs.ox.ac.uk/chaste/.

18.3 Application of the Hybrid CPM to Blood Vessel Formation

Over the past decades, a number of mathematical and computational models have been developed to propose new models for the mechanisms of embryonic development. These mechanisms span all spatial and temporal scales encompassed by this complex process, ranging from the molecular level all the way to the organismal level and its environment, and ranging from microseconds (chemical reactions) up to years (homeostasis, aging, cancer). Although at present "computing" a human is beyond reach (although successful first steps have been taken for much simpler multicellular creatures [77]), cellular Potts modeling has been applied to the understanding of relatively more simple mechanisms, including the formation of blood vessels.

During embryogenesis, vascular networks (blood vessels) are formed from initially dispersed endothelial cells (ECs), a process called vasculogenesis. Once the vasculature is established, capillary sprouts can branch off from the preexisting vasculature in response to externally supplied angiogenic stimuli, a process called angiogenesis. The new sprouts provide tissues and organs with oxygen and nutrients, and remove metabolic waste. Angiogenesis takes place in physiological situations, such as embryonic development, wound healing and reproduction [18]. The healthy body controls angiogenesis by balancing pro- and anti-angiogenic factors [19]. This balance, though, is sometimes disrupted and angiogenesis also appears in many pathologies, like diabetes [55], rheumatoid arthritis [43], cardiovascular ischemic complications [16], proliferative retinopathy [30], and cancer [29].

Sprouting angiogenesis typically starts from hypoxic tissues or cells (e.g., retinal astrocytes [76]) upregulating their production of pro-angiogenic factors such as vascular endothelial growth factor A (VEGFA) [28]. These angiogenic factors diffuse and bind to endothelial cell receptors on nearby blood vessels. Subsequently, the extracellular matrix (ECM) and basement membrane, surrounding the ECs, are degraded locally by activated proteases (e.g., matrix metalloproteinases, MMPs) produced by ECs.

Mathematical modeling is a useful tool for understanding the mechanisms of angiogenesis and to design experiments of a predictive nature. Since vessels often consist of only a few cells, explicitly considering individual cells is essential. In most modeling frameworks, the detailed investigation of cell-level properties, such as cell shape and cell adhesion, are mathematically difficult, if not impossible to consider. Therefore, the CPM with the advantage of representing cells as individual entities with a particular shape is an appropriate framework to study blood vessel formation. Over the past two decades, a plethora of mathematical and computational models have been developed to study aspects of angiogenesis. For a comprehensive review of mathematical and computational models in angiogenesis see [65], and references therein.

Typical modeling studies investigate how growth factors and receptors promote endothelial cell proliferation, how groups of endothelial cells assemble into individual vessels, and how tumors recruit the ingrowth of whole microvascular networks.

Here we briefly review cellular Potts approaches to the analysis of blood vessel formation, describing the required extensions to the CPM in technical detail.

18.4 Modeling Collective Cell Behavior During *de Novo* Vasculogenesis

Vasculogenesis is an embryogenic process during which endothelial cells organize into vascular networks. Computational modeling has been used in the search for generic cell behaviors that drive vasculogenesis. In this section, we will discuss four CPM-based models in which the individual behavior of initially dispersed cells collectively results in their organization into vascular networks.

18.4.1 Chemotactic Cell Aggregation

Serini et al. [78] showed in a continuum model that dispersed cells self-organize into polygonal patterns, when these cells secrete a chemoattractant to which all cells respond. However, Merks et al. [59] showed in a CPM-based model that with these assumptions on a longer time scale dispersed cells form rounded aggregates (see 18.2.2 and Fig. 18.2) rather than polygonal patterns. Merks et al. [59] suggested additional model assumptions to explain vascular network formation: (1) endothelial cells (ECs) adhere to one another with VE-cadherins and (2) VE-cadherin-binding inhibits VEGF signaling by interacting with the VEGF receptor 2, and (3) the ECs secrete a chemoattractant (e.g., VEGF [22]) that attracts other ECs. These assumption were simplified in the model as follows: the ECs only responded to the chemoattractant at regions of their membrane adjacent to the ECM, whereas at cell–cell interfaces the chemotaxis was inhibited, a mechanism called contact-inhibited chemotaxis. Indeed,

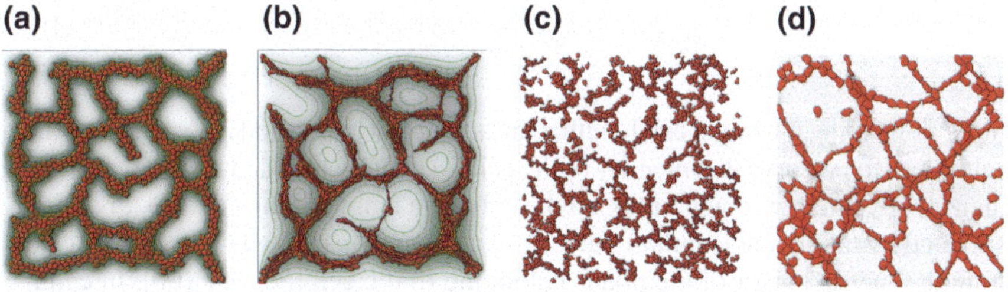

Fig. 18.3 Vasculogenesis models. Simulations of vasculogenesis driven by various mechanisms: **a** contact-inhibited chemotaxis [59], **b** cell elongation with chemotaxis [56], **c** preferential adhesion to elongated cells [89], and **d** mechanical cell-matrix interactions [92]. Panel C reprinted from Ref. [89]; copyright (2008), with permission from the Biophysical Society

the simulations showed self-organization of endothelial cells into vascular networks (Fig. 18.3a).

VE-cadherin mediated contact-inhibited chemotaxis was implemented in the cellular Potts model by only letting cells chemotact at cell–medium interfaces of their membrane, and not at cell–cell interfaces, resulting in $\lambda_{chem} = 0$ at cell–cell interfaces and $\lambda_{chem} > 0$ for cell–medium interfaces using chemotaxis as described in Sect. 18.2.2. The complete Hamiltonian of the model depends on a volume constraint for the cells, adhesion between the cells, and contact-inhibited chemotaxis. The concentration field of VEGF is described in Sect. 18.2.2.

Although VE-cadherin mediated contact-inhibited chemotaxis reproduces vasculogenesis, Köhn-Luque et al. [44] note that the diffusion speed assumed for VEGF by Merks et al. [59] is much lower than reported for most VEGF isoforms. Köhn-Luque et al. [44] propose an alternative CPM-based model for vascularization in which VEGF, containing ECM-binding domains, is secreted by the underlying endoderm. Endothelial cells scavenge VEGF by the secretion of ECM and subsequently chemotact more strongly to ECM-bound VEGF than to soluble VEGF, resulting in network formation.

18.4.2 Cell Elongation

Endothelial cells are often seen to elongate during network formation. Palm and Merks [63] showed with a CPM-based model that elongated, adhesive cells can self-organize into vascular structures. Cells aggregate into elongated structures that can only rotate very slowly, while connected in the branch points. If the model would run for infinity, the cells would form a spheroid, but this process is so slow that the cells dynamically arrest in a network-like pattern. Addition of chemotaxis to an auto-secreted chemoattractant to this cell elongation model [56] stabilizes network formation and speeds up the patterning process (Fig. 18.3b). Cell elongation is modeled by the addition of a length constraint to the Hamiltonian:

$$H'_{length} = \lambda_{length} \sum_{\sigma>0} (l(\sigma) - L(\sigma))^2, \tag{18.7}$$

with L describing the target cell length, l the actual cell length and λ_{length} the Lagrange multiplier. To preserve the integrity of the cells, a penalty is added to the Hamiltonian when a copy attempt would break up the cell. The length of the cell, $l(\sigma)$, is usually estimated by taking the length of the long axis of an ellipse fitted to the cell by calculating the inertia tensor of the pixels belonging to the cell [56, 99, 100]. In cellular Potts implementations, the inertia tensors of the cells can be efficiently calculated by keeping track of the first and second order raw momenta of the cellular coordinates, in addition to the cellular area or volume [56]. The length constraint requires an additional connectivity constraint to prevent cells from splitting up into two disconnected patches; the connectivity constraint prevents updates that would split a cell in two patches [56]. Chemotaxis is implemented as described in Sect. 18.2.2.

18.4.3 Preferential Attraction to Elongated Structures

Similarly to the work of Palm and Merks [63], a mechanism for vascular patterning excluding external factors has been proposed by Szabó et al. [88, 89]. The reason for a mechanism not involving chemical and mechanical forces originates from experiments showing vascular patterning under normal tissue conditions on a solid substrate [88]. Based on experimental results of [89] a new hypothesis for vascular network formation was proposed. These experiments showed elevated cell motility within the presence of elongated structures and cells were observed to migrate faster within narrow sprouts, while cells in wider sprouts have a decreased motility. Furthermore, the width of more elongated sprouts increases with a faster rate. The led to authors to propose that cells are highly attracted to elongated structures. Szabó et al. [88] implemented this attraction to elongated structures in a particle based method in previous work [88], where cell shape was not resolved.

In their follow-up work, Szabó and coworkers [89] added their mechanism to the CPM by adding a bias at the time of copying (*cf.* Eq. (18.5)), as

$$\Delta H_{\text{pref\_attr}} = \lambda_{\text{pref\_attr}} \left(1 - \delta(\mathbf{x}', 0)) - (1 - \delta(\mathbf{x}, 0))\right) \sum_{\{\mathbf{y} | \mathbf{y} \in \text{NB}(\mathbf{x}) \wedge \sigma(\mathbf{y}) \notin \{0, \sigma(\mathbf{x}), \sigma(\mathbf{x}')\}\}} \theta(\sigma(\mathbf{y})), \tag{18.8}$$

where $\theta(\sigma) = \left(\frac{\mu(\sigma)}{\nu(\sigma)}\right)^{\frac{1}{2}} - 1$ with $\mu(\sigma) \geq \nu(\sigma)$ the two eigenvalues of the cellular inertia tensor, representing the long and short axis of the cell. Thus, $\theta(\sigma)$ is a measure of the eccentricity of a cell with spin equal to σ. The summation in Eq. (18.8) only goes over the neighboring sites of $\mathbf{x}'$ that belong to cells other than $\sigma(\mathbf{x})$, $\sigma(\mathbf{x}')$. The term $1 - \delta(\mathbf{x}, 0)) - (1 - \delta(\mathbf{x}', 0))$ ensures that the medium is not influenced by this preference, and that copies at cell–cell interfaces are independent of this preference, such that no cell has more advantage than an other cell. If Eq. (18.8) is rewritten, see [89], it becomes clear that it can be considered as an asymmetric extension of the adhesive energy J. Thus, the attraction to elongated structures is actually a preferential adhesion to elongated structures.

Simulations of this model with initially dispersed patches of connected cells show network formation (Fig. 18.3c). Subsequent to an initial budding of one cell from a connected patch, other cells are attracted to the base of the sprout and follow the leading elongated cell. Migration of the sprout continues until a branch is established and stabilized. Due to surface tension, branches can break up, whereas new branches form continuously. The resulting networks are thus quasi-stable, the networks change continuously, while overall the statistical properties (number of branches, wavelength, and so forth) of the pattern are stable. A biological basis for the preferential adhesion to elongated structures is not yet established. Szabó et al. [89] propose that the preference can arise from mechanical tension of elongated structures, which cells can respond to by VE-cadherin based mechanosensing.

18.4.4 Mechanical Cell-ECM Interactions

The models described above explained vascular network formation based on chemical interactions between cells. The models by Palm and Merks [63] and Szabó et al. [89] suggested that cells are able to form network-like structures in absence of a substrate to transmit the chemical signal by elongating or by preferentially adhering to elongated structures, respectively. Another explanation for network formation can be found when considering the mechanical environment of cells. The extracellular matrix (ECM), a network of extracellular proteins that surround most cells in tissues, dictates the mechanical environment. The rigidity of the ECM influences cell behavior; cells have been observed to migrate in the direction of higher stiffness [48], and orient to the direction of stretch [36]. Further, focal adhesions, macromolecular assemblies by which the cytoskeleton connects to the ECM, stabilize under mechanical force [71] or on rigid substrates [66]. Cells do not only respond to the mechanical properties of the ECM, but also actively deform it [35, 97]. By applying traction forces, induced by stress fibers within the cell, cells can locally orient [93] and stiffen [97] the substrate they adhere to. This allows for cells to mechanically communicate with each other [70, 97]. Califano et al. [15] have shown that on polyacrylamide gels of sufficient compliance, cells self-organize into vascular-like networks, while they are unable to do so on very rigid substrates [14].

Some previous cell-based modeling has already been dedicated to mechanical cell-ECM interactions [7, 20], where cells contract the matrix and in response align to each other. In [51], a continuum model where cell and ECM density dynamics are regulated by chemical and mechanical forces is presented that leads to network formation. However, in this model, strains in the matrix did not significantly influence network formation. To further investigate the influence of mechanical cell-ECM interactions via strains in the ECM, van Oers et al. [92] have developed a hybrid CPM and Finite Element Model to study network formation. The traction forces that the cells apply to the ECM are described with a model proposed by Lemmon and Romer [46]. This experimentally validated model treats the cytoskeleton as a single cohesive unit, as a result of which the cell forces that cells generate at each point depend on the local cell shape. The strains that are generated in the ECM are calculated using finite elements, where the finite elements correspond to the lattice sites of the CPM. Subsequently, the cells respond to the strains in the matrix. It is assumed that cells preferentially protrude in the direction of higher strain. This was implemented by adding the following bias to the Hamiltonian at the time of copying,

$$\Delta H_{\text{mech}} = -g(\mathbf{x}, \mathbf{x}')\lambda_{\text{durotaxis}} \left(h(E(\epsilon_1))(\mathbf{v}_1 \cdot \mathbf{v}_m)^2 + h(E(\epsilon_2))(\mathbf{v}_2 \cdot \mathbf{v}_m)^2 \right), \quad (18.9)$$

with $g(\mathbf{x}, \mathbf{x}') = 1$ for extensions and $g(\mathbf{x}, \mathbf{x}') = -1$ for retractions, $\lambda_{\text{durotaxis}}$ is a parameter that describes the mechanical sensitivity of cells. $\mathbf{v}_m = \widehat{\mathbf{x} - \mathbf{x}'}$, is the direction of copying, and ϵ_1 and ϵ_2, and $\mathbf{v}_1$ and $\mathbf{v}_2$ are the eigenvalues and eigenvectors of $\underline{\epsilon}$ that represent the principal strains and strain orientation. Thus, extension in the direction of higher strain are promoted and likewise retractions are inhibited. The sigmoid function $h(E) = 1/(1 + \exp(-\beta(E - E_\theta)))$, starts at zero, goes up when there is sufficient stiffness, and eventually reaches a maximum. This means that a certain level of stiffness, due to strain stiffening, $E(\epsilon) = E_0(1 + (\epsilon/\epsilon_{st})1_{\epsilon \geq 0})$ is needed to cause a cell to spread.

On a single cell level, this model predicts that a single cell elongates due to a positive feedback loop of increasing traction and strain stiffening, as previously suggested by Winer et al [97]. Further, two cells in each others vicinity locally align. On a collective scale, these cell-level dynamics lead to vascular-like network formation (Fig. 18.3d); Cells are seen to elongate and locally align to form connected patches of aligned cells. Notably, network formation only occurs on substrates of intermediate stiffness and the simulated networks continuously remodel. Bridging events occur, where two groups of cells penetrate an existing lacuna, forming two lacunae. The paths that cells follow to divide a lacuna is directed by strain lines. Such bridging events have been observed in experimental conditions as well [92].

18.5 Modeling Sprouting During Angiogenesis

So far we have seen four independent mechanism that can lead to vascular-like network formation and thus give different explanations of the mechanisms of *de novo* vasculogenesis. A natural question then is whether these mechanism can also give rise to sprouting angiogenesis? In this section, we will explain how these four CPM-based models can drive sprouting from spheroids. In the next section, we will discuss how other CPM studies have contributed to investigating sprouting angiogenesis.

18.5.1 Sprouting-Like Behavior of Cells in de Novo *models*

Merks et al. [56, 59] showed that cells that secrete and chemotact toward a chemoattractant sprout from a spheroid when either cells elongate or exhibit contact-inhibition chemotaxis. With just plain chemotaxis, sprouting was merely possible for a small range of diffusion constants or strong cell–cell adhesion. So, what gives these two mechanisms, contact-inhibited chemotaxis and cell elongation, the ability to drive chemotactic cells to sprout? The initiation of sprouts at the surface of the spheroid are thought to occur due to a buckling instability; cells in the core of the cluster are compressed due to the pressure the cells on the surface of the spheroid apply inward due to chemotaxis toward the increasing chemoattractant concentration inside the

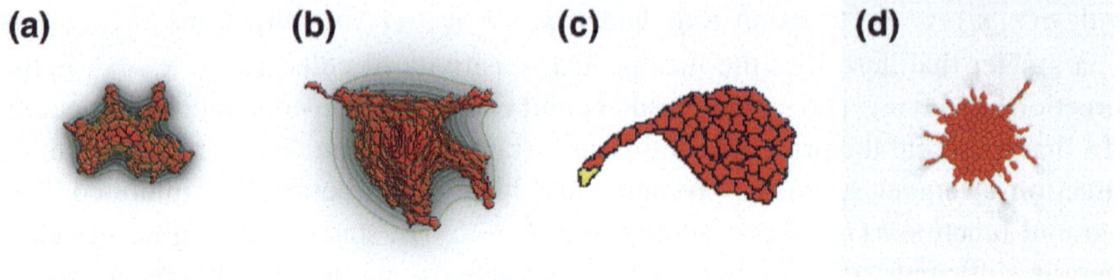

Fig. 18.4 Angiogenesis models. Simulation results of angiogenesis driven by **a** contact-inhibited chemotaxis [59], **b** cell elongation with chemotaxis [56], **c** preferential adhesion to elongated cells [87], and **d** by mechanical cell-matrix interactions [92]

spheroid. Then, for the case of contact-inhibited chemotaxis, chemoattractant gradients at convex regions of the cell aggregate are more shallow than the gradients at concave regions. This makes it more likely for cells to protrude from convex regions of the surface, i.e., at the tips of sprouts (Fig. 18.4b). In spheroids of elongated cells, sprouts start to extend due to local alignment of cells (Fig. 18.4b).

Szabó and Czirók [87] also investigated under what conditions sprouting from a spheroid occurs. It turns out that the assumed mechanism of preferential adhesion to elongated structures suffices for the cells to sprout. Sprouting also occurs by adhesion only, but sprouts can quickly break down. Thus, this attraction stabilizes sprout extensions. Finally, it is argued that the inclusion of leader cells, that polarize and have a persistence in migratory direction, is required to obtain sprouting dynamics that are more similar to experimental results (Fig. 18.4c).

The mechanical model by Van Oers et al. [92] is also able to reproduce sprouting from a spheroid (Fig. 18.4d). Similar to vascularization, only sprouts are formed on substrates of intermediate stiffness. Due to random motility, one cell protruding from the spheroid increases the strain in front of it and subsequently follows it. This instigates a positive feedback loop of strain development and cells extending from the surface that are guided by the strain lines; forming the sprout.

18.5.2 CPM Models of Sprouting Angiogenesis

In the works by Bauer et al. [5, 6] and Daub and Merks [23], VEGF and the ECM are incorporated to study how gradients of VEGF and properties of the ECM can influence sprout formation. The fibrous nature of the ECM influences cell migration in various ways. Fiber orientation directs cell migration, by contact guidance. Further, cells exhibit haptotaxis; migration toward higher ECM densities and haptokinesis; increased movement on intermediate ECM densities. In both models, sprout formation is investigated in the context of sprouting from a blood vessel toward a tumor secreting VEGF.

Bauer et al. [5, 6] model the ECM geometry by including ECM fibers and interstitial fluids, as CPM pixels and frozen tissue-specific cells. The endothelial cells preferentially adhere to the ECM fibers and migrate toward the tumor. Endothelial cells interact with other tissue cells via adhesion. The vessel and tumor are located at opposite sides of the domain. The tumor secretes VEGF that diffuses and is degraded in the model domain and is taken up by the endothelial cells. Endothelial cells interact with the ECM by uptaking and degrading it, and chemotact toward VEGF. Haptotaxis is incorporated by high cell-fiber adhesion. In addition, a distinction between tip and stalk cells is made, by letting tip cells perform chemotaxis and degrade the ECM. Sprout migration is then made possible as tip cells degrade the matrix and stalk cells follow by means of haptotaxis. The model shows that speed, direction and branching of sprouting is dependent on ECM fiber density and composition.

Daub and Merks [23] investigated the effects of ECM densities on sprout morphology and branching by coupling the CPM with a PDE describing ECM density dynamics. The model set-up resembles the one used by Anderson and Chaplain [2], who used a stochastic discrete PCA-like model based on PDE discretization, to describe sprouting toward a VEGF secreting tumor. In [23], a VEGF gradient is presented to the CPM cells, to which cells respond by chemotaxis. Furthermore, VEGF induces the cellular secretion of proteolytic enzymes that degrade the ECM. Cells in turn respond to the ECM by haptokinesis and haptotaxis. Haptokinesis promotes the formation of branches and increases the sprout velocity on intermediate ECM densities. The degree of sprouting is most influenced by the haptotaxis parameter. Again, this work has showed the importance of cells interacting with the ECM properties to sprout formation.

18.6 Multiscale Models

The above CPM models of blood vessel formation asked how a single, stereotypic set of cell behaviors results in multicellular patterns. This can be an accurate representation of the situation in some *in vitro* cell cultures, but in realistic situations, i.e., in actual organisms, the situation is usually much more complex. Blood vessels typically consist of a number of cell types, including endothelial cells, pericytes, and smooth muscle cells, each of which require their own description using the CPM. Usually this is done by assigning each cell type a different value of τ (see Eq. (18.3)), and assigning different parameters to each cell type (e.g., by giving a different value of J to each combination of cell types; cf. Sect. 18.2.1).

The situation becomes more complicated if the cells change type depending on the signals they receive from adjacent cells, a common situation in biology. In angiogenesis two phenotypes of endothelial cells are distinguished, the tip cell that has many protrusions and is highly migratory but rarely divide, and the stalk cell that has few cellular protrusion and can proliferate. The differentiation into two types is mediated by the cell–cell contact signaling (e.g., through the Notch pathway [39]). To model this situation, the CPM is often extended with sets of coupled ODEs, where each cell (or spin σ) obtains its own set of ODEs. The ODEs can then also be coupled with the ODEs of adjacent cells, to model chemical signaling, or with sets of PDEs, e.g., to describe the diffusion and reaction dynamics of chemoattractants. To model cell differentiation, the CPM parameters of the cells (i.e., the target areas $A(\sigma)$, the cell rigidity, $\lambda(\sigma)$, the interfacial tension parameters J, and so forth) can be replaced for functions of the intracellular ODEs. As a result, the dynamics of the ODE can lead to changes in the behaviors and positions of the cells in the CPM, which can in turn affect the ODE, resulting in interesting multiscale dynamics. An example of this approach is given in Sect. 18.6.1.

In these examples, the cells in the CPM are still treated as homogenous structures, whereas in actual organisms the internal structure, e.g., the cytoskeleton, affect the behavior of cells in the tissue. The regulatory networks simply regulate the parameters of the cells. In a number of problems, it becomes important to describe the internal structure of the cells in more detail. To explain the dynamics of a type of highly motile skin cell, the keratocyte, Marée et al. [54] extended the CPM with an intracellular, dynamic model of the actin cytoskeleton, an approach that was later generalized to study the response speed to chemotactic cues in eukaryotic cells [53]. This work made use of an intracellular set of PDE's to describe the polymerization and orientation of actin filaments and of the enzymes regulating the polymerization rates. The polymerization model then biased the extensions and retractions in the CPM by modifying ΔH during the copy attempt (cf. Eqs. (18.5) and (18.9)). Another approach to include internal structure is the compartmental cellular Potts model. In this approach multiple spin domains are bundled together to form one biological cell. This approach was used by Boas and Merks [10] to model the formation of the lumen, i.e., the hollowing out of new blood vessels such that blood can flow through. Section 18.6.2 will briefly review this approach.

18.6.1 Sprouting Morphogenesis with Tip Cell Selection

A suitable example problem to illustrate the structure and dynamics of multiscale models that include a model of cell differentiation, is the selection of tip and stalk cells via the Delta-Notch molecular signaling pathway. Delta-Notch signaling acts as a lateral-inhibition mechanism, where a high expression of Delta activates the expression Notch in adjacent cells, which in turn suppresses the activity of Delta. Delta-Notch signaling is involved in a variety of processes in developmental biology, including the formation of body segments: somitogenesis [26], asymmetric cell division [4, 72], neuronal plasticity [1, 47]), and the initiation of angiogenesis [8]. Delta-Notch is mediated by interactions between Notch receptors and Delta/Serrate/LAG-2 (DSL) ligands [13]. In angiogenesis, extracellular VEGF has been shown to initiate the endothelial Delta-Notch signaling leading to the dynamic stalk-tip cell selection [8] (see Fig. 18.5).

Prokopiou and coworkers [68, 69] have introduced a detailed, multiscale model of sprouting angiogenesis based on the CPM. In this model, each cell includes an ODE model of the Delta-Notch-VEGF signaling pathway, which acts to regulate task

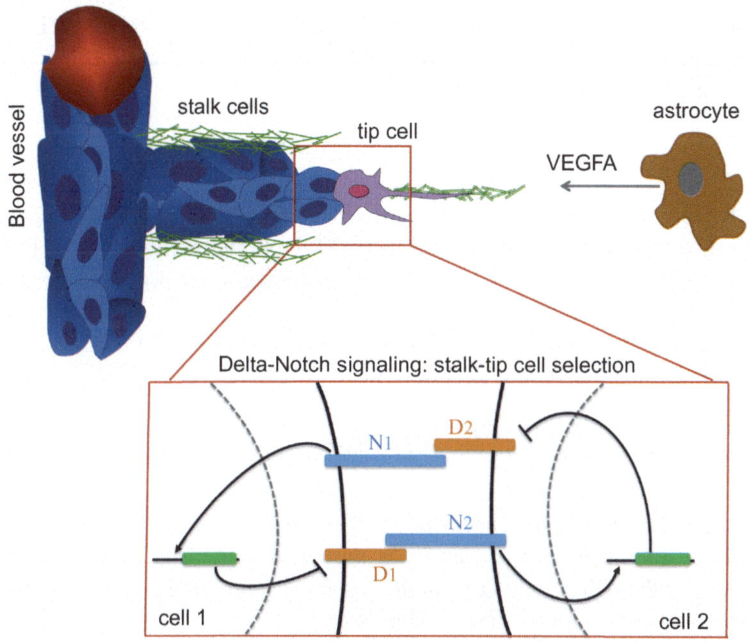

Fig. 18.5 Stalk-tip cell selection in angiogenesis. Extracellular VEGF stimulates Delta-Notch signaling pathway in endothelial cells of a nearby blood vessel. The Delta-Notch signaling pathway is responsible for (stalk-tip) cell fate decisions. In particular, high Delta (low Notch) leads to a tip cell phenotype, and low Delta (high Notch) leads to a stalk cell phenotype. N_i: Notch (cell i); D_i: Delta (cell i)

division between adjacent cells. We will review their model in detail here; similar approaches have been taken recently by Palm et al. [62] and by Boas and Merks [11].

18.6.1.1 Growth Factor and Extracellular Matrix Fields

Prokopiou's model considers the interaction between endothelial cells and the extracellular matrix (ECM). In the standard CPM (see, e.g., Sect. 18.2) the substrate, or medium, is represented as a homogenous cell covering the whole computational domain. Such a homogenous material can also be a suitable description of the ECM if the spatial inhomogeneity of ECM does not affect the problem under study. However in the case of EC migration, empirical evidence (e.g., in the developmental retina) showed that ECM form fiber bundle networks that are highly inhomogeneous and that ECs follow the tracks of fiber bundles. It is therefore necessary to model the ECM as a discrete field. In order to model such a non-homogenous ECM, the ECM fibers were distributed randomly. In particular, these fibers are modeled as a static field in the numerical domain. Each pixel in the numerical domain occupied by an ECM fiber is given a non zero (=1) value (and zero elsewhere). Cell tracking along the fibers is modeled as preferential adherence to the fibers. Thus, haptotaxis, the directional migration of cells up the ECM density field, is incorporated as an additional mechanism. In the CPM, haptotaxis can be implemented similar to chemotaxis (Eq. 18.5), with the main difference that the ECM field does not diffuse. During a copy attempt, the following term is added to ΔH,

$$\Delta H_{hapt} = \lambda_{\mathrm{ECM}} \left(\mathrm{ECM}(\mathbf{x}) - \mathrm{ECM}(\mathbf{x}') \right), \tag{18.10}$$

where, $\mathrm{ECM} \in \{0, 1\}$ is the presence or absence of ECM at site **x**, and λ_{ECM} is the strength of the preferential attachment to ECM. Note that remodeling of ECM by endothelial cells, stiffness of ECM, and ECM degradation are not considered in this model.

We assume that a source secretes VEGF. The dynamics of VEGF is described by the following equation, similar to (Eq. (18.6)):

$$eq : ODEmodel \frac{\partial [VEGF]}{\partial t} = D\nabla^2 [VEGF] + s - \epsilon [VEGF], \tag{18.11}$$

where $[VEGF]$ is the VEGF concentration, D is the diffusion coefficient, s represent the secretion rate at the source, and ϵ is the decay rate. This equation is solved numerically with no flux boundary conditions at the simulation domain. Chemotaxis is incorporated as previously described in Sect. 18.2.2.

18.6.1.2 Subcellular Level: Modeling Lateral-Inhibition

At the subcellular level, tip cell differentiation is regulated via the Delta-Notch signaling pathway, which is activated by VEGF. The contact lateral-inhibition effect for the exchange of the endothelial (stalk-tip) phenotype is implemented using a modification of a well mathematical model proposed by Collier et al. [21], where a system of coupled ODEs describes the dynamic processes of Delta and Notch activation and inhibition between cells that are in contact with each other.

Motivated by the experimental work of Lobov et al. [49], which showed that VEGF induces Delta in the retinal vasculature, Prokopiou and coworkers [68, 69] extended the model of Collier et al. [21] to incorporate the contribution of VEGF (as defined in Eq. (18.11)), the non-dimensionalized form is:

$$\begin{aligned}
\text{Delta}: \quad & \frac{dD_j}{dt} = v\left(\alpha \frac{[VEGF_j]}{VEGF_h + [VEGF_j]} \frac{1}{1 + bN_j^2} - D_j \right), \\
\text{Notch}: \quad & \frac{dN_j}{dt} = \frac{\bar{D}_j^{\,2}}{a + \bar{D}_j^{\,2}} - N_j \,, \\
\text{trans-Delta}: \quad & \bar{D}_j = \sum_i \frac{D_i P_{ij}}{P_j} \,.
\end{aligned} \tag{18.12}$$

where D_j, N_j, represent the levels of Delta and Notch expression in cell j. $[VEGF_j] = (1/a_j)\sum_i^{\omega} VEGF_{ji}$ is the average VEGF in a cell j; that is, the sum of VEGF at each pixel i inside cell j over the cell area, a_j, where ω is the total number of pixels in cell j. $VEGF_h$ is the VEGF level at which the production rate of Delta is half maximal. The trans-Delta ($\bar{D}_j$) is taken to be the sum over the immediate (contacting) neighbors i of cell j. P_j is the perimeter of cell j, and P_{ij} is the common area of cell j with its neighbor cells i, which is defined as

$$P_{ij} = \sum_{(\mathbf{x},\mathbf{x}')} \left(1 - \delta_{\sigma(\mathbf{x}),\sigma(\mathbf{x}')}\right) \left(1 - \delta_{\sigma(\mathbf{x}'),0}\right) \delta_{\sigma(\mathbf{x}),i}. \tag{18.13}$$

The summation is over all pairs of adjacent sites in the lattice.

Equation (18.12) describe (i) the activation of Notch production within each cell as a function of the levels of (trans-) Delta expressed by neighboring cells, (ii) the inhibition of Delta expression by Notch, and (iii) the activation of Delta production by extracellular VEGF. In the absence of VEGF signaling, there is no up-regulation of Delta and, therefore, no tip cell activation. Equation (18.12) was implemented using the Systems Biology Workbench and integrated within the CompuCell3D framework.

18.6.1.3 Coupling of ODE Model to CPM

To simulate the effect of the regulation by the signaling network on cell behavior, we let the level of Delta in the ODEs (Eq. (18.12)) determine the cell type $\tau \in \{\text{tip, stalk}\}$ (cf., Eq. (18.3)): if $D(\sigma) > \theta_{\text{tip}}$, the cell type becomes $\tau = $ tip, or else the cell type becomes $\tau = $ stalk. Each cell type is associated with a prescribed set of properties. The tip cells have a higher chemotactic coefficient than stalk cells ($\lambda_{\text{chem}}(\text{tip}) > \lambda_{\text{chem}}(\text{stalk})$); the stalk cells if they are adjacent to tip cells, can grow by gradually increasing their target areas. The latter property is to implement the assumption that only stalk cells adjacent to tip cells proliferate, because the proliferation of all stalk cells would lead to a thick/swollen sprout and parent vessel.

To avoid any predefined or probabilistic rules of cell growth and division, we assign to each cell a clock $\phi(\sigma)$ that progresses only for stalk cells adjacent to tip cells. The clock progresses at a rate $a = 0.01$ h per MCS. In addition, considering that we want a cell to divide after doubling in size, the target volume of a cell (V; cf. Eq. (18.3)) should grow by one initial target volume, $V(t = 0)$ during one cell cycle of 17 hours, yielding a growth rate,

$$\mu = \frac{V(t=0)}{t_{\text{cell-cycle}}} = \frac{25}{17} = 1.47 \text{ pixels/h} = 1.47 \times 10^{-2} \text{ pixels/MCS}. \tag{18.14}$$

Thus, a stalk cell divides if two conditions are satisfied: (1) its clock reaches the cell cycle duration, $t_{\text{cell-cycle}} = 17$ h, and (2) its cell area has doubled. At $t = 0$ the clock phase is drawn at random from a uniform distribution $\phi(0) \in [0 - 17]$ h, and when $\phi(t) = 17$ h the clock is reset to zero. Cell division is implemented by defining a

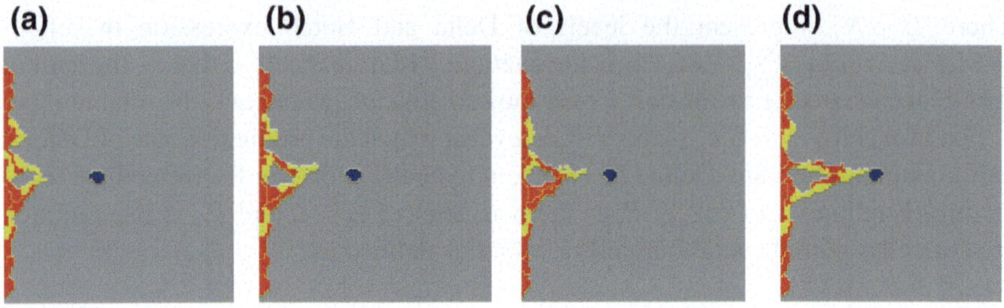

Fig. 18.6 Sprout anastomosis due to Notch-Delta pathway. Representative simulation snapshots of sprout evolution showing anastomosis. **a**, **b** Tip cell fusion (two adjacent tip cells) in 18 and 24 h, **c** one of the two tip cells becomes a stalk cell (lateral-inhibition effect from Delta-Notch signaling) in 26 h, and **d** the leading tip cell moves up the astrocyte-derived VEGF gradients in ~38 h. Key: stalk cells (*red*), tip cells (*yellow*), astrocyte (*blue*)

division plane (usually the short axis of an ellipse fitted to the cell), and assigning a new spin σ to half of the cell. Finally, only the tip cells and the stalk cells adjacent to the tip cells elongate, according to Eq. (18.7).

18.6.1.4 Simulation Results

The Notch-Delta pathway, the VEGF source, and the ECM heterogeneity all work in concert to sprouting angiogenesis. We first analyzed Eq. (18.12) to find the parameter values for α such that the homogeneous steady state become unstable. For a string of cells, the solution is a dynamic 'salt and pepper' pattern of cells with alternating high and low Delta values; for a 2D sheet of cells, the solution is a dynamic 'checker board' pattern. When we define a tip cell as a cell with a Delta-level above a threshold, the solutions then lead to a dynamic interchange of phenotypes between stalk and tip cells. Hence in our simulations, the phenotype distribution of ECs along the capillary sprout is determined by two main mechanisms: the astrocyte-derived VEGF that activates the Delta activity in each cell, and the Notch-Delta signaling pathway that yields the 'salt-pepper' pattern. Following the tip/stalk selection, the ECs then migrate chemotactically to VEGF distribution and haptotactically to ECM distribution. Figure 18.6 shows the development of multiple tip cells, each can potentially lead the formation of a sprout; when the head tip cell of two growing sprouts meet, the Notch-Delta pathway re-establishes the tip, resulting in the apparent fusion of two sprouts into one, in a process termed anastomosis.

We summarize the effect of different VEGF and ECM profiles (Table 18.1) on the resulting morphology of the capillary sprouts. Figure 18.7 shows representative snapshots of sprout evolution in each scenario.

No VEGF Gradient (Scenarios 1 and 2)

In scenarios 1 and 2, there is no VEGF gradient. A sufficiently high level of VEGF activates the Notch-Delta pathway, and leads to selection of tip cells (yellow) and

Table 18.1 Different scenarios regarding VEGF and ECM profiles presented in Fig. 18.7

Scenarios	
1.	Homogeneous VEGF and homogeneous ECM
2.	Homogeneous VEGF and heterogeneous ECM
3.	Static VEGF gradients and homogeneous ECM
4.	Static VEGF gradients and heterogeneous ECM
5.	Heterogeneous VEGF and homogeneous ECM
6.	Heterogeneous VEGF and heterogeneous ECM

stalk cells (red). However the cells do not receive directional guidance from a gradient of VEGF, resulting in a much reduced migration. Figure 18.7 shows that cell proliferation and elongation are undirected and, therefore, stalk and tip cells evenly fill the space. This morphology was observed in experiments [31, 60], where a spatial gradient in VEGF was removed in the retina, by increasing expression levels of VEGFA in transgenic mouse models. In scenario 2, the addition of a non-uniform ECM has a weak effect, because the ECM does not offer an overall gradient.

Static VEGF Gradient (Scenarios 3 and **4)**

In these two scenarios, we incorporated static VEGF gradients, which eventually lead to either a swollen (scenario 3) or a thin (scenario 4) sprout formation. The results of scenarios 1–4 look quite similar up to approximately 12 h. The sprouts are dominated by single, elongated tip cells. However, differences become visible at later time points. Particularly, in scenario 4, cell proliferation is focused onto a single sprout as a result of the VEGF gradients and the heterogeneous ECM.

Dynamic VEGF from Single Source (Scenarios 5 and **6)**

Here, a fixed astrocyte (VEGF source) is responsible for the VEGF gradients. Figure 18.7 (scenarios 5 and 6) demonstrates the model's ability to reproduce realistic capillary sprout morphologies (up to $\sim$38 h). Scenario 5 (with homogeneous ECM) can give a polarized sprout, but the emerged sprout in scenario 6 (with heterogeneous ECM) has the right extension speed ($\sim$1.6 μm/h) as it was evaluated from our experimental (unpublished) data. Therefore, we suggest that scenario 6 provides a close approximation to a growing vascular sprout. However, since the astrocyte cannot move away, scenario 6 does not allow for the formation of longer sprouts, because at late time points (85–100 h) a mass of cells starts surrounding the astrocyte.

18.6.2 Lumen Formation

The work by Scianna [74] and the model by Boas and Merks [10] illustrate the use of the so called compartmental CPM [74, 82, 86] to treat subcellular structures during angiogenesis. In the compartmental CPM, the Potts domains (clusters of

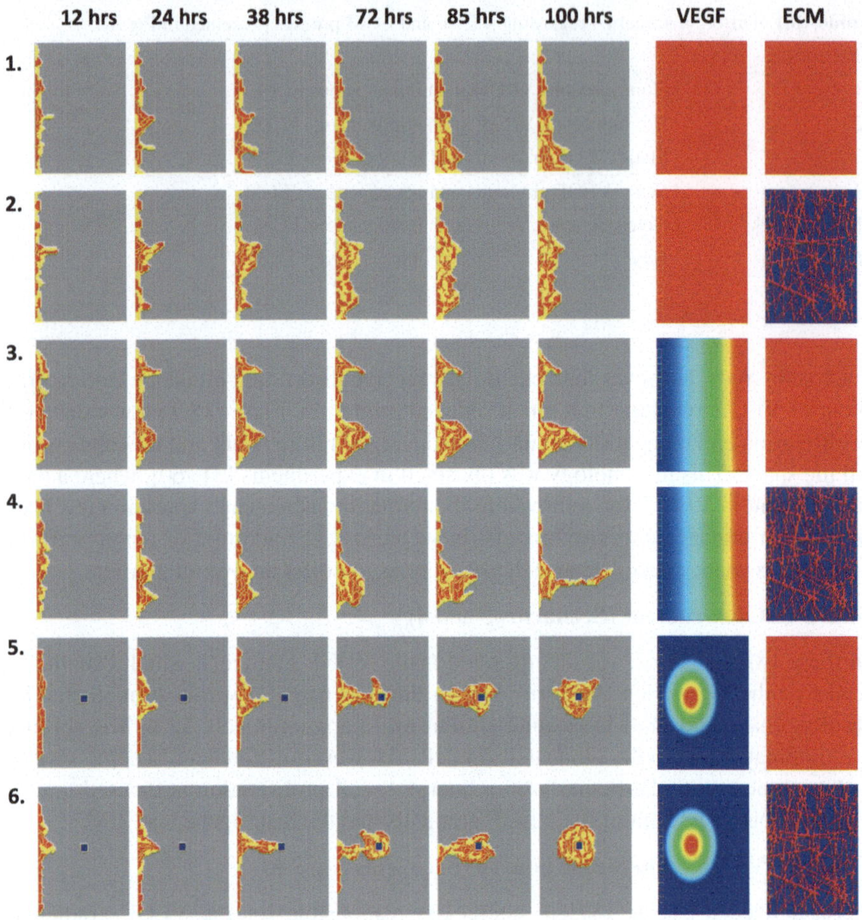

Fig. 18.7 Sprout evolution in different VEGF and ECM profiles. Representative simulation snapshots (from 10 simulations) of sprout evolution for the six scenarios outlined in Table 18.1. Scenario 6 gives the right extension speed ($\sim$1.6 μm/h) as it was evaluated from our experimental (unpublished) data. Key: stalk cells (*red*), tip cells (*yellow*), fixed astrocyte (*blue*)

spins, σ) represent parts of cells, rather than individual cells. The compartments are then bundled together to represent one biological cell using a cluster identifier, ξ. All spins belonging to the same cell then have the same cluster identifier, $\xi(\sigma)$. Additional constraints can be imposed on the whole cell (see, e.g., Ref. [10]), on individual components [75] or both [10]. Example applications include a model of the nuclei of endothelial cells [74], in particular the way how the nuclei can slow down the migration of the cells in the ECM if the nuclei are larger than the typical pore size [75], and the model of lumen formation that we will review in more detail here.

Once new blood vessels are formed, they must hollow out to allow the perfusion of blood. The mechanisms of hollowing or lumen formation have been debated for centuries. Experimental research has led to two main hypotheses: vacuolation [9, 25, 41, 96] and cell–cell repulsion [85]. During vacuolation, vacuoles are suggested to form by the fusion of pinocytotic vesicles. Initially, lumens were thought to form intracellularly by spanning the cell with a large vacuole that then fuses to the cell membrane on both sides of the cell [25, 41]. Later, lumens were also suggested to form extracellularly by the secretion of vacuoles between cells [9, 96]. During cell–cell repulsion, cell membranes of adjacent cells are suggested to repulse each other to form an extracellular lumen between the cells [85]. Both hypotheses are supported by strong experimental evidence, leaving the debate unresolved. To address this debate, Boas and Merks [10] developed a computational model of lumen formation that can represent both hypotheses.

The lumen formation model is initialized with twelve endothelial cells in a branched blood vessel, surrounded by immobile extracellular matrix (ECM). Each cell is modeled as a cluster of CPM compartments (σ) with the same cluster identifier (ξ) to allow for polarization of the cell membrane and for the formation of vesicles and vacuoles within the cell. The cell polarizes into two cell membrane compartments and a cytosol compartment upon contact with the ECM, representing cell membrane polarization by integrin signaling from the ECM. All membrane pixels that are in contact with the ECM form a membrane compartment of type $\tau(\sigma)$ = basolateral. The adjacent second neighbor order membrane pixels hereof are added to this compartment to represent tight junctions between cells, and the rest of the membrane becomes the second membrane compartment of type τ = apical. The membrane is repolarized every other time step. To mimic cell–cell repulsion, apical membranes of opposing cells are assigned a high adhesion energy.

During vacuolation, membrane pixels that internalize into the cytosol compartment have a probability to become cell compartments of type vesicle to represent pinocytosis. These single-pixel vesicles move through the cell following a biased random walk, by swapping the position of a vesicle with a neighboring pixel. Acceptance of a swap depends on a constant probability P_A multiplied by a Boltzmann probability $P_{\text{Boltzmann}}(\Delta H)$, with ΔH the change in effective energy resulting from changes in adhesion energy between compartments due to the swap. Vesicles prefer to adhere to other vesicles and vacuoles. Once vesicles meet, they fuse together into a single compartment of type vacuole, which moves by regular CPM dynamics. Vesicles and vacuoles are secreted when in contact with the apical membrane, forming a new extracellular compartment of type luminal fluid. Upon contact, luminal fluid compartments fuse into a single lumen.

Continuous lumens can be formed in the model through the branched blood vessel by vacuolation as well as by cell–cell repulsion (Fig. 18.8). However, lumen formation is far more robust to parameter values changes when the two hypotheses are combined, suggesting that the two hypotheses work synergistically. Vacuolation can help lumen formation by cell–cell repulsion by piercing cells and by enlarging the luminal space in-between cells. The cell–cell repulsion hypothesis assists

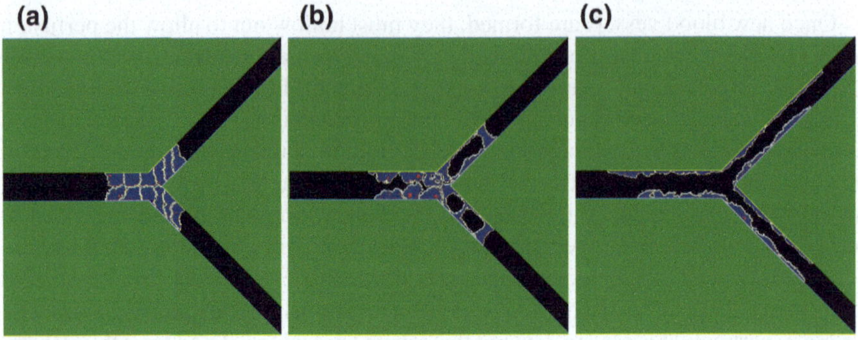

Fig. 18.8 Lumen formation. Simulation of lumen formation [10] by synergy of vacuolation and cell–cell repulsion. A branched blood vessel consists of twelve cells (*blue*) within the ECM (*green*) and fluid is colored black. The cell membrane is polarized into an basolateral membrane (*gray*) and an apical membrane (*yellow*). Vesicles and vacuoles are colored *red*. Reproduced from Ref. [10] under the terms of the Creative Commons License http://creativecommons.org/licenses/by/3.0

lumen formation by the vacuolation hypothesis by preventing collapse of the formed extracellular lumens and by extension of them.

One may question synergy of the two hypotheses as experimentalists mostly find evidence for one or the other hypothesis. It is important to realize that lumen formation by vacuolation is mostly studied in small intersegmental vessels (ISV) of zebra fish, while cell–cell repulsion is mostly studied in aortae of mice. Interestingly, when lumen formation by synergy of the two hypotheses is performed in the model initialized with a one-cell thick vessel, the resulting lumen formation visually resembles vacuolation. In contrast, when lumen formation by synergy of the two hypotheses is performed in the model initialized with a multicell thick vessel, the resulting lumen formation visually resembles cell–cell repulsion. In conclusion, the computational model of lumen formation suggests that vacuolation and cell–cell repulsion work synergistically and that the discrepancy between observations of different experimental groups might be explained by the vessel sizes they are studying.

18.6.3 Integrating Angiogenesis Models into CPM Models of Organogenesis

Angiogenesis is a key process in many developmental and pathological processes. For this reason, the simple models of endothelial cell–cell interactions have been integrated in larger models of organ development and tumor growth. Shirinifard et al. [79] have integrated an angiogenesis model similar to the one proposed by [59] with a tumor growth model, where the growth of tumor cells was made dependent on the availability of oxygen. Kleinstreuer et al. [42] integrated a cellular Potts model of in vitro angiogenesis with a large dataset of the US Environmental Protection

Agency (EPA) of pesticides and their effects on vascular morphogenesis. By linking the adverse effects of pesticides to individual cell behaviors of the cell types involved in vasculogenesis, they could construct a first toxicological, predictive model based on the cellular Potts model.

A further example of how cellular Potts models of angiogenesis can be integrated into larger models of tissue development is on age-related macular degeneration, by Shirinifard et al. [80]. Age-related macular degeneration (AMD) is the main source of vision loss in the elderly and a looming epidemic for our aging society. There are two basic forms of AMD, the "dry" form and the "wet" form. In dry AMD, the layer of retinal pigment epithelial cells (RPE) in the macula degenerate and die (atrophy). These RPE cells support the light sensitive photoreceptor cells that are critical to vision. Dry AMD can progress slowly and culminate with the more advanced stage called Geographic Atrophy, where a patch of photoreceptor cells die off. The wet AMD is due to the abnormal blood vessels (known as choroidal neovascularization or CNV) growing under the retina and macula. These new blood vessels may then bleed and leak fluid, causing the macula to bulge or lift up from its normally flat position, thus distorting or destroying central vision. Under these circumstances, vision loss may be rapid and severe.

In CNV, after capillaries initially penetrate basement membrane under the RPE (called the Bruch's membrane or BrM), invading vessels may either regress or expand. Clinically, during early and late CNV, the expanding vasculature usually spreads in one of three distinct patterns: in a layer between BrM and the retinal pigment epithelium (sub-RPE or Type 1 CNV), in a layer between the RPE and the photoreceptors (sub- retinal or Type 2 CNV) or in both loci simultaneously (combined pattern or Type 3 CNV). Most previous studies hypothesized that CNV primarily results from growth-factor effects or holes in BrM, but failed to explain the initiation nor progression patterns of CNV. Shirinifard et al. [80] used 3D CPM of the normal and pathological maculae to recapitulate these three growth patterns (Fig. 18.9). The key feature of these tissue models are the adhesions within and between different tissue layers: BrM, RPE, and photoreceptor outer segment (POS) (Fig. 18.9a), in addition to endothelial cell dynamics, VEGF dynamics and MMP degradation of ECM. These models aimed to test the hypothesis that CNV results from combinations of impairment of adhesion, in particular: RPE-RPE epithelial junctional adhesion, adhesion of the RPE basement membrane complex to BrM (RPE-BrM adhesion), and adhesion of the RPE to the photoreceptor outer segments (RPE-POS adhesion). Figure 18.9b shows a time sequence of snapshots from a typical simulation of Type 3 CNV, where new blood vessel invades both under and above the RPE layer. Results from all combinations of adhesion parameters were summarized into tables and risk maps. Figure 18.9c is an example of the risk map for the three main adhesion parameters. Key findings from the simulations are that when an endothelial tip cell penetrates BrM: (1) RPE with normal epithelial junctions, basal attachment to BrM and apical attachment to POS resists CNV. (2) Small holes in BrM do not, by themselves, initiate CNV. (3) RPE with normal epithelial junctions and normal apical RPE-POS adhesion, but weak adhesion to BrM (e.g., due to lipid accumulation in BrM) results in Early sub-RPE CNV. (4) Normal adhesion of RPE to BrM, but reduced

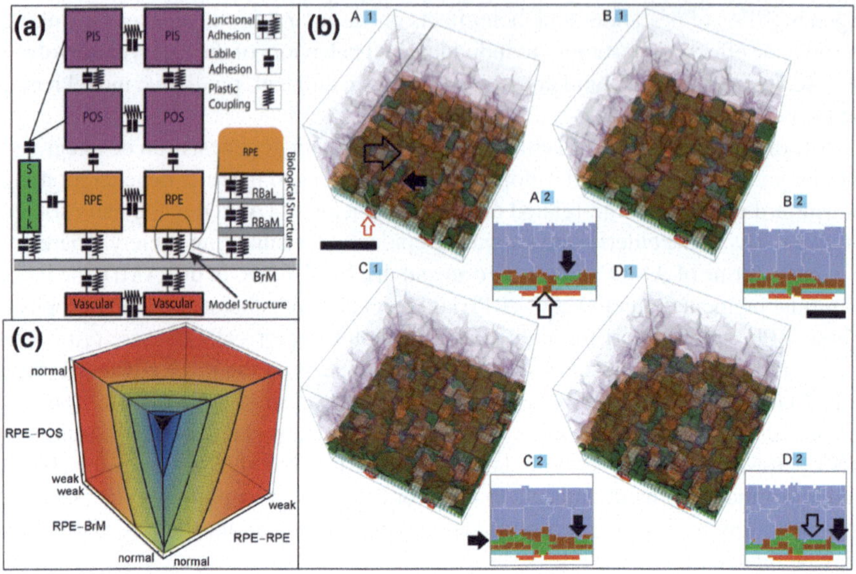

Fig. 18.9 Angiogenesis in age-related macular degeneration. **a** Schematic of the adhesive interactions in the model macula, both labile and junctional adhesions are modeled. **b** Time sequence snapshots of a sample simulation of the development of a combined sub-RPE and above-RPE CNV or Type 3 CNV. **c** The probability of CNV initiation as a function of three key adhesion mechanisms, from zero (*black*) when all adhesion strengths are normal to 1 (*red*) when each of the adhesion strength is weak. Figure modified from Ref. [80] under the terms of the Creative Commons Attribution License

apical RPE-POS or epithelial RPE-RPE adhesion (e.g., due to inflammation) results in Early sub-retinal CNV. (5) Simultaneous reduction in RPE-RPE epithelial binding and RPE-BrM adhesion results in either sub-RPE or sub-retinal CNV which often progresses to combined pattern CNV. These findings suggest that defects in adhesion dominate CNV initiation and progression. This conclusion is both novel and surprising, but coherently explain the heterogeneous range of CNV growth patterns and dynamics.

18.7 Conclusion

In this chapter, we have introduced the cellular Potts model and discussed how it can be seen as a special case of PCA. In contrast to the formal definition of PCA, the cellular Potts model is asynchronous, and its rules are not strictly local. The dynamics, as guided by the Hamiltonian (Eq. (18.3)), depend on the local neighborhood of the lattice sites, as well as on the properties of the whole biological "cell," i.e., the set of all lattice sites $\mathbf{x}$ that have the same state, or *spin* $\sigma(\mathbf{x})$. Examples of such non-local dependencies include the volume constraint Eq. (18.3) and the length constraint (Eq. (18.7)). As an advantage of this approach relative to traditional PCA that represent biological 'cells' with individual lattice sites, cells in the CPM can assume arbitrary shapes, which can be given by the model (see e.g., Sect. 18.4.2) or change dynamically during the simulations. The resulting simulation images and movies are often perceived by biological researchers as "realistic," allowing for one on one visual and quantitative comparison with microscopic data.

Of course, such flexibility comes at a cost. With the current speed of serial processors, typical simulations of the CPM are fast enough that large scale parameter studies can be performed [64], although full three-dimensional simulations can be limited to at most several million cells for individual simulations. What is currently largely out of reach are formal mathematical analyses of the CPM similar to those performed for PCA, making it practically impossible to generalize or proof any insights obtained with the CPM beyond what was tested numerically for individual parameter sets. Fortunately first attempts to formalize the CPM have been made, as shown in Chap. 19 (see also Ref. [94]). Apart from its non-locality, another mathematical limitation of the CPM is its required asynchronicity. Apart from complicating formal treatment, it hinders its implementation on graphical processing units (GPUs). To reach optimal speedup, GPUs rely heavily on the synchronicity and the locality of the algorithm. Although GPU-implementations of the CPM have been proposed [90, 98], a fully synchronous, local reformulation of the CPM would help dramatically speedup CPM-simulations.

After introducing hybrid CPMs, in which the CPM dynamics affects the kinetics of a PDE model and vice versa, we illustrated the applicability of the CPM to biomedical problems. Here we focused on the modeling and simulation of blood vessel growth: angiogenesis and vasculogenesis. After discussing in Sects. 18.4 and 18.5 how CPMs have been instrumental in proposing and analyzing new hypothesis for the cell behavior that is responsible for the formation of blood vessel like structures, Sect. 18.6 showed how such models can be incorporated into more complete, multiscale models. Such multiscale models typically contain more detailed models of the intracellular kinetics, implemented using ODEs or using a compartmental CPM. Finally, we have shown in Sect. 18.6.3 how these can be incorporated into larger scale models of organ development or disease progression. Mathematically, such models can become complicated: we have coupled systems of PDEs, ODEs and compartmental CPMs, where several of such models might operate at of the system.

In such cases, model validation might become a serious concern. First, the behavior of the system becomes difficult to determine. Parameter sweeps are key tools for determining the behavior simulation models [64], but they can only be performed starting from one or a few sets of nominal parameter values, as a result of which some interesting or false behavior of the model might be missed. To get better insight into the whole parameter space, useful methodology includes global sensitivity analyses, which have recently been tested on a simple CPM of vascular morphogenesis [12], Second, among the plethora of potential biological mechanisms represented by our models, the ones that best describe the actual mechanism must of course be selected. Sections 18.4 and 18.5 showed that a range of different mechanism is able to describe vasculogenesis and angiogenesis (see also Ref [58]). Since these models all result in sprout and branch formation reminiscing the experimental data, they are all plausible explanations for these phenomena. To this end, experiments should be designed in order to further validate these models in order to rule out some of the different hypotheses. We must however keep in mind that it is possible that different mechanisms operate in different tissues or time periods in development. Or, most likely, different mechanisms work together in order to effectively create and stabilize vessels. In order to gain a better understanding of angiogenesis, we must figure out how and when certain mechanisms play a role and how they influence each other. Computational modeling using the CPM serves as a good starting point to get insight into the roles and interactions of alternative mechanisms of vasculogenesis in a combined model, driving the development of new, testable hypotheses.

Acknowledgements We thank Indiana University and the Biocomplexity Institute for providing the CompuCell3D modeling environment and SURFsara (www.surfsara.nl) for the support in using the Lisa Compute Cluster. The investigations were in part supported by the Division for Earth and Life Sciences (ALW) with financial aid from the Netherlands Organization for Scientific Research (NWO) through Vidi grant 864.10.009. YJ was supported partially by the National Institute of Health grant U01CA143069.

References

1. Alberi, L., Liu, S., Wang, Y., Badie, R., Smith-Hicks, C., Wu, J., Pierfelice, T.J., Abazyan, B., Mattson, M.P., Kuhl, D., Pletnikov, M., Worley, P.F., Gaiano, N.: Activity-induced notch signaling in neurons requires arc/arg3. 1 and is essential for synaptic plasticity in hippocampal networks. Neuron **69**, 437–444 (2011)
2. Anderson, A.R.A., Chaplain, M.A.J.: Continuous and discrete mathematical models of tumor-induced angiogenesis. Bull. Math. Biol. **60**(5), 857–899 (1998)
3. Anderson, A.R.A., Quaranta, V.: Integrative mathematical oncology. Nat. Rev. Cancer **8**(3), 227–234 (2008)
4. Bardin, A.J., Le Borgne, R., Schweisguth, F.: Asymmetric localization and function of cell-fate determinants: a fly's view. Curr. Opin. Neurobiol. **14**, 6–14 (2004)
5. Bauer, A.L., Jackson, T.L., Jiang, Y.: A cell-based model exhibiting branching and anastomosis during tumor-induced angiogenesis. Biophys. J. **92**, 3105–3121 (2007)
6. Bauer, A.L., Jackson, T.L., Jiang, Y.: Topography of extracellular matrix mediates vascular morphogenesis and migration speeds in angiogenesis. PLoS Comput. Biol. **5**(7), e1000,445 (2009)
7. Bischofs, I.B., Schwarz, U.S.: Cell organization in soft media due to active mechanosensing. Proc. Natl. Acad. Sci. U.S.A. **100**(16), 9274–9279 (2003)
8. Blanco, R., Gerhardt, H.: VEGF and Notch in tip and stalk cell selection. Cold Spring Harbor Perspectives in Medicine **3** (2013)
9. Blum, Y., Belting, H.G., Ellertsdottir, E., Herwig, L., Lüders, F., Affolter, M.: Complex cell rearrangements during intersegmental vessel sprouting and vessel fusion in the zebrafish embryo. Dev. Biol. **316**(2), 312–322 (2008)
10. Boas, S.E.M., Merks, R.M.H.: Synergy of cell-cell repulsion and vacuolation in a computational model of lumen formation. J. R. Soc. Interface **11**(92), e20131049 (2014)
11. Boas, S.E.M., Merks, R.M.H.: Tip cell overtaking occurs as a side effect of sprouting in computational models of angiogenesis. BMC Syst. Biol. **9**, 86 (2015)

12. Boas, S.E.M., Navarro Jimenez, M.I., Merks, R.M.H., Blom, J.G.: A global sensitivity analysis approach for morphogenesis models. BMC Syst. Biol. **9**, 85 (2015)
13. Bray, S.: Notch signalling: a simple pathway becomes complex. Nat. Rev. Mol. Cell Biol. **7**, 678–689 (2011)
14. Califano, J., Reinhart-King, C.: A balance of substrate mechanics and matrix chemistry regulates endothelial cell network assembly. Cell. Molec. Bioeng. **1**(2), 122–132 (2008)
15. Califano, J.P., Reinhart-King, C.A.: Exogenous and endogenous force regulation of endothelial cell behavior. J. Biomech. **43**(1), 79–86 (2010)
16. Cao, Y., Hong, A., Schulten, H., Post, M.J.: Update on therapeutic neovascularization. Cardiov. Res. **65**, 639–648 (2005)
17. Carlier, A., Geris, L., Bentley, K., Carmeliet, G., Carmeliet, P., Van Oosterwyck, H.: MOSAIC: a multiscale model of osteogenesis and sprouting angiogenesis with lateral inhibition of endothelial cells. PLoS Comput. Biol. **8**(10), e1002,724 (2012)
18. Carmeliet, P.: Angiogenesis in life, disease and medicine. Nature **438**, 932–936 (2005)
19. Carmeliet, P., Jain, R.K.: Angiogenesis in cancer and other diseases. Nature **407**, 249–257 (2000)
20. Checa, S., Rausch, M.K., Petersen, A., Kuhl, E., Duda, G.N.: The emergence of extracellular matrix mechanics and cell traction forces as important regulators of cellular self-organization. Biomech. Model. Mechanobiol. (2014)
21. Collier, J.R., Monk, N.A., Maini, P.K., Lewis, J.H.: Pattern formation by lateral inhibition with feedback: a mathematical model of delta-notch intercellular signalling. J. Theor. Biol. **183**, 429–446 (1996)
22. Coultas, L., Chawengsaksophak, K., Rossant, J.: Endothelial cells and vegf in vascular development. Nature **438**, 937–945 (2005)
23. Daub, J.T., Merks, R.M.H.: A cell-based model of extracellular-matrix-guided endothelial cell migration during angiogenesis. Bull. Math. Biol. **75**, 1377–1399 (2013)
24. Daub, J.T., Merks, R.M.H.: Cell-based computational modeling of vascular morphogenesis using tissue simulation toolkit. In: Ribatti, D. (ed.) Vascular Morphogenesis: Methods and Protocols, Methods in Molecular Biology, vol. 1214, pp. 67–127. Springer, New York, NY (2015)
25. Davis, G.E., Bayless, K.J.: An integrin and rho gtpase-dependent pinocytic vacuole mechanism controls capillary lumen formation in collagen and fibrin matrices. Microcirculation **10**(1), 27–44 (2003)
26. Dequeant, M.L., Pourquie, O.: Segmental patterning of the vertebrate embryonic axis. Nat. Rev. Genet. **9**, 370–382 (2008)
27. Enderling, H., Hlatky, L., Hahnfeldt, P.: Migration rules: tumours are conglomerates of self-metastases. Br. J. Cancer **100**(12), 1917–1925 (2009)
28. Folkman, J.: Tumour angiogenesis: therapeutic implications. New Engl. J. Medic. **285**, 1182–1186 (1971)
29. Folkman, J.: Tumor angiogenesis: a possible control point in tumor growth. Ann. Intern. Med. **82**, 96–100 (1975)
30. Forrester, J.V., Chapman, A., Kerr, C., Roberts, J., Lee, W.R., Lackie, J.M.: Bovine retinal explants cultured in collagen gels. a model system for the study of proliferative retinopathy. Arch. Ophthalmol. **108**, 415–420 (1990)
31. Gerhardt, H., Golding, M., Fruttiger, M., Ruhrberg, C., Lundkvist, A., Abramsson, A., Jeltsch, M., Mitchell, C., Alitalo, K., Shima, D., Betsholtz, C.: VEGF guides angiogenic sprouting utilizing endothelial tip cell filopodia. J. Cell Biol. **161**, 1163–1177 (2003)
32. Glazier, J.A., Balter, A., Popławski, N.J.: Magnetization to morphogenesis: a brief history of the Glazier-Graner-Hogeweg model. In: Anderson, A.R.A., Rejniak, K.A. (eds.) Single Cell-Based Models in Biology and Medicine, series Mathematics and Biosciences in Interaction, 3-28. Birkhaüser Verlag, Basel/Switzerland, pp. 79–106 (2007)
33. Glazier, J.A., Graner, F.: Simulation of the differential adhesion driven rearrangement of biological cells. Phys. Rev. E **47**(3), 2128–2154 (1993)

34. Graner, F., Glazier, J.A.: Simulation of biological cell sorting using a two-dimensional extended potts model. Phys. Rev. Lett. **69**, 2013–2016 (1992)
35. Harris, A.K., Wild, P.P., Stopak, D.D.: Silicone rubber substrata: a new wrinkle in the study of cell locomotion. Science (New York, NY) **208**(4440), 177–179 (1980)
36. Haston, W.S., Shields, J.M., Wilkinson, P.C.: The orientation of fibroblasts and neutrophils on elastic substrata. Exper. Cell Res. **146**(1), 117–126 (1983)
37. Hogeweg, P.: Cellular automata as a paradigm for ecological modeling. Appl. Math. Comput. **27**, 81–100 (1988)
38. Holtfreter, J.: Experimental studies on the development of the pronephros. Rev. Can. Biol. **3**, 220–250 (1994)
39. Jakobsson, L., Franco, C.A., Bentley, K., Collins, R.T., Ponsioen, B., Aspalter, I.M., Rosewell, I., Busse, M., Thurston, G., Medvinsky, A., Schulte-Merker, S., Gerhardt, H.: Endothelial cells dynamically compete for the tip cell position during angiogenic sprouting. Nat. Cell Biol. **12**(10), 943–953 (2010)
40. Käfer, J., Hayashi, T., Marée, A.F.M., Carthew, R.W., Graner, F.: Cell adhesion and cortex contractility determine cell patterning in the Drosophila retina. P. Natl. Acad. Sci. USA **104**(47), 18549–18554 (2007)
41. Kamei, M., Saunders, W.B., Bayless, K.J., Dye, L., Davis, G.E., M., W.B.: Endothelial tubes assemble from intracellular vacuoles in vivo. Nature **27**(442), 453–456 (2006)
42. Kleinstreuer, N., Dix, D., Rountree, M., Baker, N., Sipes, N., Reif, D., Spencer, R., Knudsen, T.: A computational model predicting disruption of blood vessel development. PLoS Comput. Biol. **9**(4), e1002,996 (2013)
43. Koch, A.E.: Angiogenesis as a target in rheumatoid arthritis. Ann. Rheum. Dis. **62**(Suppl 2), 60–67 (2003)
44. Köhn-Luque, A., de Back, W., Starruss, J., Mattiotti, A., Deutsch, A.e.a.: Early embryonic vascular patterning by matrix-mediated paracrine signalling: A mathematical model study. PLoS One **6**(9), e24,175 (2011)
45. Krieg, M., Arboleda-Estudillo, Y., Puech, P.H., Käfer, J., Graner, F., Müller, D.J., Heisenberg, C.P.: Tensile forces govern germ-layer organization in zebrafish. Nat. Cell Biol. **10**, 429–436 (2008)
46. Lemmon, C.A., Romer, L.H.: A predictive model of cell traction forces based on cell geometry. Biophys. J. **99**(9), L78–L80 (2010)
47. Lieber, T., Kidd, S., Struhl, G.: DSL-notch signaling in the drosophila brain in response to olfactory stimulation. Neuron **69**, 468–481 (2004)
48. Lo, C., Wang, H., Dembo, M., Wang, Y.L.: Cell movement is guided by the rigidity of the substrate. Biophys **79**(1), 144–152 (2000)
49. Lobov, I.B., Renard, R.A., Papadopoulos, N., Gale, N.W., Thurston, G., Yancopoulos, G.D., Wiegand, S.J.: Delta-like ligand 4 (Dll4) is induced by VEGF as a negative regulator of angiogenic sprouting. PNAS **104**, 3219–3224 (2007)
50. Lushnikov, P.M., Chen, N., Alber, M.: Macroscopic dynamics of biological cells interacting via chemotaxis and direct contact. Phys. Rev. E **78**(6), 061,904 (2008)
51. Manoussaki, D., Lubkin, S.R., Vernon, R.B., Murray, J.D.: A mechanical model for the formation of vascular networks in vitro. Acta Biotheoretica **44**(3–4), 271–282 (1996)
52. Marée, A., Grieneisen, V., Hogeweg, P.: The cellular potts model and biophysical properties of cells, tissues and morphogenesis. In: Single Cell-Based Models in Biology and Medicine pp. 107–136 (2007)
53. Marée, A.F.M., Grieneisen, V.A., Edelstein-Keshet, L.: How cells integrate complex stimuli: The effect of feedback from phosphoinositides and cell shape on cell polarization and motility. PLoS Comput. Biol. **8**(3), e1002,402 (2012)
54. Marée, A.F.M., Jilkine, A., Dawes, A., Grieneisen, V.A., Edelstein-Keshet, L.: Polarization and movement of keratocytes: a multiscale modelling approach. B. Math. Biol. **68**(5), 1169–1211 (2006)
55. Martin, A., Komada, M.R., Sane, D.C.: Abnormal angiogenesis in diabetes mellitus. Med. Res. Rev. **23**, 117–145 (2003)

56. Merks, R.M.H., Brodsky, S., Goligorksy, M., Newman, S., Glazier, J.A.: Cell elongation is key to in silico replication of in vitro vasculogenesis and subsequent remodeling. Devel. Biol. **289**, 44–54 (2006)
57. Merks, R.M.H., Glazier, J.A.: A cell-centered approach to developmental biology. Physica A **352**(1), 113–130 (2005)
58. Merks, R.M.H., Koolwijk, P.: Modeling morphogenesis in silico and in vitro: towards quantitative, predictive cell-based modeling. Math. Model. Nat. Pheno. **4**(5), 149–171 (2009)
59. Merks, R.M.H., Perryn, E.D., Shirinifard, A., Glazier, J.A.: Contact-inhibited chemotaxis in de novo and sprouting blood-vessel growth. PLoS Comput. Biol **4**(9), e1000,163 (2008)
60. Mitchell, C., Rutland, C., Walker, M., Nasir, M., Foss, A., Stewart, C., Gerhardt, H., Konerding, M., Risau, W., Drexler, H.: Unique vascular phenotypes following over-expression of individual VEGFA isoforms from the developing lens. Angiogenesis **9**(4), 209–224 (2006)
61. Ouchi, N.B., Glazier, J.A., Rieu, J.P., Upadhyaya, A., Sawada, Y.: Improving the realism of the cellular Potts model in simulations of biological cells. Phys. A Stat. Mech. Appl. **329**(3–4), 451–458 (2003)
62. Palm, M.M., Dallinga, M.G., van Dijk, E., Klaassen, I., Schlingemann, R.O., Merks, R.M.H.: Computational screening of angiogenesis model variants predicts that differential chemotaxis helps tip cells move to the sprout tip and accelerates sprouting. PLoS ONE **11**(11), e0159478 (2016)
63. Palm, M.M., Merks, R.M.H.: Vascular networks due to dynamically arrested crystalline ordering of elongated cells. Phys. Rev. E **87**, e012,725 (2013)
64. Palm, M.M., Merks, R.M.H.: Large-scale parameter studies of cell-based models of tissue morphogenesis using CompuCell 3D or Virtual Leaf. In: Tissue Morphogenesis. Methods in Molecular Biology, vol. 1189, pp. 301–322. Springer, New York (2014)
65. Peirce, S.M.: Computational and mathematical modeling of angiogenesis. Microcirculation **15**(8), 739–751 (2008)
66. Pelham, R.J., Wang, Y.L.: Cell locomotion and focal adhesions are regulated by substrate flexibility. Proc. Natl. Acad. Sci. U.S.A. **94**(25), 13661–13665 (1997)
67. Pitt-Francis, J., Pathmanathan, P., Bernabeu, M.O., Bordas, R., Cooper, J., Fletcher, A.G., Mirams, G.R., Murray, P., Osborne, J.M., Walter, A., Chapman, S.J., Garny, A., van Leeuwen, I.M.M., Maini, P.K., Rodriguez, B., Waters, S.L., Whiteley, J.P., Byrne, H.M., Gavaghan, D.J.: Chaste: A test-driven approach to software development for biological modelling. Comput. Phys. Commun. **180**(12), 2452–2471 (2009)
68. Prokopiou, S.A.: Integrative modelling of angiogenesis in the bovine corpus luteum. Ph.D. thesis, University of Nottingham (2013)
69. Prokopiou, S.A., Owen, M.R., Byrne, H.M., Ziyad, S., Domigan, C., Iruela-Arispe, M.L., Jiang, Y.: Integrative modeling of sprout formation in angiogenesis: coupling the VEGFA-Notch signaling in a dynamic stalk-tip cell selection. ArXiv e-prints (2016)
70. Reinhart-King, C.A., Dembo, M., Hammer, D.A.: Cell-cell mechanical communication through compliant substrates. Biophys. J. **95**(12), 6044–6051 (2008)
71. Riveline, D.D., Zamir, E.E., Balaban, N.Q., Schwarz, U.S., Ishizaki, T.T., Narumiya, S.S., Kam, Z.Z., Geiger, B.B., Bershadsky, A.D.: Focal contacts as mechanosensors: externally applied local mechanical force induces growth of focal contacts by an mDia1-dependent and ROCK-independent mechanism. J. Cell Biol. **153**(6), 1175–1186 (2001)
72. Roegiers, F., Jan, Y.N.: Asymmetric cell division. Curr. Opin. Cell Biol. **16**, 195–205 (2004)
73. Savill, N.J., Hogeweg, P.: Modelling morphogenesis: From single cells to crawling slugs. J. Theor. Biol. **184**, 229–235 (1997)
74. Scianna, M.: A multiscale hybrid model for pro-angiogenic calcium signals in a vascular endothelial cell. J. Math. Biol. **74**(6), 1253–1291 (2012)
75. Scianna, M., Preziosi, L.: Modeling the influence of nucleus elasticity on cell invasion in fiber networks and microchannels. J. Theor. Biol. **317**, 394–406 (2013)
76. Scott, A., Powner, M.B., Gandhi, P., Clarkin, C., Gutmann, D.H.e.a.: Astrocyte-derived vascular endothelial growth factor stabilizes vessels in the developing retinal vasculature. PLoS One **5**(7), e11,863 (2010)

77. Segel, L.A.: Computing an organism. PNAS **98**(7), 3639–3640 (2001)
78. Serini, G., Ambrosi, D., Giraudo, E., Gamba, A., Preziosi, L., Bussolino, F.: Modeling the early stages of vascular network assembly. EMBO J. **22**(8), 1771–1779 (2003)
79. Shirinifard, A., Gens, J.S., Zaitlen, B.L., Popławski, N.J., Swat, M., Glazier, J.A.: 3D multi-cell simulation of tumor growth and angiogenesis. PLoS ONE **4**(10), e7190 (2009)
80. Shirinifard, A., Glazier, J.A., Swat, M., Gens, J.S., Family, F., Jiang, Y., Grossniklaus, H.E.: Adhesion failures determine the pattern of choroidal neovascularization in the eye: a computer simulation study. PLoS Comput. Biol. **8**(5), e1002,440 (2012)
81. Sozinova, O., Jiang, Y., Kaiser, D., Alber, M.: A three-dimensional model of myxobacterial fruiting-body formation. P. Natl. Acad. Sci. USA **103**(46), 17255–17259 (2006)
82. Starruß, J., Bley, T., Søgaard-Andersen, L., Deutsch, A.: A new mechanism for collective migration in myxococcus xanthus. J. Stat. Phys. **128**(1–2), 269–286 (2007)
83. Starruß, J., De Back, W., Brusch, L., Deutsch, A.: Morpheus: a user-friendly modeling environment for multiscale and multicellular systems biology. Bioinformatics (2014)
84. Steinberg, M.S.: Reconstruction of tissues by dissociated cells. Some morphogenetic tissue movements and the sorting out of embryonic cells may have a common explanation. Science **141**, 401–408 (1963)
85. Strilić, B., Eglinger, J., Krieg, M., Zeeb, M., Axnick, J., Babál, P., Müller, D.J., Lammert, E.: Electrostatic cell-surface repulsion initiates lumen formation in developing blood vessels. Curr. Biol. **20**(22), 2003–2009 (2010)
86. Swat, M.H., Thomas, G.L., Belmonte, J.M., Shirinifard, A., Hmeljak, D., Glazier, J.A.: Multi-Scale Modeling of Tissues Using CompuCell3D, vol. 110. Elsevier Inc. (2012)
87. Szabó, A., Czirók, A.: The role of cell-cell adhesion in the formation of multicellular sprouts. Math. Model Nat. Phenom. **5**(1), 106–122 (2010)
88. Szabó, A., Erica, D.P., Czirók, A.: Network formation of tissue cells via preferential attraction to elongated structures. Phys. Rev. Lett. **98**(3), 038,102 (2007)
89. Szabó, A., Mehes, E., Kosa, E., Czirók, A.: Multicellular sprouting in vitro. Biophys. J. **95**(6), 2702–2710 (2008)
90. Tapia, J.J., D'souza, R.M.: Parallelizing the cellular potts model on graphics processing units. Comput. Phys. Commun. **182**(4), 857–865 (2011)
91. Turner, S., Sherratt, J.A.: Intercellular adhesion and cancer invasion: a discrete simulation using the extended potts model. J. Theor. Biol. **216**, 85–100 (2002)
92. Vernon, R.B., Sage, E.H.: Between molecules and morphology. Extracellular matrix and creation of vascular form. Am. J. Pathol. **147**(4), 873–883 (1995)
93. van Oers, R.F.M., Rens, E.G., LaValley, D.J., Reinhart-King, C.A., Merks, R.M.H.: Mechanical cell-matrix feedback explains pairwise and collective endothelial cell behavior in vitro. PLoS Comput Biol **10**(8), e1003,774 (2014)
94. Voss-Böhme, A.: Multi-scale modeling in morphogenesis: a critical analysis of the cellular potts model. PLoS ONE **7**(9), e42, 852 (2012)
95. Vroomans, R.M.A., Marée, A.F.M., de Boer, R.J., Beltman, J.B.: Chemotactic migration of T cells towards dendritic cells promotes the detection of rare antigens. PLoS Comput. Biol. **8**(11), e1002,763 (2012)
96. Wang, Y., Kaiser, M.S., Larson, J.D., Nasevicius, A., Clark, K.J., Wadman, S.A., Roberg-Perez, S.E., Ekker, S.C., Hackett, P.B., McGrail, M., Essner, J.J.: Moesin1 and VE-cadherin are required in endothelial cells during in vivo tubulogenesis. Development **137**, 3119–3128 (2010)
97. Winer, J.P., Oake, S., Janmey, P.A.: Non-linear elasticity of extracellular matrices enables contractile cells to communicate local position and orientation. PLoS ONE **4**(7), e6382 (2009)
98. Yu, C., Yang, B.: Parallelizing the cellular potts model on gpu and multi-core cpu: An opencl cross-platform study. In: 2014 11th International Joint Conference on Computer Science and Software Engineering (JCSSE), pp. 117–122 (2014)
99. Zajac, M., Jones, G., Glazier, J.: Simulating convergent extension by way of anisotropic differential adhesion. J. Theor. Biol. **222**(2), 247–259 (2003)
100. Zajac, M., Jones, G.L., Glazier, J.A.: Model of convergent extension in animal morphogenesis. Phys. Rev. Lett. **85**, 2022–2025 (2000)

Chapter 19
Cellular Potts Models for Interacting Cell Populations: Mathematical Foundation, Challenges, and Future Prospects

Anja Voss-Böhme

Abstract Cellular Potts models (CPMs) are extensions of asynchronous probabilistic cellular automata (PCA) developed specifically to model interacting cell populations. They constitute a modeling framework for the field of cell and tissue biology that is particularly useful when the details of intercellular interaction are essentially determined by the shape and the size of the individual cells as well as the length of the contact area between neighboring cells. In this chapter, the mathematical foundation of CPMs and their relation to PCA as well as to standard Markov chains are reviewed. On the basis of their mathematical properties, the challenges of applying CPMs for studying tissue organization from the cell-based approach are explained. In conclusion, future prospects and necessary developments are discussed from the mathematical and the modeling point of view.

19.1 Introduction

Probabilistic cellular automata (PCA) are frequently applied as stochastic models for the temporal evolution of spatially extended systems. While they are already approved for numerically studying equilibrium behavior in statistical physics by means of Markov chain Monte Carlo methods [11], they prove to be more and more beneficial in another field of application: biological systems, which are by nature open, nonequilibrium systems where noise is ubiquitous. In particular, when studying tissue organization, PCA can be utilized as cell-based models for the analysis of emergent collective behavior that results from local interactions of biological cells.

The behavior of individual biological cells is the result of complex intra- and extracellular processes at different spatial and temporal scales. The intercellular

A. Voss-Böhme (✉)
Hochschule für Technik und Wirtschaft Dresden, University of Applied Sciences, Friedrich-List-Platz 1, 01069 Dresden, Germany
e-mail: anja.voss-boehme@htw-dresden.de; anja.voss-boehme@tu-dresden.de

A. Voss-Böhme
Zentrum für Informationsdienste und Hochleistungsrechnen, Technische Universität Dresden, 01062 Dresden, Germany

P.-Y. Louis and F.R. Nardi (eds.), *Probabilistic Cellular Automata*, Emergence, Complexity and Computation 27, https://doi.org/10.1007/978-3-319-65558-1_19

coordination and regulation of these processes lead to structure, form, and function at the tissue scale. Many intracellular processes become manifest at the cellular level only as effective behavior, which remains almost unchanged for a range of microscopic variations and small perturbations in the contributing mechanisms. A *cell-based modeling approach* focuses on effective cell behaviors and interactions, such as cell migration, cell polarization, cell proliferation and death, as well as cell adhesion and alignment of cellular orientations, and studies the consequences of specific cellular interaction mechanisms for the tissue-scale behavior of the whole cell population.

To exploit PCA as cell-based models for biological tissue formation and maintenance, it is essential to describe cell migration in the PCA. If the biological cells have approximately equal sizes and shapes, the frameworks of interacting particle systems as well as that of lattice-gas cellular automata have been suggested as suitable PCA extensions for the cell-based modeling of interacting and migrating biological cell populations [5, 26]. Another approach to model tissue organization by means of PCA is to use a cellular Potts model as proposed by [7] in the context of cell sorting.

Cellular Potts models (CPMs) are asynchronous probabilistic cellular automata developed specifically to model interacting cell populations. They constitute a modeling framework for the field of cell and tissue biology that is particularly useful when the details of intercellular interaction are essentially determined by the shape and the size of the individual cells as well as the length of the contact area between neighboring cells. Synonyms for the name *cellular Potts model* are *Glazier-Graner-Hogeweg model* or *extended Potts model* [6]. In a CPM, individual cells are represented by simply connected domains of nodes. The dynamics evolves by updating one node at a time based on probabilistic rules such that one cell shrinks in size by one lattice site and a neighboring cell increases in size by occupying this site resulting in shifts of the two cells' centers of mass. These dynamics are interpreted to resemble cell surface fluctuations. The transition rules follow a *modified Metropolis algorithm* defined with respect to a *Hamiltonian* that reflects the assumed interdependence structure between the biological cells.

In this chapter, the mathematical foundation of CPMs and their relation to PCA as well as to standard Markov chains is reviewed. The challenges of applying CPMs if studying tissue organization from a cell-based perspective are explained on the basis of their mathematical properties. In conclusion, future prospects and necessary developments are discussed from the mathematical and the modeling point of view.

19.2 Mathematical Foundations

19.2.1 CPM State Space

A CPM assigns a state $\eta(x)$ from a set $W = \{0, 1, ..., \kappa\}$ to each site x of a finite set S, cp. Fig. 19.1. The set S resembles the discretized spatial domain and is chosen as

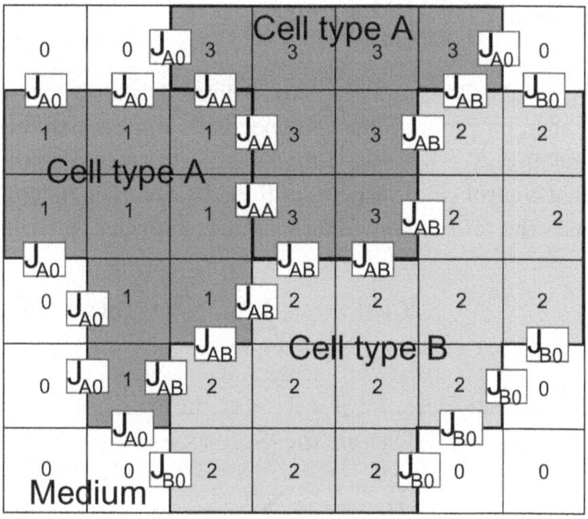

Fig. 19.1 Cell surface interaction in the cellular Potts model. Three cells, each one covering several lattice sites, interact with each other at the cell surfaces. The strengths J_{AA}, J_{AB}, J_{BB} of the interactions depend on the cell types, type A depicted in *dark gray*, type B in *light gray*. Interactions between the cells and the medium (*white*) have strength J_{A0} and J_{B0}. Interactions at the boundary of the domain are not shown

a two- or three-dimensional regular lattice. The set $W = \{0, 1, ..., \kappa\}$ contains the elementary states, called *cell indices* in the context of CPMs, where $\kappa \in \mathbb{N}$ is the absolute number of cells that are considered in the model. The state of the system as a whole is described by *configurations* $\eta \in \mathbb{X} = W^S$. Given a configuration $\eta \in \mathbb{X}$, a *cell* is the set of all points in S with the same cell index,

$$\text{cell}_w := \{x \in S : \eta(x) = w\}, \quad w \in W \backslash \{0\}.$$

The value 0 is assigned to a given node if this node is not occupied by a cell. These spatial positions are interpreted to be filled by extracellular medium. Each cell is of a certain *cell type*, which determines the migration and interaction properties of the cell. The set of all possible cell types is denoted by Λ. Denote by $\tau : W \to \Lambda$ the map that assigns each cell its cell type. A cell with index $w \in W$ has *volume*[1]

$$V_w(\eta) := \sum_{x \in S} \delta(w, \eta(x)),$$

and *surface length*

$$M_w(\eta) := \frac{1}{2} \sum_{\text{interfaces}(x,y)} \delta(w, \eta(x)).$$

The sum in the last term is taken over all *interfaces* of a given configuration η, that are all pairs $(x, y) \in S^2$ with $|x - y| = 1$ of lattice neighbors which do not belong to the same cell, that is for which $\eta(x) \neq \eta(y)$.

[1] For the Kronecker symbol δ, it holds that $\delta(u, v) = 1$ if $u = v$ and $\delta(u, v) = 0$ otherwise.

19.2.2 CPM Dynamics

A cellular Potts model is a time-discrete Markov chain with state space $\mathbb{X}$, where the transition probabilities are specified with the help of a *Hamiltonian*. The latter is a function $H : \mathbb{X} \to \mathbb{R}$ with a special structure. Usually, it is the sum of several terms that control single aspects of the cells' interdependence structure. The standard CPM uses the following two terms. First, a *surface interaction term*

$$H_s(\eta) = \sum_{\text{interfaces}(x,y)} \frac{1}{2} J(\tau(\eta(x)), \tau(\eta(y))), \quad \eta \in \mathbb{X}, \tag{19.1}$$

is specified. Here, $J : \Lambda \times \Lambda \to \mathbb{R}$, the matrix of so-called *surface energy coefficients*, is assumed to be symmetric. Second, a *volume constraint*

$$H_\upsilon(\eta) = \sum_{w \in W \setminus \{0\}} \lambda_{\tau(w)} \left(V_w(\eta) - \upsilon_{\tau(w)}\right)^2, \quad \eta \in \mathbb{X}. \tag{19.2}$$

is used. Here υ_τ, the target volume, and λ_τ, the strength of the volume constraint, are cell-type-specific parameters, $\tau \in \Lambda$. Depending on the phenomenon under investigation, further summands can be included. For instance, a constraint can be imposed on the surface length [18],

$$H_m(\eta) = \sum_{w \in W \setminus \{0\}} \alpha_{\tau(w)} \left(M_w(\eta) - m_{\tau(w)}\right)^2, \quad \eta \in \mathbb{X}. \tag{19.3}$$

Again m_τ, the target surface length, and α_τ, the strength of the surface constraint, are parameters, $\tau \in \Lambda$. Thus, the typical structure of a CPM Hamiltonian is

$$H = H_s + H_\upsilon + H_0, \tag{19.4}$$

where H_s, H_υ are given in (19.1) and (19.2) and $H_0 : \mathbb{X} \to \mathbb{R}$ is a model-specific addend. See Sect. 19.2.3 below for specific examples of H_0.

Transitions from one configuration to another follow a special rule which is called *modified Metropolis algorithm*. It requires two additional parameters. First, a *temperature* $T > 0$ is specified. It is a measure for the intensity of cell surface fluctuations in the CPM and is, therefore, interpreted as a biological analogue to thermal fluctuations in statistical physics. Second, a *transition threshold* h is introduced which helps to calibrate the dynamical behavior and to avoid artificial temporal oscillations [18, 20]. Then, the following algorithm is performed:

(0) Start with configuration η.
(1) Pick a target site $x \in S$ at random with uniform distribution on S.
(2) Pick a neighbor y of x at random with uniform distribution among all nearest lattice neighbors of x.
(3) Calculate the energetic difference

$$\Delta H_x^y := \Delta H_x^y(\eta) := H(\eta_x^y) - H(\eta) \tag{19.5}$$

of a transition $\eta \to \eta_x^y$, where

$$\eta_x^y(z) := \begin{cases} \eta(y), & \text{if } z = x, \\ \eta(z) & \text{otherwise.} \end{cases}$$

is a *trial configuration.*
(4) Accept the transition by setting $\eta := \eta_x^y$ with probability $p(\Delta H_x^y)$, or ignore the transition with probability $1 - p(\Delta H_x^y)$, where

$$p(\Delta H_x^y) = \begin{cases} 1, & \text{if } \Delta H_x^y < h, \\ \exp\left\{-\dfrac{\Delta H_x^y - h}{T}\right\}, & \text{otherwise.} \end{cases}$$

(5) Go to (1) or end the algorithm.

Under the described dynamics, only such transitions are possible where the index of at most one lattice site is changed. The new assignment to this lattice site is chosen from the cell indices of the neighboring lattice sites. Thus, one cell increases in size by one lattice node, while another one shrinks in size by one node. These dynamics are interpreted as cell surface fluctuations. They result in shifts of the CPM cells' centers of mass.

To complete the model, appropriate boundary conditions must be specified. If the influence of the boundary shall be neglected, periodic boundary conditions are used. This means that the space can be thought of as being mapped onto a torus. However, fixed boundary conditions, where the interaction between cell surfaces and confining environment is explicitly modeled, can be defined as well.

19.2.3 CPM Extensions

The CPM model formalism has been extended by several problem-specific add-ons. In general, this is done by including additional terms in the Hamiltonian, as described in (19.4). These extensions are called *energetic extensions*, since they affect the Hamiltonian, or 'energy,' describing the cells' interdependence structure. In some cases, however, additional terms are added directly to the energetic difference ΔH_x^y calculated in (19.5) thereby changing the weights for the acceptance of a proposed transition in the modified Metropolis algorithm. The latter extensions are called *kinetic extensions*, since they directly affect the transition rates. Below, several examples are outlined which demonstrate how CPM extensions can be used to adapt the CPM framework to specific applications.

As already explained above, *cell motility* emerges in the CPM implicitly from the fluctuations of the cells' centers of mass. To explicitly model physical characteristics of cell motility such as cell persistence and inertia, additional terms that constrain the cell displacement per time step can be added to the difference ΔH_x^y in (19.5) as kinetic CPM extensions. Inertia, for example, has been modeled via the term

$$\Delta H_{\text{inertia}}(t) = \sum_{w \in W \setminus \{0\}} \lambda_{\text{inertia}}(w) \left\| \overrightarrow{\text{vel}}(w,t) - \overrightarrow{\text{vel}}(w,t-\Delta t) \right\|^2, \tag{19.6}$$

where $\overrightarrow{\text{vel}}(w,t)$ denotes the instantaneous center-of-mass velocity of the cell w at time t, $\lambda_{\text{inertia}}(w)$ is a cell-specific parameter, and Δt is the length of one or more Monte Carlo time steps [2]. Notice that formula (19.6) is formulated with explicit time parameter to refer the velocity increment between two time points but it is nevertheless a temporally homogeneous rule.

Cell shapes arise in the CPM implicitly via the volume constraint. In the two-dimensional CPM, cells adopt approximately rotation-symmetrical shapes, producing a space tiling pattern comparable to epithelial tissues. Elongated cell shapes can be modeled via energetic CPM extensions by imposing a cell length constraint which renders the major axis of the ellipsoidal approximation of the cell's shape to be close to a predefined target length or ratio [29]. Rodlike cell shapes with particular stiffness have been modeled using a compartmentalized cell concept, where each cell consists of a row of standard CPM cells [21].

Chemotactic response to some field $c : S \to [0,\infty)$ of signals can be modeled in the simplest form by an energetic addend

$$H_{chemo} = \sum_{w \in W \setminus \{0\}} \lambda_{chemo}(w) \sum_{x \in \text{cell}_w} c(x)$$

to the Hamiltonian, where λ_{chemo} is a possibly cell-type-specific chemotactic response parameter [6]. If $\lambda_{chemo} < 0$, the cells prefer to move up a chemotactic gradient; for $\lambda_{chemo} > 0$, they prefer to move down such a gradient. There have been several more refined extensions to the CPM that model chemotaxis [6]. One example is the following kinetic extension used by [20] where the positions of the target spin x and the trial spin y in a proposed transition $\eta \to \eta_x^y$ are taken into account,

$$\Delta H_{chemo} = \lambda_{chemo}(\eta(y))(c(y) - c(x)). \tag{19.7}$$

The CPM can be coupled to non-lattice formalisms as well, typically to systems of differential equations. Such *hybrid approaches* enable *multi-scale modeling* in which molecular species are represented as continuous quantities, and cells are treated as discrete entities. For instance, CPM parameters pertaining to cellular properties can be under the control of ordinary differential equations representing subcellular processes such as gene regulation. Furthermore, CPM cell behavior can be linked, for instance via chemotaxis, to lattice-based reaction-diffusion systems. Such approach has been adopted, for example, to model the intracellular biochemistry that exerts influence on the protrusions and retractions in the CPM by kinetic modulation of the transition probabilities [14]. There exist extended simulation platforms on the basis of the hybrid CPM framework that provide environments for 'in silico' experiments with interacting cell populations [22, 24].

19.2.4 CPM Properties

Since the probability for the next transition in a given CPM is determined solely by the present configuration but not by the past ones, the temporal evolution of a CPM is a Markov chain [3, Definition 2.1.1]. The dynamics of the latter is completely characterized by a transition matrix $p = (p(\eta, \zeta)_{\eta,\zeta \in \mathbb{X}}$, where $p(\eta, \zeta)$ is the probability of a transition $\eta \to \zeta$ by one step of the modified Metropolis algorithm, $\eta, \zeta \in \mathbb{X}$. Thus, although CPM dynamics are customarily described in algorithmic form as above, they can be defined in conventional mathematical terms as well. To this end, *transition kernels* $p_x : \mathbb{X} \times W \to [0, 1]$ are specified for each node $x \in S$ via

$$p_x(\eta, \upsilon) = \begin{cases} \dfrac{1}{|\mathcal{N}|} \sum\limits_{z \in \mathcal{N}_1(x)} \delta(\eta(z), \upsilon) \min\left\{\exp\left\{-\dfrac{\Delta H_x^{\upsilon}(\eta) - h}{T}\right\}, 1\right\}, & \upsilon \neq \eta(x), \\ 1 - \sum_{\upsilon \in W, \upsilon \neq \eta(x)} p_x(\eta, \upsilon), & \upsilon = \eta(x), \end{cases}$$

where $\mathcal{N}(x)$ is the set of lattice neighbors of x and $|\mathcal{N}(x)|$ is their number. The value $p_x(\eta, \upsilon)$ determines the probability to change the state at node x into state υ by following the modified Metropolis algorithm given that the present configuration is η. Since CPM transitions are performed in an asynchronous way, during one transition at most on node is changed. Therefore, the transition probability $p(\eta, \zeta)$ for a transition from configuration $\eta \in \mathbb{X}$ to a configuration $\zeta \in \mathbb{X}$ under the CPM dynamics is given by

$$p(\eta, \zeta) := \begin{cases} \frac{1}{|S|} p_x(\eta, \zeta(x)), & \text{if } \eta(z) = \zeta(z), z \neq x, \text{ and } \eta(x) \neq \zeta(x); \\ \frac{1}{|S|} \sum_{x \in S} p_x(\eta, \eta(x)), & \text{if } \eta = \zeta; \\ 0, & \text{if } |\{x \in S : \eta(x) \neq \zeta(x)\}| \geq 2. \end{cases} \tag{19.8}$$

Although CPMs are usually applied to model local cell–cell contact-dependent interactions, the transition probabilities in the CPM model depend on the present configuration in a nonlocal manner. The latter means that not only the states of neighboring nodes but those of the whole lattice must be evaluated to determine the next state at a given position. This is due to the volume constraint, which is a necessary term for each CPM Hamiltonian since otherwise cell sizes cannot be constrained under the dynamics. Notice that standard PCA are defined by local transitions where the update probabilities for a given node depend only on the states of the neighboring nodes. Therefore, CPMs must be referred to as generalized PCA. The non-locality of the transition rates in CPMs hinders the extension of CPM models to infinite lattices and thus the mathematical rigorous application of spatiotemporal limit procedures for the analysis of the emergent macroscopic behavior [27]. See Sect. 19.3 for a discussion of this issue.

The *standard Metropolis algorithm* [13] and the modified Metropolis algorithm described above differ with respect to the choice of the trial cell index in steps (2) and (3). Indeed, for the standard Metropolis algorithm, the trial state υ at node x is chosen uniformly among all elements of W, while it is chosen uniformly among all states of the nearest lattice neighbors of x in the modified algorithm. Then, the energetic difference $\Delta H_x^\upsilon := H(\eta_x^\upsilon) - H(\eta)$ of the Hamiltonians evaluated for both the trial configuration $\eta_x^\upsilon \in \mathbb{X}$,

$$\eta_x^\upsilon(z) := \begin{cases} \eta(z), & z \neq x, \\ \upsilon, & z = x, \end{cases} \quad x \in S, \upsilon \in W,$$

and the present configuration $\eta \in \mathbb{X}$ is calculated, and the transition is performed as described in step (4). The modification of the Metropolis algorithm is necessary if modeling interacting cell populations to ensure that cells remain simply connected domains under the dynamics and to prevent heterogeneous nucleation [6], that is the spontaneous emergence of islands of foreign cell indices within a given cell. However, this alteration of the update algorithm has far-reaching consequences for the CPM dynamics in comparison with systems with standard Metropolis dynamics [27]. The standard Metropolis algorithm has been constructed in such a way that the corresponding Markov chain converges to a stationary behavior, which is typically the actual object of interest. In contrast, the modified Metropolis algorithm leads to a Markov chain with absorbing states. Indeed, all those configurations that consist of only one cell covering all nodes of the lattice cannot be left under the CPM dynamics. Therefore, any CPM is eventually trapped in one of these absorbing states regardless of the special structure of its Hamiltonian. Although the time until absorption can be long compared to biologically realistic timescales [27], this means nevertheless that transient Markov chain behavior is studied when using CPMs. See Sect. 19.3.1 for a detailed discussion of this issue.

19.3 Challenges of CPM Applications

The starting point of CPM modeling is the cellular level, that is, the activity and interaction of biological cells. The multitude of complex intracellular mechanisms is incorporated into the model by their effects on cell contact-dependent interaction mechanisms. The aim of this cell-based approach is to find out which mechanisms at the cellular level are characteristic for certain phenomena at the tissue level. There are two basic modeling lines that are pursued within this approach.

The first one, called *mechanistic modeling* here, focuses on simplifying the complex interaction patterns present in interacting cell populations with the aim to identify major organization principles of tissue organization. Here, the focus is on three typical questions:

(i) Identification of potential effective cellular interactions that could explain an observed behavior at the tissue level (inverse question);
(ii) Comparison of different mechanisms at the cellular level, in terms of similarities and differences in the resulting emergent behavior at the tissue level (direct question);
(iii) Assessment of the robustness and sensitivity of the mechanism with respect to small changes in specific process parameters (robustness and sensitivity question).

Question (i) is primarily studied to find a possible explanation for an observed biological phenomenon. Using studies of type (ii), it can be examined which out of several hypothetical processes is more appropriate to match the biological observations at the tissue level, or whether a particular mechanism can be excluded as explanation. Since biological behavior is impressively robust with respect to small disturbances and simultaneously sensitive to regulatory stimuli, a hypothesized mechanism that is proposed to explain a biological observation must have these two properties as well. At the same time, it must be ensured that the necessary simplifications that lead to the mathematical model are in fact of secondary importance. In this respect, question (iii) completes the analysis, although it is often particularly challenging. Notice that, for mechanistic models, the matching between model and experimental data is often sought initially on a qualitative basis since one quests primarily for mechanisms that are possible in principle.

In contrast, the second modeling line, called *phenomenological modeling* here, is targeted at reproducing observed biological behavior as exactly as possible by a mathematical simulation model. It is characterized by repeated modeling-experiment-modeling cycles to refine the model and adapt the parameters. The resulting model is often very complex and merely amenable to numerical analysis. Therefore, it provides only to some extent insight into abstract organization principles. However, it constitutes a basis for 'in silico' experimentation, that is, computer-based predictions of system behaviors under biologically realistic conditions.

19.3.1 Exploiting CPMs as Mechanistic Models for Tissue Organization

The mechanistic construction of CPMs is displayed schematically in Fig. 19.2. Central for CPM dynamics is the Hamiltonian, which codes both the considered rules of intercellular interaction and the assumed cellular characteristics. Together with potential kinetic extensions, it regulates the intensity of CPM cells' surface fluctuations. However, the actual impact of the Hamiltonian on the CPM dynamics is diluted since the Hamiltonian determines only transient behavior of CPM dynamics but not the stationary states. Sooner or later, any CPM is trapped in one of the absorbing states, independently of the special form of the Hamiltonian. Thus, the modeler's control of the CPM dynamics is constrained.

In addition, the parametrization of CPMs is intertwined. The surface fluctuations, which drive the dynamics of CPMs, determine simultaneously the actual behavior of a CPM at the cellular scale, the specifics of intercellular interaction, and the behavior at the tissue scale. Single aspects of cellular properties in the model, for instance cell shape flexibility, mean square cell displacement per unit time or cells' surface roughness, and of the intercellular interaction, like the strength of intercellular adhesion, cannot be controlled individually but are interlinked with each other. Likewise, the control of purely technical model behavior such as the maintenance of cellular cohesion, that is the property of CPM cells to span over connected, essentially convex lattice domains, is coupled indirectly to effects on biologically relevant cellular and intercellular properties like the strength of intercellular adhesion. Therefore, since it is hard to describe the causal relationships in CPM models, the utilization of CPMs as mechanistic models is limited.

However, despite the described drawbacks in mechanistic construction of CPMs, there are only a few model classes besides the CPM that allow to study interacting cell populations with non-isotropic and type-specific cell morphologies. The vertex model [10, 15, 28] and the subcellular element model [17, 19], for instance, which are both spatially continuous models, operate with a similar spatial resolution and encounter similar mechanistic challenges. Since CPMs allow to incorporate type-specific cell shapes and sizes, they are very valuable if the details of intercellular interaction are essentially determined by the shape and the size of the individual cells as well as the length of the contact area between neighboring cells [27]. In applications where all cells have approximately equal shapes and sizes, it is often sufficient to choose a model class with lower spatial resolution, such as interacting particle systems or probabilistic cellular automata with one node representing at least one cell.

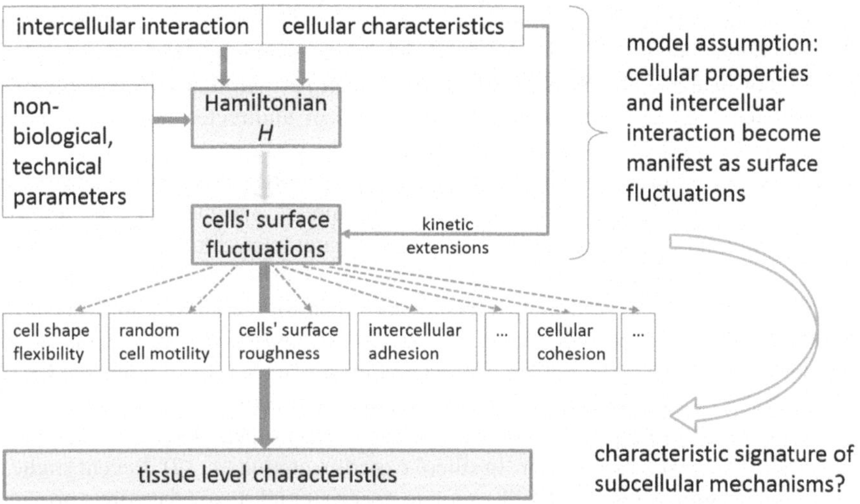

Fig. 19.2 Mechanistic structure of CPMs. Both the rules of intercellular interaction and the considered cellular characteristics are eventually coded via the Hamiltonian or by kinetic extensions into expressions that regulate the intensity of CPM cells' surface fluctuations. Additional technical parameters are integrated into the Hamiltonian to suppress phenomenologically unrealistic behavior. The actual impact of the Hamiltonian on the CPM dynamics is diluted since it determines only transient behavior of CPM dynamics but not the stationary states. The surface fluctuations drive simultaneously the actual behavior of a CPM at the cellular scale, the specifics of intercellular interaction, and the behavior at the tissue scale. Single aspects of the cellular properties in the model, for instance cell shape flexibility, the magnitude of random cell displacements or cells' surface roughness, and of the intercellular interaction, like the strength of intercellular adhesion, cannot be controlled individually but are interlinked with each other. Likewise, the control of purely technical model behavior such as the maintenance of cellular cohesion, that is the property of CPM cells to span over connected, essentially convex lattice domains, is coupled indirectly to effects on biologically relevant cellular and intercellular properties. The emerging tissue-scale behavior is solely rooted in the specified characteristics of the CPM cells' surface fluctuations and not linked directly to cellular and intracellular specifics. The figure is reprinted with modifications from [27]

19.3.2 CPM-Based Simulation Platforms

Exploiting CPMs as phenomenological models for interacting and migrating cell populations requires to define the principle structure of the Hamiltonian and of potential kinetic extensions as well as to estimate and adjust the model parameters. The structure of the Hamiltonian is derived from the cellular properties and cell–cell interactions which are considered essential for the biological process at hand. There is, however, considerable ambiguity in the choice of the model parameters. Most importantly, several CPM parameters such as the temperature T, the transition threshold h but also the strength of the volume constraint, are not directly related to biological observables, while standard biological observables do not directly translate into model parameters [27]. For instance, cell migration is modeled indirectly in standard

CPMs via cell surface fluctuations. As a result, the cell migration parameters in a CPM are determined by a multitude of model factors which interfere in a complex manner. For instance, the mean square displacement per unit time of a model cell's center of mass, a key characteristic for the extent of undirected random motility, emerges in the model from the specific choices of the surface energy coefficients J, the strength λ_τ of the volume constraints, the target number of nodes that constitute a cell, coded in the target volume V, the temperature T, and the transition threshold h but cannot be controlled directly by a model parameter. Similarly, the effective interactions between cells cannot be described by clearly identifiable parameters but emerge from the overall system's behavior. Therefore, a CPM's adjustment to hypothetical or observed biological cell behavior depends to a large extent on the modeler's experience and intuition. The state of the art consists of adapting the CPM parameters to biological data by repeated simulation sweeps over varied parameter ranges. If the CPM performs sufficiently well for certain parameter combinations, these values are taken as basis for 'in silico' experimentation [4, 12]. Recent studies automatically import into the CPM structure experimental image data, thus providing realistic blueprints for the spatial arrangement of temporarily fluctuating cells which then constitute the spatial basis for describing coupled processes at the intracellular scale, such as reaction-diffusion systems in irregular and fluctuating cellular domains [22].

19.3.3 Robustness of CPMs

The challenges concerning the robustness and sensitivity of CPMs with respect to small perturbations of the system parameters shall be illustrated using the example of a simple growth model for migrating cells. Cell proliferation can be modeled in a CPM by allowing cell size growth and by implementing cell division if a cell size threshold is exceeded [9]. Cell migration in a CPM results indirectly by shifts of the cells' center of mass caused by the cell surface fluctuations. The latter is determined by the Hamiltonian and potential kinetic extensions. As argued above, the actual cell motility in a CPM cannot be adjusted directly but arises from the interrelation of all model parameters. For given model parameters, the motility of a CPM cell can be measured by the mean square displacement (MSD) per unit time of that cell. If different CPMs with the same proliferation rate and the same MSD per unit time are compared, one might expect that the cell number increase of the total population is identical for equal initial conditions. Actually, if one starts with a few cells concentrated in the center of the spatial domain, a transition from exponential to linear growth of the total cell number is observed in all CPMs after some characteristic time span. However, the time until the transition to linear growth occurs depends strongly on the chosen parametrization of the different CPMs [8]. Thus, the model predictions are reliable in qualitative terms but quantitatively less trustable.

An understanding of the robustness and sensitivity of CPMs with respect to the system parameters is even more relevant if additional intracellular mechanisms shall be incorporated in the model. Intracellular processes usually occur on other spatial and temporal scales than cellular processes. Therefore, the effect of added intracellular processes in a CPM might be observed at the cellular scale only by slight deviations of the cellular behavior and by only quantitative alterations of tissue-scale properties. To causally attribute the observed behavior at the tissue scale to specific intracellular processes, it is essential to demonstrate that the effects of alternative CPM parameterizations are negligible in comparison with the effects of the considered intracellular processes.

19.4 Future Prospects

The CPM framework was initially developed by [7] to explore the tissue-scale consequences of the differential adhesion hypothesis [23] which holds that cell-type-dependent disparities in the expression of molecules that regulate intercellular adhesion are responsible for cell sorting. Since then, CPMs have been applied and elaborated to study a wide range of tissue-organization problems. Although a CPM's resolution is below the cellular scale, CPMs are referred to as cell-based models here. This is due to the fact that the rules for CPM dynamics are cell-based. Intracellular processes, such as cytoskeletal reorganization or spatially heterogeneous expression of certain cell surface molecules, are incorporated via their effective, cell-type-specific impact on the cell surface fluctuations at the cellular level. To extent CPMs to model also subcellular processes, it would be important to link the subcellular dynamics of cell size and cell shape adaption in the model to intracellular core processes of the considered biological system. In particular, the basic mechanisms of cytoskeletal and cell surface reorganization existent in biological cells should be reflected in the dynamics of such a model, which then actually describes mechanisms on the subcellular scale. From the modeling point of view, it remains a challenge to realistically model the collective movement of several physically connected subcellular units which potentially interact with each other.

Future developments concerning the utilization of CPMs as phenomenological models will benefit from a greater degree of standardization. For well-defined benchmark problems, where the biological conditions are stated statistically well founded and experimentally reproducible, sample CPM systems should be developed which can quantitatively reproduce these biological experiments as closely as possible. By model adjustment to biological data, standard practice concerning the handling of the following problems can be established.

- Which biological experiment shall be reproduced within which parameter ranges?
- Which degree of accordance shall be achieved with respect to which observables? Which statistical data is necessary to measure this accordance?
- Which CPM specifications are chosen for which reasons? Which alternative specifications are rejected for which reasons?

Based on several successful benchmark models, a kind of protocol for the adjustment of CPMs to new experimental situations could be developed which defines routines for proper CPM modeling.

From a mathematical point of view, two main topics are interesting. First, it is important to develop methods which relate CPMs to other model frameworks by suitable coarse graining or limiting procedures. These include moment closure methods for the derivation of related differential equation systems as well as the derivation of conditions under which CPMs can be converted into cell-based probabilistic cellular automata or interacting particle systems with one node representing one cell. The methods that are available for CPM analysis so far comprise essentially numerical simulation studies, such as [16, 18], and heuristic approximations as in [1, 25], for instance. Second, all questions concerning the robustness and sensitivity of CPMs with respect to variations in the technical as well as in the biologically interpretable model parameters are essentially mathematical questions. Here, it is crucial to distinguish, at least for some simplified, paradigmatic CPMs, which parts of the CPM dynamics are robust with respect to varying model factors and which parts react very sensible to those changes. The consideration of additional intracellular processes in CPMs, as it is done in hybrid CPM studies, is reasonable only if one is able to justify that the effects described by the intracellular dynamics dominate compared to the consequences of the necessary model abstractions and compromises.

Acknowledgements The author thanks Andreas Deutsch for discussions and comments.

References

1. Alber, M., Chen, N., Lushnikov, P.M., Newman, S.A.: Continuous macroscopic limit of a discrete stochastic model for interaction of living cells. Phys. Rev. Lett. **99**, 168102 (2007)
2. Balter, A., Merks, R.M.H., Poplawski, N.J., Swat, M., Glazier, J.A.: The Glazier-Graner-Hogeweg model: extensions, future directions, and opportunities for further study. In: Anderson, A.R.A., Chaplain, M.A.J., Rejniak, K.A. (eds.) Single Cell-Based Models in Biology and Medicine. Mathematics and Biosciences in Interaction, pp. 151–167. Springer, Berlin (2007)
3. Brémaud, P.: Markov chains, Gibbs Fields, Monte Carlo Simulation and Queues. Springer, Berlin (1999)
4. Czirok, A., Varga, K., Mehes, E., Szabo, A.: Collective cell streams in epithelial monolayers depend on cell adhesion. New J. Phys. **15**(7), 075006 (2013)
5. Deutsch, A., Dormann, S.: Cellular Automaton Modeling of Biological Pattern Formation: Characterization, Applications, and Analysis. Birkhäuser, Boston (2005)
6. Glazier, J.A., Balter, A., Poplawski, N.J.: Magnetization to morphogenesis: a brief history of the Glazier-Graner-Hogeweg model. In: Anderson, A.R.A., Chaplain, M.A.J., Rejniak, K.A.

(eds.) Single Cell-Based Models in Biology and Medicine. Mathematics and Biosciences in Interaction, pp. 79–106. Springer, Berlin (2007)
7. Glazier, J.A., Graner, F.: Simulation of the differential adhesion driven rearrangement of biological cells. Phys. Rev. E. **47**(3), 2128–2154 (1993)
8. Guzzetti, F.: Stochastic lattice-based model: a biological application for contact inhibition. Master thesis, Universita degli Studi di Milano (2012)
9. Hogeweg, P.: Evolving mechanisms of morphogenesis: on the interplay between differential adhesion and cell differentiation. J. Theor. Biol. **203**(4), 317–333 (2000)
10. Landsberg, K.P., Farhadifar, R., Ranft, J., Umetsu, D., Widmann, T.J., Bittig, T., Said, A., Jülicher, F., Dahmann, C.: Increased cell bond tension governs cell sorting at the drosophila anteroposterior compartment boundary. Curr. Biol. **19**(22), 1950–1955 (2009)
11. Lebowitz, J.L., Maes, C., Speer, E.R.: Statistical mechanics of probabilistic cellular automata. J. Stat. Phys. **59**(1–2), 117–170 (1990)
12. Li, J.F., Lowengrub, J.: The effects of cell compressibility, motility and contact inhibition on the growth of tumor cell clusters using the cellular potts model. J. Theor. Biol. **343**, 79–91 (2014)
13. Madras, N.N.: Lectures on Monte Carlo methods. Fields Institute Monographs. American Mathematical Society (2002)
14. Marée, A.F.M., Jilkine, A., Dawes, A., Grieneisen, V.A., Edelstein-Keshet, L.: Polarization and movement of keratocytes: a multiscale modelling approach. Bull. Math. Biol. **68**(5), 1169–1211 (2006)
15. Nagai, T., Kawasaki, K., Nakamura, K.: Vertex dynamics of two-dimensional cellular patterns. J. Phys. Soc. Jpn **57**(7), 2221–2224 (1988)
16. Nakajima, A., Ishihara, S.: Kinetics of the cellular Potts model revisited. New J. Phys. **13**(3), 033035 (2011)
17. Newman, T.J.: Modeling multicellular systems using subcellular elements. Math. Biosci. Eng. **2**(3), 611–622 (2005)
18. Ouchi, N.B., Glazier, J.A., Rieu, J., Upadhyaya, A., Sawada, Y.: Improving the realism of the cellular Potts model in simulations of biological cells. Physica A **329**(3–4), 451–458 (2003)
19. Sandersius, S.A., Weijer, C.J., Newman, T.J.: Emergent cell and tissue dynamics from subcellular modeling of active biomechanical processes. Phys. Biol. **8**(4), 045007 (2011)
20. Savill, N.J., Hogeweg, P.: Modelling morphogenesis: from single cells to crawling slugs. J. Theor. Biol. **184**(3), 229–235 (1997)
21. Starruss, J., Bley, T., Sogaard-Andersen, L., Deutsch, A.: A new mechanism for collective migration in Myxococcus xanthus. J. Stat. Phys. **128**(1–2), 269–286 (2007)
22. Starruß, J., de Back, W., Brusch, L., Deutsch, A.: Morpheus: a user-friendly modeling environment for multiscale and multicellular systems biology. Bioinformatics **30**(9), 1331–1332 (2014)
23. Steinberg, M.S.: Reconstruction of tissues by dissociated cells. Science **141**, 401–408 (1963)
24. Swat, M.H., Thomas, G.L., Belmonte, J.M., Shirinifard, A., Hmeljak, D., Glazier, J.A.: Multiscale modeling of tissues using CompuCell3d. In: Asthagiri, A.R., Arkin, A.P. (eds.) Computational Methods in Cell Biology. Methods in Cell Biology, pp. 325–366. Academic Press, Dublin (2012)
25. Turner, S., Sherratt, J.A., Painter, K.J., Savill, N.J.: From a discrete to a continuous model of biological cell movement. Phys. Rev. E **69**(2), 021910 (2004)
26. Voss-Böhme, A., Deutsch, A.: On the cellular basis of cell sorting kinetics. J. Theor. Biol. **263**(4), 419–436 (2010)
27. Voss-Böhme, A.: Multi-scale modeling in morphogenesis: a critical analysis of the cellular Potts model. PLoS ONE **7**(9), e42852 (2012)
28. Weliky, M., Oster, G.: The mechanical basis of cell rearrangement. Development **109**(2), 373–386 (1990)
29. Zajac, M., Jones, G.L., Glazier, J.A.: Simulating convergent extension by way of anisotropic differential adhesion. J. Theor. Biol. **222**, 247–259 (2003)

Chapter 20
Cellular Automata for Clouds and Convection

Daan Crommelin

Abstract Numerical models of the global atmosphere have spatial resolutions that are much too coarse to resolve clouds and convection processes explicitly. Because these processes play an important role in the atmosphere and climate system, they are included in numerical models by means of simplified representations, so-called parameterizations. Traditional parameterization schemes for atmospheric convection are deterministic. To overcome the limitations of these deterministic schemes, stochastic parameterizations are being developed. The use of probabilistic cellular automata (PCA) for this application is very new and can provide a way to generate spatial patterns of convection as observed in the atmosphere. It is approached from two directions, both briefly reviewed here. In one approach, convection and other sub-grid-scale processes are represented with deterministic CA. In recent work, this is extended to PCA. In the other approach, convection is represented by means of discrete stochastic processes (finite state Markov chains) on a lattice. In most studies in this direction, there is no direct coupling between neighboring lattice nodes, however recently such couplings are considered as well. To illustrate the concept of parameterization, a frequently used test model (the L96 model) is discussed as well in this chapter. Parameterization of atmospheric convection and clouds with PCA has several interesting mathematical aspects. One is the interactive (two-way) coupling of the PCA to a partial differential equation for large-scale atmospheric flow. The state of the PCA couples to the time evolution of the flow, and in turn the PCA rules (transition probabilities) depend on the flow state. Furthermore, for convection it is natural to consider N-state PCAs with $N > 2$ rather than a binary ($N = 2$) PCA. Finally, statistical inference can be a fruitful approach to construct the PCA rules or transition probabilities for convection. The PCA dependence on the time-evolving atmospheric flow and the large number of configurations for PCAs with $N > 2$ provide interesting challenges for such inference.

D. Crommelin (✉)
Centrum Wiskunde & Informatica (CWI), Amsterdam, The Netherlands
e-mail: Daan.Crommelin@cwi.nl

D. Crommelin
Korteweg-de Vries Institute for Mathematics, University of Amsterdam, Amsterdam, The Netherlands

P.-Y. Louis and F.R. Nardi (eds.), *Probabilistic Cellular Automata*, Emergence, Complexity and Computation 27, https://doi.org/10.1007/978-3-319-65558-1_20

20.1 Introduction

The representation of clouds and convection processes in numerical models of climate and atmosphere is of great importance. Atmospheric convection is the vertical motion of moist air and is a key element in the transportation of moisture through the atmosphere and in the hydrological cycle of the climate system. If water vapor in rising air condensates, the resulting microscopic water droplets form clouds. Further thermodynamical and physical processes such as evaporation, freezing, and precipitation add to the complexity and richness of convection and cloud dynamics. The interaction of clouds with incoming solar radiation and outgoing infrared radiation (e.g., reflection) is important in the context of climate change through the mechanism of the so-called cloud-climate feedback [1].

Despite their importance, the spatial resolutions of numerical models for climate and weather prediction are too coarse to resolve clouds and convection processes explicitly [2, 3]. This is due to computational limitations: current state-of-the-art global (i.e., covering the entire earth) operational weather forecasting models can afford spatial (horizontal) resolutions on the order of 10 km, whereas the atmospheric components of global climate models have even coarser resolutions (50–100 km) because they are used for simulations over much longer time intervals. The consequence is that clouds and convection must be represented in a simplified way in these global numerical models.

In atmosphere–ocean science, such simplified representations are known under the name *parameterizations*. The state of the atmosphere that can be resolved by the global numerical model is given as input to a parametrization module, which returns a contribution from convection to the overall rate of change of the model atmosphere. Let $\Psi(x, y, z, t)$ denote the state of the atmosphere at the geographical location (x, y, z) at time t. Typically, x stands for longitude, y for latitude, and z for elevation above the earth surface. In the most commonly used models, the state Ψ consists of five variables: wind velocities in three directions, temperature, and moisture. For ease of exposition, we assume that the time evolution of Ψ is governed by a nonlinear partial differential equation (PDE) (in practice, there are additional algebraic equations):

$$\frac{\partial}{\partial t}\Psi = \mathcal{N}(\Psi, \nabla\Psi) + R \tag{20.1}$$

This nonlinear PDE is derived from the Navier–Stokes equation. The variable $R(x, y, z, t)$ denotes the contribution from unresolved physical processes such as convection. Thus, it is assumed that the nonlinear differential operator $\mathcal{N}(\Psi, \nabla\Psi)$ accounts for physical and dynamical processes that can be adequately resolved in the global numerical model. As mentioned, the contributions from processes that cannot be resolved in the numerical model are collected in R. In the rest of this chapter, we will focus on convection, although in practice other unresolved processes are also parameterized in global models (e.g., atmospheric gravity waves, interactions with the underlying land or ocean surface, ...).

In order to close the system, a model for R is required. Traditionally, parameterizations are set up in a deterministic fashion, so that R is effectively a function of Ψ, without any randomness or uncertainty involved. Furthermore, it is common practice to assume that R is determined by Ψ locally in x and y (but not in z). By this we mean the following: let (x_i, y_j) be the (x, y) coordinates of the node (i, j) of the horizontal grid (or lattice) of the numerical model. We define $R_{i,j}(z, t) := R(x_i, y_j, z, t)$ and similarly for $\Psi_{i,j}(z, t)$. The "locality" assumption means that $R_{i,j} = f(\Psi_{i,j})$, i.e., R at node (i, j) is determined by Ψ at the same node (and at the same time), but not by Ψ or R at other nodes. The assumption does not involve the vertical coordinate z; the full vertical profile of $\Psi_{i,j}$ determines the full vertical profile of $R_{i,j}$. For convection, vertical nonlocal effects can be important.

Traditional convection parameterization schemes are based on physical reasoning and intuition, and they are effectively deterministic mappings $\Psi_{i,j} \mapsto R_{i,j}$ (although they are usually not formulated in such explicit manner). To overcome the limitations of these traditional schemes, stochastic parameterization schemes started to receive a lot of attention in the last 10–15 years. In these schemes, the deterministic mapping from $\Psi_{i,j}$ to $R_{i,j}$ is effectively replaced by a probabilistic one. This reflects the uncertainty about sub-grid-scale processes that is inevitable in numerical models with finite resolution

Although much work has been done on developing stochastic convection parameterization schemes in the last 10–15 years, a still outstanding challenge is how such schemes can generate realistic spatial patterns for convection, with appropriate spatial correlations. The "locality" assumption discussed above translates into conditional independence of R at different grid nodes, e.g., $R_{i,j}|\Psi_{i,j}$ and $R_{i,j+1}|\Psi_{i,j+1}$ are assumed to be uncorrelated. This is a limitation, because convection, although it is a physical process at small spatial scales, can organize into larger-scale structures (sometimes dubbed meso-scale structures), with clusters (and clusters-of-clusters) of convective elements spreading out over multiple horizontal grid nodes. Such structures are difficult to capture with parameterization schemes operating under the locality assumption [4]. Cellular automata (CA) can provide a way to generate these spatial patterns.

In this chapter, we discuss the relevance and prospects of representing clouds and convection processes in atmosphere models using probabilistic cellular automata (PCA). The use of PCA for this application is very new and is approached from two different angles. In one line of research, discussed in Sect. 20.2, convection and other sub-grid-scale processes are represented with the help of deterministic CA. In recent work, this is extended to include PCA. In the other approach, reviewed in Sect. 20.3, convection is represented by means of discrete stochastic processes (finite state Markov chains) on a horizontal lattice. In most studies in this direction, there is no direct coupling between neighboring lattice nodes, however recently such couplings are considered as well, see Sect. 20.5. To clarify these ideas, in Sect. 20.4 a test model is presented that is often used for designing and testing new methods for sub-grid modeling. Furthermore, in Sect. 20.6 it is discussed how statistical inference can contribute to determine the rules (cell transition probabilities) of a PCA for convection.

20.2 Convection Parameterization with Cellular Automata

The proposal to use cellular automata (CA) for parameterizing the feedback from unresolved scales in numerical models of the atmosphere goes back at least to the late 1990s [5, 6]. The idea was taken up for the purpose of parameterizing the so-called backscatter of kinetic energy from unresolved scales [7–9]. A CA is used to generate dynamically evolving spatial patterns that determine patterns of kinetic energy input from unresolved scales. More specifically, if R in (20.1) is a kinetic energy source term, it is modeled as $R_{i,j}(z,t) = K(\Psi_{i,j}(z,t))\, S_{i,j}(t)$, i.e., as the product of an appropriate function K of Ψ and a time-evolving pattern S generated by a CA (see e.g., [7]).

The CA in these studies is a deterministic, synchronous CA, with a layer of memory (or history) added to it; a cell that "comes to life" has multiple lives $L_{\max}$ (in the above-mentioned studies, $L_{\max} = 32$). Each time a cell does not meet the rules for survival, it loses one of its lives. Only neighboring cells that have the maximum amount of lives are relevant for determining whether a cell comes to life or survives (see [8] for more details). To be more precise, let us denote by $L_{i,j}(t)$ the number of lives of a cell with lattice index (i, j) at time t, so $0 \leq L_{i,j} \leq L_{\max}$. If $L_{i,j} = L_{\max}$, cell (i, j) is called "fertile". Let $M_{i,j}(t)$ be the number of fertile cells in the Moore neighborhood of (i, j), excluding (i, j) itself. Clearly, $0 \leq M_{i,j} \leq 8$. If $L_{i,j}(t) = 0$ and $M_{i,j}(t) \in \{2, 3\}$, then $L_{i,j}(t+1) = L_{\max}$ ("birth"). If $L_{i,j}(t) > 0$ and $M_{i,j}(t) \in \{3, 4, 5\}$ then $L_{i,j}(t+1) = L_{i,j}(t)$ ("survival"). In all other cases, $L_{i,j}(t+1) = \max(L_{i,j}(t) - 1, 0)$ ("death").

The "raw" CA state or pattern at time t is determined by the pattern of the $L_{i,j}(t)$. To arrive at the pattern $S(t)$, the raw CA pattern is spatially smoothed. An example is shown in Fig. 20.1. Note that the rules of this CA are deterministic. The rules were chosen heuristically, not inferred or derived in a rigorous way. Also, the CA evolves independently of the large-scale atmospheric state, i.e., there is no coupling of Ψ back to S.

In several studies (e.g., [10–12]), the use of CA for convection parameterization is explored, with a setup quite similar to the kinetic energy backscatter CA schemes mentioned above. The CA with memory added (as described above) is the starting point, however in recent papers feedback from Ψ to S has been introduced by making the CA rules also dependent on Ψ [10, 11]. Furthermore, the CA pattern can be advected (transported horizontally) by the large-scale flow determined by Ψ. As already mentioned, the CA rules are deterministic, although elements of randomness are introduced in several of these papers by initializing the CA randomly or by adding randomly located live cells at each time step. A probabilistic version of the CA rules has also been considered [11]: if a cell meets the rule for either birth or survival, it will come to (or remain in) life with a probability smaller than one (with the deterministic rule, the probability equals one). This probability can be fixed or made dependent on advection (i.e., on Ψ). It is reported [11] that the probabilistic, advection-dependent rule generates patterns that look more like convection than those generated with the deterministic rule.

Fig. 20.1 Example of a pattern generated by the deterministic, synchronous CA with memory as used in e.g., [7]. The *top panel* shows the raw pattern (number of lives) in a 720×360 CA. The *bottom panel* shows the pattern after coarse-graining, smoothing and normalizing

20.3 Markov Chains on a Lattice

The CA for convection parameterization discussed in the previous section were primarily deterministic. Although some recent studies consider probabilistic extensions, the starting point is a deterministic CA. In a different line of research, the problem is approached from almost the opposite perspective: convection is parameterized using discrete stochastic processes (finite state Markov chains) on a lattice, but mostly without direct interaction or coupling between the Markov chains at neighboring lattice nodes. Thus, in this approach the starting point is stochastic and relies on the locality assumption discussed earlier.

The model contributions $R_{i,j}$ due to convection are functions of the vertical coordinate z, hence they live in an infinite-dimensional space. To make stochastic modeling of $R_{i,j}$ more tractable, in a number of studies this function space is effectively discretized, so that the time evolution of $R_{i,j}$ can be modeled as a finite state Markov chain (e.g., [13–17]). In these studies, a small number of convective states or cloud states are chosen, and the Markov chain determines the transition probabilities of switching between the states. To reflect the dependence on $\Psi_{i,j}$, the transition probabilities are conditional on (specific functions of) $\Psi_{i,j}$.

An important quantity for convection parameterization is the so-called *convective area fraction* (CAF). Every horizontal grid node (i, j) of the global numerical model has a surface area associated with it (roughly equal to $\Delta x_i \, \Delta y_j$, with $\Delta x_i = x_{i+1} - x_i$

and $\Delta y_j = y_{j+1} - y_j$). The CAF is the fraction of this area that is in a convective state. In conventional, deterministic parameterizations, the CAF is fixed (e.g., at 0.03). Many stochastic approaches focus on stochastic modeling of the CAF; the resulting CAF is then used as input to calculate $R_{i,j}$ in much the same way as it is done in deterministic schemes (making use of so-called mass flux parameterization methods).

The convective/cloud states can be defined on a "microscopic" lattice, with many micro-lattice nodes pertaining to a single node of the global model grid (the "macroscopic" lattice) [13, 15, 17, 18]. The CAF for the macro-node (i, j) is given by the fraction of micro-nodes, associated with (i, j), that are in an appropriately convective state. In the simplest form, there are two possible states at each node (convective and non-convective); a more complicated setup involves more than two states. For example, in the multicloud model from [13] there are four states (three cloud states, one of which is strongly convective and a clear sky state), and in [15] this multicloud model is extended to five states.

To formalize this, let (k, l) be the node index of the micro-grid. For every macro-node (i, j), k and l range from 1 to K and L, respectively. We define by $b(i, j, k, l)$ the state at node (k, l) of the micro-grid associated with macro-node (i, j). This state takes values in a finite set of states, $b(i, j, k, l) \in S := \{c_1, \ldots, c_N\}$, where $c_1, c_2, \ldots$ denote convective/cloud states. We denote by $\sigma_n(i, j)$ the area fractions of the various states:

$$\sigma_n(i, j) = \frac{1}{K\,L} \sum_{k=1}^{K} \sum_{l=1}^{L} \mathbf{1}\{b(i, j, k, l) = c_n\} \tag{20.2}$$

where $\mathbf{1}\{.\}$ is the indicator function. Suppose that c_N is a strongly convective state, the only one that contributes to the CAF. Then we simply have that the CAF for macro-node (i, j) is given by $\sigma_N(i, j)$. A mass flux parameterization scheme then takes $\sigma_N(i, j)$ as input, together with $\Psi_{i,j}$, to determine $R_{i,j}$. In this approach, the only information about the sub-grid-scale convection processes that enters the global numerical model (i.e., the macro-model) is $\sigma_N(i, j)$. We note that the states $b(i, j, k, l)$ evolve in time, in accordance with the Markov chain transition probabilities that are conditioned on $\Psi_{i,j}$. As a consequence, $\sigma_N(i, j)$ also changes in time.

In an alternative setup, it is the CAF itself that is modeled with a Markov chain [16]. The CAF is discretized in multiples of 0.01 (including zero), and there is no micro-lattice involved. Furthermore, in [14] there is not even a CAF involved. The $R_{i,j}$ themselves are discretized, using a clustering algorithm, hence the states of the Markov chain correspond to entire functions of the vertical coordinate.

The transition probabilities for the Markov chain can be obtained in various ways. One approach is to rely on physical reasoning, as in [13, 17]. An alternative approach is to make use of available datasets on convection (stemming from high-resolution models or from observations) to obtain transition probabilities through statistical inference [14–16, 18, 19].

20.4 Test Case: The L96 Model

The Lorenz '96 (L96) model [20] is an idealized model of atmospheric flow. It is used frequently as a test bed for developing new ideas and algorithms for parameterization and predictability, e.g., the Markov chain approach discussed in the previous section [19]. The model consists of a set of coupled nonlinear ordinary differential equations (ODEs), and although it was not derived from a PDE it is commonly interpreted as having spatial extent, describing an atmosphere-like dynamical system on a one-dimensional lattice of constant latitude. The model ODEs are as follows:

$$\frac{d}{dt} X_i = X_{i-1}(X_{i+1} - X_{i-2}) - X_i + F + R_i \,, \tag{20.3a}$$

$$\frac{d}{dt} Y_{i,k} = \frac{1}{\varepsilon}\big(Y_{i,k+1}(Y_{i,k-1} - Y_{i,k+2}) - Y_{i,k} + h_y X_i\big) \,, \tag{20.3b}$$

$$R_i = \frac{h_x}{K} \sum_{k=1}^{K} Y_{i,k} \,. \tag{20.3c}$$

The variables $X_i(t)$ are interpreted as describing the system at large spatial scales, the $Y_{i,k}(t)$ as variables of small-scale processes. The $i \in \{1, \ldots, I\}$ and $k \in \{1, \ldots, K\}$ are interpreted as one-dimensional lattice indices (i on a macro-lattice, k on a micro-lattice). The lattice has periodic boundary conditions, so that $X_i = X_{i+I}$, $Y_{i,k} = Y_{i+I,k}$ and $Y_{i,k+K} = Y_{i+1,k}$ (we note that the use of indices here differs somewhat from the convention used in e.g., [19, 20], this is done for the sake of consistency with other sections in this chapter).

The $Y_{i,k}$ evolve on a faster timescale than the X_k. The time scale separation is controlled by the parameter ε: with $\varepsilon \ll 1$ there is large scale separation, with $\varepsilon \approx 1$ there is no scale separation. Other parameters in (20.3) are the coupling strengths h_x and h_y and the forcing F. For further discussion and interpretation of these parameters, we refer to [19, 20] and the references therein. In what follows we use $\varepsilon = 0.5$, $h_x = -1$, $h_y = 1$, and $F = 10$, as in [19]. Finally, the total number of variables (X_i and $Y_{i,k}$) is $I + I \times K$, examples of settings are $I = 36$, $K = 10$ [20] and $I = 18$, $K = 20$ [19].

The goal of sub-grid-scale parameterization, in the context of the L96 model, is to simulate the dynamics of X as generated by (20.3) as well as possible without having to simulate Y explicitly (here, X denotes the vector $(X_1, \ldots, X_K)$ and similarly for Y). The analogy with realistic atmosphere models is that in such models it is computationally much too expensive to resolve all relevant small-scale variables (Y), it is only feasible to resolve the large-scale variables (X). For the L96 model, this requires a parameterization of the $R_i(t)$ in terms of the $X_i(t)$. The $R_i(t)$ are the quantities that provide the feedback from the small scales to the large scales, see (20.3). The parameterization (or model) for R together with the ODEs for X in (20.3a) form a system with $2\,I$ degrees of freedom, a large reduction compared to the $(K+1)I$ degrees of freedom in the full L96 model (20.3).

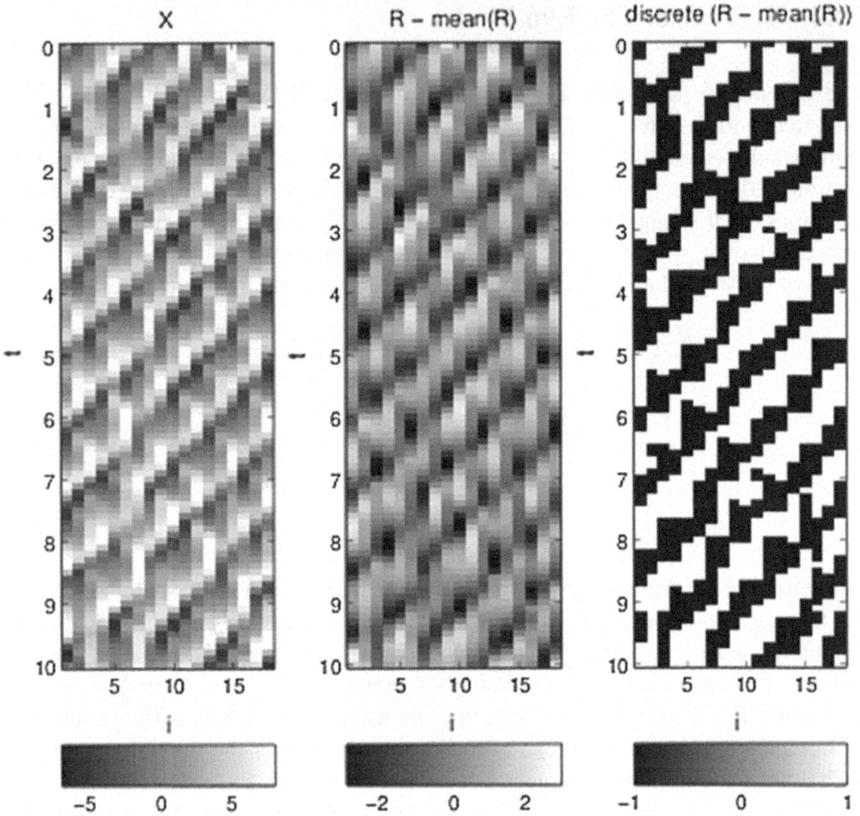

Fig. 20.2 Example timeseries generated by the full L96 model (20.3) with parameter settings from [19]. The *left panel* shows $X(t)$, the *middle panel* shows $R(t)$ with its mean subtracted. In the *right panel*, $R(t)$-mean(R) is discretized into two states: -1 for negative values, $+1$ for positive values. In all panels, time runs from *top* to *bottom* in increments of 0.1 and the spatial index i is on the horizontal axis

Modeling R is far from straightforward. The dynamics of Y, and hence of R, is dependent on the state of X, see (20.3b). Also, Y has its own chaotic dynamics and is not simply "slaved" to X. In case of large scale separation, i.e., $\varepsilon \ll 1$, asymptotic methods such as averaging and homogenization [21] may be used to derive a reduced model for X, however in realistic atmosphere models there is no clear scale separation between resolved-scale flow and convection.

Figure 20.2 shows an example of data (timeseries) generated by numerical integration of the full L96 model (20.3), i.e., generated by simulating Y as well as X. The parameter settings $(I, K, \varepsilon, h_x, h_y, F) = (18, 20, 0.5, -1, 1, 10)$ are those from [19]. The left panel shows the time evolution of the vector $X(t)$, the middle panel that of $R(t)$ with its mean subtracted. A simple two-state discretization, in which each $R_i(t)$-mean(R) is mapped to either $+1$ or -1 depending on its sign, is shown in the right panel.

Although the behavior shown in Fig. 20.2 is chaotic, wave-like structures can be seen to travel through the spatial domain, not only in X but also in R. A parameterization should capture these noisy space-time patterns of R and their dependence on the patterns of X. Under the locality assumption discussed earlier, a stochastic parameterization for R consists of I copies of a scalar stochastic process for R_i conditioned on X_i. In the Markov chain approach, the parameterized R_i can take on only a finite number of values.

20.5 From Markov Chains to PCA

The conditional Markov chain (CMC) lattice models described in the previous sections do not involve direct interactions between Markov chains at neighboring lattice nodes. However, the CMC states at neighboring nodes are not independent, due to the coupling to $\Psi_{i,j}$. In the case of a micro-lattice, the chains governing the time evolution of $b(i, j, k, l)$ and $b(i, j, k \pm 1, l \pm 1)$ are conditioned on the same $\Psi_{i,j}$. If the Markov chains are only defined at the level of the macro-lattice, then $b(i, j)$ and $b(i \pm 1, j \pm 1)$ are correlated because $\Psi_{i,j}$ and $\Psi_{i\pm1,j\pm1}$ are coupled through the PDE model (20.1).

Notwithstanding these indirect couplings between Markov chains at neighboring lattice nodes, there may be reason to couple them more directly. For example, it was demonstrated [15] that such direct coupling can strongly enhance the variance of the area fractions $\sigma_n(t)$. Recently, a detailed investigation into the limitations of the locality assumption for capturing large-scale coherence in convection modeling was presented [4].

Let $\{k, l\}$ denote the neighborhood of the micro-lattice node (k, l) (e.g., the Moore neighborhood or the von Neumann neighborhood). If we generalize the CMC model to include dependencies on the neighborhood, while also retaining the (local) dependence on the macrostate Ψ, we arrive at a model characterized by the following transition probabilities for the cloud states $b(i, j, k, l, t)$:

$$b(i, j, k, l, t + \Delta t) \mid b(i, j, k, l, t),\ \Psi(i, j, t),\ b(i, j, \{k, l\}, t) \tag{20.4}$$

Clearly, this "conditional PCA" is a rich model with many possible scenarios. There are N possible states for $b(i, j, k, l, t)$, so that with the Moore neighborhood there are N^9 different configurations for $b(i, j, k, l, t)$ and $b(i, j, \{k, l\}, t)$ together in case of a two-dimensional lattice. The dependence on $\Psi(i, j, t)$ makes the number of possible configurations even higher. To control these possibilities, it is nearly inevitable to impose certain structures on the model. How to do this is largely an open question. In [15], some ad-hoc choices were made to control the number of parameters that determine the transition probabilities for the conditional PCA. Controlling the parameters in a more systematic way is still a challenge.

The CMC lattice model from [13] was recently generalized to include interactions between neighboring cells on the micro-lattice [22]. The transition probabilities (PCA rules) are designed and motivated from physical intuition, similar to [13]. Energies (or interaction potentials) are assigned to all possible combinations of two neighboring cell states. For a given configuration of the lattice model, the sum of the potentials of all interactions present in that configuration determines a Hamiltonian energy. The transition rates for the individual cells are functions of this Hamiltonian. As the system state (configuration) evolves over time, so do the Hamiltonian and the transition rates. Only nearest neighbors are taken into account (i.e., Moore or von Neumann neighborhood), as it is argued that these are physically the most relevant [22].

20.6 Statistical Inference for PCA

To obtain the rules or transition probabilities of a PCA for clouds and convection, several of the papers mentioned previously rely on physical intuition and heuristics, e.g., [11, 22]. An alternative approach is to infer these rules from available datasets. Such data can come from two sources: observations/measurements of the real physical atmosphere and numerical simulations with high-resolution models. Regarding the latter, we note that it is possible to do fairly realistic simulations of convection processes (although the detailed physics of e.g., the involved phase changes (ice –water vapor– liquid water) and ice microphysics are still challenging). However, these simulations require extremely high model resolution, so that they are restricted in practice to limited spatial domains and short time intervals (e.g., 24 h on a 100 km by 100 km horizontal domain). It will be many years before such high-resolution simulations become feasible for the global atmosphere on climate timescales (years, decades, and longer). Hence, the need to parameterize convection will persist for many more years.

With a dataset of sufficient spatial and temporal resolution, a pre-processing step is needed to assign a discrete state to all the lattice nodes at every time step. In previous sections, it was discussed how clouds and convection can be modeled by defining a few cloud states (e.g., deep, stratiform, clear sky). The step of classifying the states at the lattice nodes, i.e., of deciding in what cloud or convective state a cell is, is nontrivial. However, we will not discuss it here further as it is primarily a matter of physical insight.

Once the space-time patterns of the discrete states are extracted from the dataset, one can attempt to fit a PCA to these patterns by means of statistical inference of the PCA rules or transition probabilities. There are two aspects to this inference task: selecting the neighborhood and identifying the rules. In previous sections, we have mainly focused on neighborhoods that are one step deep in space and time (e.g., the neighborhood for $R_i(t + \Delta t)$ consisting of $\{R_{i-1}(t), R_i(t), R_{i+1}(t)\}$ in case of a one-dimensional lattice). Larger neighborhoods, either in space or time, may give better results but can also lead to over-fitting, hence selecting the neighborhood is part of the inference problem. Furthermore, inferring the PCA rules with a given neighborhood is equally nontrivial.

Various methods have been developed for neighborhood selection and rule identification, see e.g., [23–26]. The focus in these studies is mostly on binary systems, i.e., CA with two states. However, for PCA modeling of convection more than two states are typically used, as discussed in previous sections. A method for N-state systems proposed in [27] has not yet been used for convection PCA identification.

A major complication for inferring a PCA for convection (or other sub-grid processes) from data is the influence of the large-scale state. As already discussed, Ψ and R in (20.1) are two-way coupled, so the behavior of R is dependent on Ψ. How to infer a PCA for R that is dependent on Ψ, with R and Ψ both evolving in time, is an open question and a challenging research topic. It is assumed here that time series data of both R and Ψ are available to infer the PCA. It may be fruitful to consider Ψ as a time-dependent covariate for R, although strictly speaking, Ψ does not evolve independently of R, see (20.1).

In the Markov chain approach discussed in Sect. 20.3, the dependence on the large-scale state is considered in several papers. In [19], the inference of transition probabilities for R conditional on X from L96 model data is carried out through a straightforward extension of maximum likelihood estimation. This procedure is also used in e.g., [15, 16, 18]. In [28] a, Bayesian approach is proposed to estimate parameters of the multicloud model from [13]. It is mentioned in [28] that this approach can be extended to the multicloud model with neighbor interaction as proposed in [22].

20.7 Summary and Conclusion

Modeling of atmospheric convection and clouds is an emerging application for PCA that entails several interesting mathematical challenges. An important aspect is the interactive (two-way) coupling to a PDE for large-scale atmospheric flow, see equation (20.1). The state of the PCA for R couples to the time evolution of the large-scale flow state Ψ through (20.1), and at the same time the PCA rules (transition probabilities) depend on Ψ. Furthermore, in most studies so far where convection is modeled as a discrete process, more than two states are used, so it is natural to consider N-state PCAs with $N > 2$ rather than a binary ($N = 2$) PCA.

As discussed in Sects. 20.2 and 20.3, PCA modeling for convection emerges from two different research directions. The use of deterministic CA for modeling convection and other sub-grid processes has been pursued for more than ten years, see Sect. 20.2, however the extension of this approach to stochastic modeling (i.e., to PCA) is quite recent (e.g., [11]). Markov chain lattice models for convection, discussed in Sect. 20.3 have also been studied for a while. These models are stochastic from the outset, but they usually do not include interactions between neighboring cells. Such interactions were added recently [22].

Deriving the PCA rules or transition probabilities from first principles is very challenging for convection. Besides physical intuition and heuristics, statistical inference can be a fruitful approach to construct these rules. A major challenge for inference is the fact that a PCA for convection should be dependent on the large-scale state Ψ. Some work has been done to include this dependence in the Markov chain lattice models, but the generalization to PCA has hardly been explored yet.

In Sect. 20.4, the L96 model was discussed, an often used idealized model for experimenting with sub-grid-scale parameterizations. This would be a suitable model for testing and validating new ideas and algorithms to tackle the challenges summarized here.

Acknowledgements DC is financially supported by the Netherlands Organisation for Scientific Research (NWO) through the Vidi project *Stochastic models for unresolved scales in geophysical flows*.

References

1. Stephens, G.L.: Cloud feedbacks in the climate system: a critical review. J. Clim. **18**, 237–273 (2005)
2. Arakawa, A.: The cumulus parameterization problem: past, present, and future. J. Clim. **17**, 2493–2525 (2004)
3. Randall, D., Khairoutdinov, M., Arakawa, A., Grabowski, W.: Breaking the cloud parameterization deadlock. B. Am. Meteorol. Soc. **84**, 1547–1564 (2003)
4. Tan, J., Jakob, C., Lane, T.P.: The consequences of a local approach in statistical models of convection on its large-scale coherence. J. Geophys. Res. Atmospheres **120**, 931–944 (2015)
5. Palmer, T.N.: A nonlinear dynamical perspective on model error: a proposal for non-local stochastic-dynamic parameterization in weather and climate prediction models. Q. J. R. Meteorol. Soc. **127**, 279–304 (2001)
6. Palmer, T.N.: On parametrizing scales that are only somewhat smaller than the smallest resolved scales, with application to convection and orography. In: Proceedings of the ECMWF Workshop on New Insights and Approaches to Convective Parametrization, pp. 328–337 (1997)
7. Berner, J., Doblas-Reyes, F.J., Palmer, T.N., Shutts, G., Weisheimer, A.: Impact of a quasi-stochastic cellular automaton backscatter scheme on the systematic error and seasonal prediction skill of a global climate model. Philos. Trans. R. Soc. A **366**, 2561–2579 (2008)
8. Shutts, G.: A stochastic kinetic energy backscatter algorithm for use in ensemble prediction systems. Technical Memorandum 449, ECMWF (2004)
9. Shutts, G.: A kinetic energy backscatter algorithm for use in ensemble prediction systems. Q. J. R. Meteorol. Soc. **131**, 3079–3102 (2005)
10. Bengtsson, L., Körnich, H., Källén, E., Svensson, E.: Large-scale dynamical response to sub-grid-scale organization provided by cellular automata. J. Atmos. Sci. **68**, 3132–3144 (2011)
11. Bengtsson, L., Steinheimer, M., Bechtold, P., Geleyn, J.F.: A stochastic parametrization for deep convection using cellular automata. Q. J. R. Meteorol. Soc. **139**, 1533–1543 (2013)
12. Berner, J., Shutts, G., and Palmer, T.: Parameterising the multiscale structure of organised convection using a cellular automaton. In: ECMWF Workshop on Representation of Sub-grid Processes Using Stochastic-Dynamic Models, pp. 129–139 (2005)
13. Khouider, B., Biello, J., Majda, A.J.: A stochastic multicloud model for tropical convection. Commun. Math. Sci **8**, 187–216 (2010)
14. Dorrestijn, J., Crommelin, D.T., Siebesma, A.P., Jonker, H.J.J.: Stochastic parameterization of shallow cumulus convection estimated from high-resolution model data. Theor. Comput. Fluid Dyn. **27**, 133–148 (2013)

15. Dorrestijn, J., Crommelin, D.T., Biello, J.A., Böing, S.J.: A data-driven multicloud model for stochastic parameterization of deep convection. Philos. Trans. R. Soc. A **371**(1991), 20120374 (2013)
16. Gottwald, G.A., Peters, K., Davies, L.: A data-driven method for the stochastic parametrisation of subgrid-scale tropical convective area fraction. Q. J. R. Meteorol. Soc. **142**, 349–359 (2016)
17. Majda, A.J., Khouider, B.: Stochastic and mesoscopic models for tropical convection. Proc. Natl. Acad. Sci. **99**, 1123–1128 (2002)
18. Dorrestijn, J., Crommelin, D.T., Siebesma, A.P., Jonker, H.J.J., Jakob, C.: Stochastic parameterization of convective area fractions with a multicloud model inferred from observational data. J. Atmos. Sci. **72**, 854–869 (2015)
19. Crommelin, D., Vanden-Eijnden, E.: Subgrid-scale parameterization with conditional Markov chains. J. Atmos. Sci. **65**, 2661–2675 (2008)
20. Lorenz, E.N.: Predictability - a problem partly solved. In: Proceedings of the 1995 ECMWF Seminar on Predictability, ECMWF, Reading, UK, 118 (1996)
21. Pavliotis, G., Stuart, A.: Multiscale Methods: Averaging and Homogenization. Springer (2008)
22. Khouider, B.: A coarse grained stochastic multi-type particle interacting model for tropical convection: nearest neighbour interactions. Comm. Math. Sci **12**, 1379–1407 (2014)
23. Richards, F.C., Meyer, T.P., Packard, N.H.: Extracting cellular automaton rules directly from experimental data. Physica D **45**, 189–202 (1990)
24. Adamatzky, A.I.: Identification of Cellular Automata. CRC Press (1994)
25. Billings, S.A., Yang, Y.: Identification of probabilistic cellular automata. IEEE Trans. Syst. Man Cybern. B Cybern. **33**, 225–236 (2003)
26. Sun, X., Rosin, P.L., Martin, R.R.: Fast rule identification and neighborhood selection for cellular automata. IEEE Trans. Syst. Man Cybern. B Cybern. **41**, 749–760 (2011)
27. Guo, Y., Billings, S.A., Coca, D.: Identification of N-state spatio-temporal dynamical systems using a polynomial model. Int. J. Bifurc. Chaos **18**, 2049–2057 (2008)
28. De La Chevrotiere, M., Khouider, B., Majda, A.J.: Calibration of the stochastic multicloud model using Bayesian inference. SIAM J. Sci. Comput. **36**, B538–B560 (2014)

Participants of the 2013 Eindhoven Meeting

Pablo Arrighi (Université de Grenoble and ÉNS de Lyon, France)
Pablo.Arrighi@imag.fr

Franco Bagnoli (Università di Firenze, Italy)
franco.bagnoli@unifi.it

Sonja Boas (Centrum Wiskunde & Informatica, Amsterdam, The Netherlands)
boas@cwi.nl

Olivier Bouré (Université de Lorraine – inria, France)
olivier.boure@loria.fr

Jean Bricmont (Université de Louvain, Belgique)
jean.bricmont@uclouvain.be

Anne Briquet (Université de Lorraine, France)
anne.briquet@univ-lorraine.fr

Ana Bušić (inria Paris, France)
ana.busic@inria.fr

Philippe Chassaing (Université de Lorraine, Institut Élie Cartan, France)
chassaingph@gmail.com

Alessandro Checco (National University of Ireland, Ireland)
alessandro.checco@nuim.ie

Yadeta Chimdessa (Arba Minch University, Ethiopia)
yady.chimdessa@gmail.com

Emilio N.M. Cirillo (Sapienza Università di Roma, Italy)
emilio.cirillo@uniroma1.it

Francesca Collet (Università di Bologna, Italy)
francesca.collet@unibo.it

https://www.eurandom.tue.nl/events/workshops/2013/PCA/PCA.html

P.-Y. Louis and F.R. Nardi (eds.), *Probabilistic Cellular Automata*, Emergence, Complexity and Computation 27, https://doi.org/10.1007/978-3-319-65558-1

Daan Crommelin (Centrum Wiskunde & Informatica Amsterdam, The Netherlands)
Daan.Crommelin@cwi.nl

Paolo Dai Pra (Università di Padova, Italy)
daipra@math.unipd.it

Andreas Deutsch (Technische Universität Dresden, Deutschland)
andreas.deutsch@tu-dresden.de

Sander Dommers (Università di Bologna, Italy)
sander.dommers@unibo.it

Jesse Dorrestijn (Centrum Wiskunde & Informatica, Amsterdam, The Netherlands)
J.Dorrestijn@cwi.nl

Johan Dubbeldam (TU Delft, Nederland)
j.l.a.dubbeldam@tudelft.nl

Aernout van Enter (Rijksuniversiteit Groningen, Nederland)
a.c.d.van.enter@rug.nl

Nazim Fatès (inria – Nancy, France)
nazim.fates@inria.fr

Roberto Fernández (Universiteit Utrecht, Nederland)
R.Fernandez1@uu.nl

Lucas Gerin (Université Paris 10, France)
Lucas.Gerin@u-paris10.fr

Remco van der Hofstad (Technische Universiteit Eindhoven, Nederland)
rhofstad@win.tue.nl

Tim Hulshof (Technische Universiteit Eindhoven, Nederland)
w.j.t.hulshof@tue.nl

Yi Jiang (Georgia State University, USA)
yjiang12@gsu.edu

Thomas Geert de Jong (Technische Universiteit Eindhoven, Nederland)
t.g.d.jong@tue.nl

Carlo Lancia (Technische Universiteit Eindhoven, Nederland and Roma Tor Vergata, Italy)
c.lancia@tue.nl

Kerry Landman (University of Melbourne, Australia)
kerryl@unimelb.edu.au

Arnaud Le Ny (Université Paris–Est Créteil, France)
arnaud.le-ny@u-pec.fr

Pierre–Yves Louis (Université de Poitiers, France;
visiting researcher EURANDOM, Eindhoven, Nederland)
pierre-yves.louis@math.univ-poitiers.fr

Christian Maes (KU Leuven, België)
Christian.Maes@fys.kuleuven.be

Jean Mairesse (LIAFA, CNRS, Université Paris 7, France)
mairesse@liafa.univ-paris-diderot.fr

Danuta Makowiec (Uniwersytet Gdański, Polska)
fizdm@univ.gda.pl

Irène Marcovici (LIAFA, Université Paris Diderot – Paris 7, France)
irene.marcovici@liafa.univ-paris-diderot.fr

Nevena Marić (University of Missouri – St. Louis, USA)
maric@math.umsl.edu

Carsten Mente (Technische Universität Dresden, Deutschland)
carsten.mente@tu-dresden.de

Roeland Merks (Centrum Wiskunde & Informatica, Amsterdam;
and Mathematical Institute, Leiden University, The Netherlands)
merks@cwi.nl

Ida Minelli (Università degli Studi dell'Aquila, Italy)
ida.minelli@dm.univaq.it

Adrian Muntean (Technische Universiteit Eindhoven, Nederland)
a.muntean@tue.nl

Tobias Muller (Universiteit Utrecht, Nederland)
t.muller@uu.nl

Francesca R. Nardi (Technische Universiteit Eindhoven, Nederland)
F.R.Nardi@tue.nl

Ioana Niculescu (Universiteit Utrecht, Nederland)
ioana_niculescu@yahoo.com

Markus Owen (Nottingham, United Kingdom)
Markus.Owen@nottingham.ac.uk

Margriet Palm (Centrum Wiskunde & Informatica, Amsterdam, Nederland)
m.m.palm@cwi.nl

Fernando Peruani (Université de Nice, France)
Fernando.Peruani@unice.fr

Lise Ponselet (Université Catholique de Louvain, Belgique)
lise.ponselet@uclouvain.be

Damien Regnault (Université d'Évry Val d'Essonne, France)
damien.regnault@ibisc.fr

Wioletta Ruszel (Technische Universiteit Delft, Nederland)
w.m.ruszel@tudelft.nl

Ville Salo (University of Turku, Finland)
vosalo@utu.fi

Benedetto Scoppola (Università Roma 2, Italy)
scoppola@mat.uniroma2.it

Elisabetta Scoppola (Universita degli Studi Roma Tre, Italy)
scoppola@mat.uniroma3.it

Sylvain Sené (Université d'Évry-Val-d'Essonne, France)
sylvain.sene@ibisc.univ-evry.fr

Gordon Slade (University of British Columbia, Canada)
slade@math.ubc.ca

Piotr Słowiński (University of Warwick, United Kingdom)
p.m.slowinski@warwick.ac.uk

Cristian Spitoni (Universiteit Utrecht, Nederland)
C.Spitoni@uu.nl

Siamak Taati (Universiteit Utrecht, Nederland)
siamak.taati@gmail.com

Lorenzo Taggi (Max Planck Institute, Leipzig, Deutschland)
Lorenzo.Taggi@landis.mpg.de

Christoph Temmel (Vrije Universiteit Amsterdam, Nederland)
ctc@temmel.me

Kiamars Vafayi (Technische Universiteit Eindhoven, Nederland)
k.vafayi@tue.nl

Hanne Van Den Bosch (Université Catholique de Louvain, Belgique)
hanne.vandenbosch@student.uclouvain.be

Krishnan Vinu
vinuooty@gmail.com

Anja Voß–Böhme (Technische Universität Dresden, Deutschland)
anja.voss-boehme@tu-dresden.de

Renske Vroomans (Universiteit Utrecht, Nederland)
R.M.A.Vroomans@uu.nl

The manufacturer's authorised representative in the EU is Springer Nature Customer Service Centre GmbH, Europaplatz 3, 69115 Heidelberg, Germany. If you have any concerns regarding our products, please contact ProductSafety@springernature.com

Printed and bound by CPI Group (UK) Ltd, Croydon, CR0 4YY
05/10/2026
02228139-0015